AF838153

Coastal and Marine Geospatial Technologies

Coastal Systems and Continental Margins

VOLUME 13

Series Editor

Bilal U. Haq

Editorial Advisory Board
M. Collins, *Dept. of Oceanography, University of Southampton, U.K.*
S. Berne, *Université Perpignan Via Domitia, Perpignan, France*
K.E. Louden, *Dept. of Oceanography, Dalhousie University, Halifax, NS, Canada*
J.D. Milliman, *School of Marine Science, The College of William & Mary, Gloucester Point, VA, U.S.A.*
H.W. Posamentier, *Chevron Energy Technology Company, Houston, TX, U.S.A.*
A. Watts, *Dept. of Earth Sciences, University of Oxford, U.K.*

For further volumes:
http://www.springer.com/series/5799

Coastal and Marine Geospatial Technologies

Edited by

David R. Green
Centre for Marine and Coastal Zone Management (CMCZM),
Department of Geography and Environment, University of Aberdeen,
Scotland, UK

Springer

Editor
Dr. D.R. Green
Centre for Marine and Coastal Zone
Management (CMCZM)
Department of Geography and Environment
University of Aberdeen
St. Mary's, Elphinstone Road
Aberdeen, AB24 3UF
Scotland, UK
d.r.green@abdn.ac.uk

ISBN 978-1-4020-9719-5 e-ISBN 978-1-4020-9720-1
DOI 10.1007/978-1-4020-9720-1
Springer Dordrecht Heidelberg London New York

Library of Congress Control Number: 2009930957

© Springer Science+Business Media B.V. 2010, All rights reserved for pp. 57–64 and 73–85
No part of this work may be reproduced, stored in a retrieval system, or transmitted in any form or by any means, electronic, mechanical, photocopying, microfilming, recording or otherwise, without written permission from the Publisher, with the exception of any material supplied specifically for the purpose of being entered and executed on a computer system, for exclusive use by the purchaser of the work.

Cover illustration: Cover photographs taken by David R. Green

Printed on acid-free paper

Springer is part of Springer Science+Business Media (www.springer.com)

Preface

In 2005 the CoastGIS symposium and exhibition was once again held in Aberdeen, Scotland, in the UK, the second time that we have had the privilege host this international event in the city of Aberdeen. This was the 6th International Symposium Computer Mapping and GIS for Coastal Zone Management, a collaboration between the International Cartographic Association's (ICA) Commission on Marine Cartography, and the International Geographical Union's (IGU) Commission on Coastal Systems.

The theme for 2005 was: *Defining and Building a Marine and Coastal Spatial Data Infrastructure*. As a major coastal event, the CoastGIS series of conferences always attracts an international audience of coastal researchers, managers, and practitioners who use one or more of the geospatial technologies (e.g. GIS, GPS, digital mapping, remote sensing, databases, and the Internet) in their work.

The CoastGIS series is fundamentally an international event which over the years has gained a strong following attracting delegates from around the globe. Hosted by the University of Aberdeen – at the Aberdeen Exhibition and Conference Centre (AECC) – once again CoastGIS 2005 provided an opportunity to communicate the results of a wide range of innovative scientific research into coastal and marine applications of the geospatial technologies, including remote sensing, Geographical Information Systems (GIS), Global Positioning Systems (GPS), databases, data models, the Internet and online mapping systems. The 2005 even offered many excellent paper and poster presentations from around the World, with delegates from the UK, Norway, Spain, Italy, France, Germany, Mexico, Indonesia, Nigeria, the USA, Canada, and Australia – indeed a truly international cross-section of people!

This current book, which is the outcome of this event, is a compilation of a selection of edited chapters given in Aberdeen. Divided into six parts this edited volume categorises the chapters into parts on: (I) *Information Networks, Spatial Data Infrastructures and Information Systems*; (II) *GIS Tools*; (III) *Spatial Planning*; (IV) *Coastal Geomorphology*; (V) *The Coastal Environment; and* (VI) *Coastal Hazards and Vulnerability*. *Part* I touches upon metadata services, spatial information systems for policy support, spatial data infrastructures in Nigeria, Indonesia, and the UK, information sharing, information networks, and national databases. *Part II* considers a range of GIS tools dealing with marine pollution and aquaculture, ICZM, macro-algal weedmat monitoring, hydrodynamic models, and the spatial analysis of and risks associated with coastal traffic. In addition, topics such as photogrammetry and visualisation are also covered. *Part III* deals with aspects of coastal planning issues and the links with spatial information systems. *Part IV* focuses on the use of GIS for studies of coastal geomorphology, morphodynamics, and shoreline evolution. *Part V* examines the currently highly topical issues of climate change of the coastal landscape and the use of GIS to manage coastal defence planning. The final *Part, VI*, deals with environmental indicators, coastal vulnerability in relation to climate change, and geohazards.

As a book it offers a collection of chapters with considerable insight into the application of these rapidly evolving technologies to many different aspects of the marine and coastal environment around the world, and at many different scales.

I would like to take this opportunity to express my thanks to a number of people and organisations who were involved in the Aberdeen event and who have been instrumental in helping progress this book into print as a permanent record of the CoastGIS 2005 Symposium:

Aberdeen City Council (ACC) for making funds available to the CoastGIS 2005 event and for hosting the Civic reception in Aberdeen; NOAA (in particular Cindy Fowler and Jeff Payne) in the USA for agreeing to sponsor CoastGIS 2005 and for input to both presentations and a special workshop; to the University of Aberdeen for the Symposium Proceedings; and finally Stronachs in Aberdeen. Also to thank all the AECC staff for making the conference work on the ground, and for their input into the organization, registration, presentations and catering, and especially to Brian Horsburgh taking the risk by underwriting this event; to Petra van Steenbergen and her team (especially Hermine Vloemans) at Springer for believing in this book and for progressing this volume to publication; the ICA and IGU; and last but not least all of the people who have supported the CoastGIS event over the years – they know who they are – and in this case for encouraging the second event in Aberdeen – particularly Darius Bartlett (UCC in Ireland) – who started the whole CoastGIS series off in 1995 – and Ron Furness in Australia who was enthusiastic, supportive and encouraging throughout the event.

Scotland, UK David R. Green

Contents

Contributors

Julien Barde IRD, UMR 212 EME, CRH, IRD, 911 Av. Jean Monnet BP 171 34200 Sète Cedex France, julien.barde@ird.fr

Katarzyna Bojar 29B Battersea Rise, SW11 1HG, London, England, UK, kasiabojar@yahoo.co.uk

Jorge Brenner Harte Research Institute for Gulf of Mexico Studies, Texas A & M University – Corpus Christi, 6300 Ocean Dr., Corpus Christi, TX 78412, USA, jorge.brenner@tamucc.edu

Iain Brown The Macaulay Institute, Craigiebuckler, Aberdeen AB15 8QH, UK, i.brown@macaulay.ac.uk

Michael J.A. Butler ACZISC Secretariat, Dalhousie University, Halifax, Nova Scotia, Canada, michael.butler@dal.ca

Rosaline R. Canessa Department of Geography, University of Victoria, P.O. Box 3060, Victoria, BC, Canada, V8W 3R4, rosaline@uvic.ca

Francisco Carreño Dpto. Bioligia y Geologia, Universidad Rey Juan Carlos, C/ Tulipán s/n, 28933 Móstoles, Madrid, Spain, francisco.carreno@urjc.es

Norval Collins Communication, Environmental and Fisheries Consultants, Halifax, Nova Scotia, Canada, ncollins@cefconsultants.ns.ca

Corinne Corbau Università degli Studi di Ferrara, Dipartimento di Scienze della Terra, Via G. Saragat, 1, 44100 Ferrara, Italy, corrine.corbau@unife.it

Valerie Cummins Coastal & Marine Resources Centre, University College Cork, Naval Base, Haulbowline, Cobh, Co. Cork, Ireland, v.cummins@ucc.ie

Paolina Curro University of Messina-Dipartimento di Studi su Risorse, Impresa, Ambiente e Metodologie Quantitative-Piazza Pugliatti 1- 98124 Messina, Italy, pcurro@unime.it

I.M. Davies FRS Marine Laboratory, 375 Victoria Road, Aberdeen AB11 9 DB, UK, daviesim@marlab.ac.uk

Jean-Christophe Desconnets IRD, ROSELT, 500 rue JF Breton, 34093 Montpellier Cedex 5, France

Tobias Dolch Alfred-Wegener-Institute for Polar and Marine Research, Wadden Sea Station Sylt, Hafenstrasse 43, 25992 List, Germany, tdolch@awi-bremerhaven.de

Uwe Dornbusch 34 Down Terrace, Brighton BN2 9 ZH, UK, uwe_dornbusch@ymail.com

Declan Dunne Coastal & Marine Resources Centre, University College Cork, Naval Base, Haulbowline, Cobh, Co. Cork, Ireland, d.dunne@ucc.ie

Ned Dwyer Coastal & Marine Resources Centre, University College Cork, Naval Base, Haulbowline, Cobh, Co. Cork, Ireland, n.dwyer@ucc.ie

Jorge I. Euán Avila CINVESTAV Unidad Merida, km 6 carretera antigua a Progreso, 97128 Merida, Yucatan, Mexico, euan@mda.cinvestav.mx

N. Evelpidou Remote Sensing Laboratory, Geology Department, University of Athens, Panepistimiopolis, Zografou, 157-84, Athens, Greece, evelpidou@geol.uoa.gr

Maurizio Farina Regione Emilia Romagna, Servizio Tecnico di Bacino Po di Volano, Viale Cavour 77, 44100 Ferrara, Italy, mfarina@regione.emilia-romagna.it

K. Gaki-Papanastassiou Department of Geography & Climatology, Faculty of Geology & Geoenvironment, University of Athens, Panepistimiopolis, Zografou 15784, Athens, Greece, gaki@geol.uoa.gr

Randy Gillespie CCMC, 155 Ridge Road, St. John's, Newfoundland, Canada A1B 3NG, rgillespie@ccmc.nf.ca

Dídac Perales Í Giménez Washington State Department of Ecology, PO Box 47600, Olympia, WA 98504-7600, USA

P. Gkavakou Department of Climatology-Physical Geography, Faculty of Geology, University of Athens, Panepistimiopolis, Zografou 157 84, Athens, Greece

Leon Gosar Faculty of Civil and Geodetic Engineering, University of Ljubljana, Jamova 2, SI-1000 Ljubljana, Slovenia, leon.gosar@fgg.uni-lj.si

Th. Gournelos Department of Climatology-Physical Geography, Faculty of Geology, University of Athens, Panepistimiopolis, Zografou 157 84, Athens, Greece, gournelos@geol.uoa.gr

Simona Gragnaniello Dipartimento di Scienze della Terra, Università degli Studi di Ferrara, Via G. Saragat, 1, 44100 Ferrara, Italy

C. Greathead FRS Marine Laboratory, 375 Victoria Road, Aberdeen AB11 9DB, UK, greathead@marlab.ac.uk

David R. Green Centre for Marine and Coastal Zone Management (CMCZM), Department of Geography and Environment, University of Aberdeen, Elphinstone Road, Aberdeen AB24 3UF, UK, d.r.green@abdn.ac.uk

M. Gubbins FRS Marine Laboratory, 375 Victoria Road, Aberdeen AB11 9DB, UK, gubbinsm@marlab.ac.uk

Alfonso Cuevas Jiménez Departamento de Manejo de Zonas Costeras, Faculdad de Ciencias, UNAM, Av. Universidad 3000, Circuito Exterior S/N, Delegación Coyoacán, C.P. 04510, Ciudad Universitaria D.F., Mexico

Jose A. Jimenez Laboratori d'Enginyeria Marítima, ETSECCPB, Universitat Politècnica de Catalunya, C/ Jordi Girona, 1-3, Modul D-1, 08034, Barcelona, Spain, Jose.Jimenez@upc.edu

S. Koussouris Department of Climatology-Physical Geography, Faculty of Geology, University of Athens, Panepistimiopolis, Zografou 157 84, Athens, Greece

Claudette Leblanc ACZISC Secretariat, Dalhousie University, Halifax, Nova Scotia, Canada

Y.S. Li Department of Civil and Structural Engineering, The Hong Kong Polytechnic University, Hong Kong, P.R. China, ceysli@polyu.edu.hk

Therese Libourel LIRMM, 161 rue Ada, F-34392 Montpellier Cedex 05, France, libourel@lirmm.fr

Sarah A. Loos Department of Geography, University of Victoria, P.O. Box 3060, Victoria, BC, Canada, V8W 3R4, sarahloos@gmail.com

Lionel Loubersac IFREMER, Campus IRD, 101 Promenade Roger Laroque, BP 2059 98846 Nouméa Cedex, Nouvelle Calédonie, France, lionel.loubersac@ifremer.fr

Gerold Lűerssen Common Wadden Sea Secratariat, Virchowstrasse 1, 26382 Wilhelmshaven, Germany, luerssen@waddensea-secretariat.org

Vincenzo Maccarrone Via del Faro 3, 91020 Campobello di Mazara TP Sicilia, Consiglio Nazionale delle Ricerche – Istituto per l'ambiente marino costiero U.O di Capo Granitola, Italy, vincenzo.maccarrone@unime.it

H. Maroukian Department of Geography and Climatology, Faculty of Geology and Geoenvironment, University of Athens, Panepistimiopolis, Zografou 157 84, Athens, Greece, maroukian@geol.uoa.gr

Pierre Maurel IRD, UMR 212 EME, CRH, IRD, 911 Av. Jean Monnet BP 171 34200 Sète Cedex, France, pierre.maurel@teledetection.fr

Nabila Mazouni CEPRALMAR, 20 rue République, 34000 Montpellier, France, mazouni@cepralmar.org

Isabel Montoya Universidad Rey Juan Carlos, ESCET, C/ Tulipán s/n, 28933 Móstoles. Madrid, Spain, isabel.montoya@urjc.es

Cherith Moses Department of Geography, University of Sussex, Chichester Building, Falmer, Brighton BN1 9QJ, UK, c.moses@sussex.ac.uk

Alister D. Nairn Geography, Australian Bureau of Statistics, Alister.Nairn@abs.gov.au

Sandy M.N. Ng Centre for Water Research, University of Western Australia, ng@cwr.uwa.edu.au

P.C. Nwilo Department of Surveying and Geoinformatics, University of Lagos, Lagos, Nigeria, pcnwilo@yahoo.com

Liz O'dea Application and Data Services, Washington State Department of Ecology, PO Box 47600, Olympia, WA 98504-7600, USA, LODE461@ECY.WA.GOV

Ifeyinwa Christabel Okeke Department of Geography, University of Lagos, Akoka, Lagos, Nigeria, ifychristabelokeke@hotmail.com

Filipa S.B.F. Oliveira LNEC, Avenida do Brasil 101, 1700-066 Lisboa, Portugal, foliveira@lnec.pt

O.O. Oni Department of Surveying and Geoinformatics, University of Lagos, Lagos, Nigeria

A.O. Onojeghuo Department of Surveying and Geoinformatics, University of Lagos, Lagos, Nigeria

D.A. Osanwuta National Inland Waterways Authority, Lokoja, Nigeria, dennisosanwuta@yahoo.com

Mike Osborne SeaZone Solutions Limited, Red Lion House, Bentley, Hampshire, GU10 5HY, UK, mike.osborne@seazone.com

Joaquim Pais-Barbosa Faculty of Engineering, Hydraulics and Water Resouces Institute, Porto University, Rua Dr Roberto Frias, s/n, 4200-465 Porto, Portugal, jlpb@fe.up.pt

Ronald Pelot Industrial Engin., Dalhousie University, P.O. Box 1000, Halifax NS, Canada B3J 2X4, ronald.pelot@dal.ca

John Pepper United Kingdom Hydrographic Office, Admiralty Way, Taunton TA1 2DN, UK, john.pepper@ukho.gov.uk

Manca Plazar Municipality of Piran, Tartinijev trg 2, 6330 Piran, Slovenia

Lauren Plummer Dalhousie University, Halifax, NS, Canada B3J 2X4, lplummer@dal.ca

S.E. Poulos Department of Geography and Climatology, Faculty of Geology, University of Athens, Panepistimiopolis, Zografou 157 84, Athens, Greece, poulos@geol.uoa.gr

Gatot H. Pramono BAKOSURTANAL (National Coordinating Agency for Surveys and Mapping), Indonesia, Gatot@bakosurtanal.go.id

Tanja Preseren Faculty of Civil and Geodetic Engineering, University of Ljubljana, Jamova 2, SI-1000, Ljubljana, Slovenia

Bebas Purnawan BAKOSURTANAL (National Coordinating Agency for Surveys and Mapping), Indonesia, bebas@bakosurtanal.go.id

David A. Robinson Department of Geography, University of Sussex, Chichester Building, BN1 9QJ, UK, D.A.Robinson@sussex.ac.uk

Emanuele Roccatagliata GISIG – GIS International Group, via Piacenza 54, 16138 Genova, Italy, e.roccatagliata@gisig.it

Inmaculada Rodríguez Universidad Rey Juan Carlos, ESCET, C/ Tulipán s/n, 28933 Móstoles, Madrid, Spain, inmaculada.rodriguez@urjc.es

Michele Rotunno ENEA National Agency for New Technologies, Energy and the Environment, Environment Department, Via Don Fiammelli 2, 40129 Bologna, Italy, michele.rotunno@bologna.enea.it

Giorgio Saio GISIG – GIS International Group, via Piacenza 54, 16138 Genova, Italy, g.saio@gisig.it

María José Sánchez Universidad Rey Juan Carlos, ESCET, C/ Tulipán s/n, 28933 Móstoles, Madrid, Spain, mariajose.sanchez@urjc.es

Rafael Sarda Centre d'Estudis Avançats de Blanes (CSIC), Carrer d'Accés a la Cala Sant Francesc, 14, 17300 Blanes, Spain, sarda@ceab.csic.es

Andrew G. Sherin Natural Resources Canada, Earth Science Sector, 1 Challenger Drive, Dartmouth, NS, Canada B2Y 4A2, sherin@nrcan.gc.ca

Umberto Simeoni Università degli Studi di Ferrara, Dipartimento di Scienze della Terra, Via G. Saragat, 1, 44100 Ferrara, Italy, g23@unife.it

Franc Steinman Faculty of Civil and Geodetic Engineering, University of Ljubljana, Jamova 2, SI-1000 Ljubljana, Slovenia, fsteinma@fgg.uni-lj.si

Tim Stojanovic Marine and Coastal Environment Research Group, School of Earth and Ocean Sciences, Cardiff University, Cardiff CF10 3YE, UK, stojanovic@cf.ac.uk

Suwahyuono Wahyu BAKOSURTANAL (National Coordinating Agency for Surveys and Mapping), Indonesia, wahyulaut@yahoo.com

Francisco Taveira-Pinto Faculty of Engineering, Porto University, Hydraulics and Water Resouces Institute, Rua Dr Roberto Frias, s/n, 4200-465 Porto, Portugal, fpinto@fe.up.pt

D. Tulett FRS Marine Laboratory, 375 Victoria Road, Aberdeen AB11 9 DB, UK, d.tulett@marlab.ac.uk

D. Vaiopoulos Remote Sensing Laboratory, Geology Department, University of Athens, Panepistimiopolis, Zografou, 157-84, Athens, Greece, vaiopoulos@geol.uoa.gr

Edi Valpreda ENEA National Agency for New Technologies, Energy and the Environment, Environment Department, Via Don Fiammelli 2, 40129 Bologna, Italy, valpreda@bologna.enea.it

A. Vassilopoulos Remote Sensing Laboratory, Geology Department, University of Athens, Panepistimiopolis, Zografou, 157-84, Athens, Greece, vassilopoulos@geol.uoa.gr

Fernando Veloso-Gomes Faculty of Engineering, Hydraulics and Water Resouces Institute, Porto University, Rua Dr Roberto Frias, s/n, 4200-465 Porto, Portugal, vgomes@fe.up.pt

Onyx W.H. Wai Department of Civil and Structural Engineering, The Hong Kong Polytechnic University, Hong Kong, P.R. China, ceonyx@polyu.edu.hk

Rendel Williams Department of Geography, University of Sussex, Chichester Building, Falmer, Brighton BN1 9QJ, UK, R.B.G.Williams@sussex.ac.uk

Colin D. Woodroffe University of Wollongong, School of Earth and Environmental Sciences, Wollongong, NSW 2522, Australia, colin@uow.edu.au

Z.H. Xu Department of Civil and Structural Engineering, The Hong Kong Polytechnic University, Hong Kong, P.R. China

Part I
Information Networks, SDI and Information Systems

Chapter 1
A Metadata Service for Managing Spatial Resources of Coastal Areas

Julien Barde, Therese Libourel, Pierre Maurel, Jean-Christophe Desconnets, Nabila Mazouni, and Lionel Loubersac

Abstract Within the framework of the SYSCOLAG pluridisciplinary research programme on ICZM in the Languedoc-Roussillon region of France, we propose an ISO-19115-compliant metadata service to share informational resources. Each resource is described using a metadata form. Each form consists of textually and spatially controlled terms which are stored in a single DBMS (Postgres and Postgis). The graphical user-interfaces contain the expert knowledge (thematic and spatial ontologies) and include a mapping service (SVG and OGC-compliant WMS-WFS) that allows spatial and thematic queries to be addressed to the metadata service.

Keywords Metadata · Internet mapping · Integrated coastal zone management · Interoperability · Standards

1.1 Introduction

This chapter presents the results obtained from the SYSCOLAG (COastal and LAGoonal SYStems) pluridisciplinary research programme on Integrated Coastal Zone Management (ICZM) in the Languedoc-Roussillon (LR) region of France. The principal objective of this programme is to develop scientific tools designed for ICZM. In conformity with the recommendations of the European Union, the LR regional authority wants, in the medium term, to constitute an "observatory" to support the sustainable and integrated management of its coastal areas. Specifically from the point of view of information systems, it is necessary to optimise access, sharing and management of information and available knowledge concerning the littoral so that, in time, the processes involved in ICZM itself can be improved (Doody, 2003).

J. Barde (✉)
IRD, UMR 212 EME, CRH, IRD 911 Av. Jean Monnet BP 171 34200 Sète Cedex, France
e-mail: julien.barde@ird.fr

D.R. Green (ed.), *Coastal and Marine Geospatial Technologies*, Coastal Systems and Continental Margins 13, DOI 10.1007/978-1-4020-9720-1_1,
© Springer Science+Business Media B.V. 2010

Faced with the heterogeneous set of actors involved as well as the myriad ways of managing informational resources,[1] we have decided to implement a metadata[2] service so that these resources can be shared and be easily retrievable. ICZM problematics revolve around thematic and spatial concepts (PNUE, 1999; Hénocque and Denis, 2001). The all-pervading presence – explicit or implicit – of the spatial dimension in the field of ICZM has led us to favour this dimension in the metadata service we propose. In addition, to impart a long life to this information system at the regional level and to ensure its interoperability, over time, with other systems at the national or international levels, we have decided to conform to international standards for the different components of the system (metadata standards, recommendations of the Open GIS Consortium (OGC), thesaurus and reference geographical objects, Unified Modeling Language (UML)). In this, we follow the conclusions of the MADAME[3] project (Craglia and Evmorfopoulou, 2000) and the INSPIRE[4] directive (Smits et al., 2002).

This chapter is organised as follows: In the second section, we will explore an approach based on international standards (ISO 19115 and OGC) and the specific character of the ICZM domain. In section three, we will detail how the spatial component was incorporated into the metadata service to help the user describe and/or search for informational resources. In section four, we will see how geographical information relevant to ICZM in the LR region can be fed into this infrastructure. We will conclude by a discussion on the possibilities opened up by the development of such a tool.

1.2 Underlying Principles in Formulating the Metadata Service

Basing ourselves on the characteristics and principles of integrated management, we have opted for an information-system (IS) architecture that allows the "centralisation" of heterogeneous and distributed information sources thanks to a shared metadata service (Barde et al., 2005). This service can be accessed by a Web portal;

---

[1]The term "resource" refers to information elements such as data, documents, protocols that will be described and qualified at a higher level (metadata) using the metadata service. Depending on the organisation and the individual, a resource might be a report, a map, an image, a video clip, a dataset, a database, a model, a value measured by a sensor, an image-segmentation software application, etc.

[2]In its basic meaning, the term "metadata" signifies "data about data, or data that describes data and helps the user make relevant use of it".

[3]MADAME, Methods for Access to data and Metadata in Europe (1998–2000). The MADAME project will implement methodologies for accessing European data and metadata and, thus, has for goal the valorizaton of public data, http://www.shef.ac.uk/%7Escgisa/MADAMENew/Content.htm.

[4]INSPIRE is an EU initiative to establish an infrastructure for spatial information in Europe that will help make spatial or geographical information more accessible and interoperable for a wide range of purposes, including those supporting sustainable development, http://www.ec-gis.org/inspire/.

(Kay and Christie, 2001) outline its specific relevance for the ICZM domain. In our case, the choice was specially driven by the fact the tool we intend creating is destined not only for the scientific community but, above all, for administrators and the general public.

The system's design consists of choosing the metadata standard, structuring the geographical and semantic reference bases for indexing and searching metadata and designing the linkages to a cartographic interface.

1.2.1 Choice of a Metadata Standard

Metadata provides several advantages for information exchange (management of access rights in particular, richness and relevance of the description). The Dublin Core standard seemed to be the obvious choice for the exchange of digital data.[5] But the spatial component that is so prominent in the case of ICZM led us finally to adopt a standard that is more suitable for the geographical dimension of informational resources that have to be shared.

The TC 211 technical committee of ISO has recently established a standard, ISO 19115 (Geographic Information – Metadata) (ISO 2003), which is a combination of existing metadata standards for exchange of geographical information, such as, among others, the American FGDC and the European CEN/TC 287 (with the ENV 12657 standard). ISO 19115 enjoys today unanimous approval and is recommended to eventually replace its predecessors and thus facilitate the interoperability of metadata services. We have adopted this standard for our project.

1.2.1.1 General Introduction to the ISO 19115 Standard

The ISO 19115 standard consists of eleven headings (Fig. 1.1).

Of these eleven headings, three bear directly on the resource's spatial dimension (*Reference system*, *Spatial representation* and *Content information*) and concern only resources of type geographical information (maps, remote sensing images, etc.).

In addition, we find, in the mandatory heading *Identification,* elements for describing the geographical extent relating to the resource. Even though optional, we have chosen to make them mandatory in the profile we have decided upon (in discussion with programme partners), and, this, irrespective of the described resource. Such an approach will allow us to process spatial requests, including on metadata that does not concern geographical information. Even though we insist on keeping these elements, their operational use is very restrictive if we do not help the user in his entry and/or search operations.

[5] Dublin Core Metadata Initiative (DCMI), http://dublincore.org/.

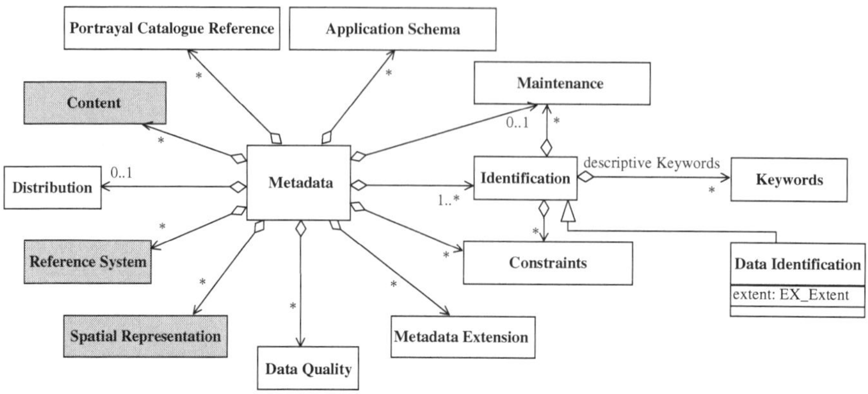

Fig. 1.1 The different heads of ISO 19115 (See also Plate 1 on page 393 in Color Plate Section)

1.2.1.2 The Three Specific Descriptors of Geographical Extent in ISO 19115

ISO permits two modes of describing the geographical extent (Fig. 1.2):

- a text mode, using toponyms (or spatial keywords): descriptor III,
- a (carto)graphic mode, itself consisting of two sub-modes: a geographical object with complex limits or its approximation in the form of a bounding box: descriptors I and II respectively.

Even though the bounding rectangle can theoretically be considered as an approximation of a more precise geometry (corresponding to a toponym), its importance cannot be denied. In fact, the relative relevance of these three descriptor modes depends on the resource types described. For some (a map of a county, for example), the ideal descriptor will be a toponym and/or the associated geographical object (the county in question). For others, the bounding rectangle will be better suited (typically satellite imagery, which is not associated with specific geographical objects but with a contiguous zone determined by the characteristics of the imaging device). This explains why we have decided to retain all the three descriptor types.

However, it is difficult to ask a user to enter the geographical extent in cartographic mode in the way required by ISO. He will have to enter the geographical coordinates in decimal degrees in text mode, identical to that used for spatial databases (a series of points delimiting the bounding rectangle or the complex geographical object).

Therefore we have decided to use a cartographic interface with which the users will be able to enter these descriptors, and, eventually, metadata searches, much more easily.

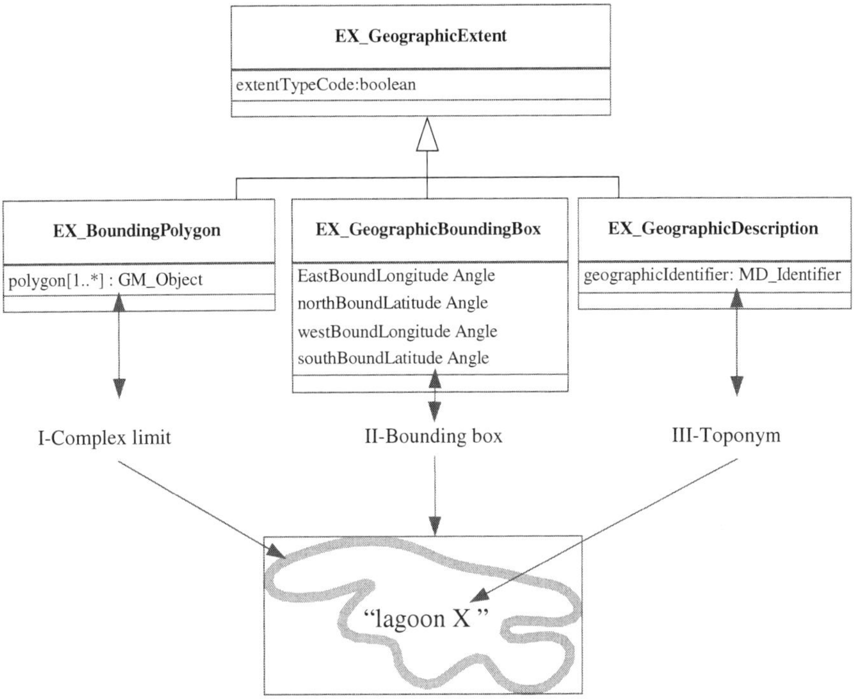

Fig. 1.2 Illustration of the three descriptor types of ISO 19115 (See also Plate 2 on page 393 in Color Plate Section)

1.2.2 Thematic and Spatial Reference Bases for the Indexing of Resources

1.2.2.1 Conceptual Model

To ensure consistency in the description of resources and the quality of metadata searches within the SYSCOLAG community, it is necessary to offer thematic and spatial reference bases to the metadata service's user so that the terms and geographical objects used can be controlled. Figure 1.3 shows how we have conceptually modelled these reference bases in the metadata base.

Experience has shown that resource indexation in our domain typically uses keywords such as: *shellfish cultivation*, *spring*, *pollution*, *nitrate*, *lagoon of Thau*. There exist different types of keywords, shown in Fig. 1.3 by specialisations of the concept class. The keywords mentioned in the example belong to these different categories: *shellfish cultivation*, *pollution* and *nitrate* are thematic keywords whereas *spring* is a temporal key word and *lagoon of Thau* is a spatial keyword (item representing the *lagoon* spatial concept) which can be represented cartographically in its role as a geographical object.

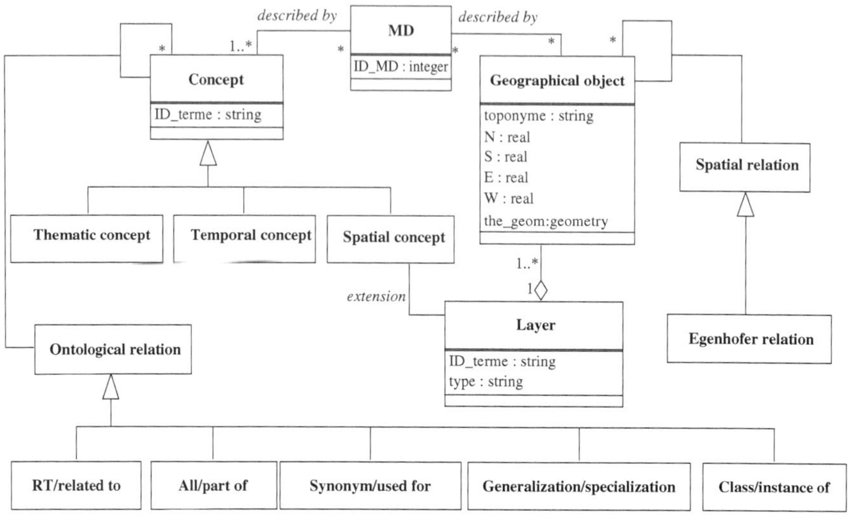

Fig. 1.3 Conceptual model of thematic and spatial reference bases

We have seen in Fig. 1.1 that, in conformity with ISO 19115, each metadata record is indexed in part by keyword and in part by different descriptors of the geographical extent, with the help of toponyms, in particular. These recommendations have been incorporated in our model in Fig. 1.3 with, respectively, the MD class, the concept class and the geographical Object class (which has as attributes the three types of geographical descriptors of ISO 19115). As a complement to ISO and to better adapt the metadata service to our context, we propose to manage semantic links that exist between these keywords by ontological relationships[6] (Barde et al., 2004a) and the use of spatial relationship(s) (according to Pullar and Egenhofer (1988) which exist between geographical objects.

In the case of our example, it is interesting to consider the specific place occupied by keywords of the type *lagoon of Thau* which we can define both as a spatial keyword (item-level concept, where the item represents the lagoon concept) and as a textual descriptor of the geographical extent (a toponym associated with an item that represents a geographical object).

To avoid this information redundancy that is possible in ISO, we have decided to reserve the use of toponyms for the description of the resource's geographical extent and not at the level of keyword elements in the form of spatial keywords. A new relationship between the spatial concepts and their graphical representation is shown, in our diagram, as a transition from the concept (spatial) to the object (geographical): i.e., from the intention, *lagoon*, to the extension, *lagoon of Thau*.

But, in GIS, usage requires that the graphical object *lagoon of Thau* be managed within a table having the name of the concept that expresses the thematic character-

[6] According to Gruber, an ontology is a specification of a conceptualization (Gruber, 1993).

istics of the object set that it groups in the form of records. We take this into account by introducing a *Layer* class whose instances are characterised by the value of the ID_term (for example, *lagoon*) and the geometric type (point, line, polygon, etc.) of the geographical Objects (such as the lagoon of Thau of type polygon) that it is composed of.

1.2.2.2 Standards Used

For the ontological relationships, we have relied on the ISO 5964 standard for developing multilingual thesauri (ISO 1985) and we have specialized the RT relationships (Barde et al., 2005). The spatial relationships used are those standardised by the OGC.

To be able to manage the textual and geographical descriptions as proposed by this model in a single RDBMS, it is necessary to use a spatial cartridge. RDBMSs that allow this type of data infrastructure range from the commercial (Oracle) to free (MySQL ver.4.1 or later with MySQL Spatial Extensions or Postgres with Postgis). We have decided to adopt the Postgres/Postgis solution.

1.3 Interaction Between Metadata and the Geographical Reference Bases Via A Cartographic Interface

Having covered some of the technical details of the implementation, we now proceed to the interactions between the user and the system that allow him to enter data or search metadata based on geographical criteria.

1.3.1 Implementation

First, a SVG (Scalable Vector Graphics) application was created with the aim of optimising functions for entry and metadata-search operations. Yet, it conforms to the OGC specifications for Web Mapping Services (WMS (ISO 2004)). Our choice of the SVG graphical format in vector mode was based on its superior rendering quality (a plus for the user) and its compactness. Moreover, as far as the graphics are concerned, the painting algorithm implemented actually improved the readability of the maps and resolved certain problems associated with the active layer by the use of transparencies. For output in SVG format, the Assvg function of Postgis works at impressive speed. The technical details of our cartographic interface's features are available in (Courtin, 2004). This application has been now replaced by MapServer and Mapbuilder.

1.3.2 Entering Geographical Descriptors in Graphics Mode

Figure 1.4 shows the graphical interface that permits the user to describe a resource's geographical extent when entering a metadata record.

One of two cases can arise:

- the user describes the geographical extent using descriptors known to the system (descriptors I or III, see Fig. 1.1), from which we can deduce descriptor II,[7]
- the user does not find appropriate descriptors I or III from those offered and prefers to describe the metadata using descriptor II. In an inverse operation of the first case, we generate[8] descriptor I corresponding to this bounding rectangle and ask the user if this new object represents a theme that is missing from our cartographic interface (the toponym corresponding to descriptor III, which will otherwise, by default, take the value Bounding rectangle).

In both cases, we have obtained all three types of ISO descriptors without a very restrictive entry requirement.

Fig. 1.4 Geographical description of a resource in ISO 19115 (See also Plate 3 on Page 394 in Color Plate Section)

[7]Using the Postgis spatial function of the type YMAX(the_geom).

[8]Using the GeometryFromText(text,[<SRID>]) Postgis function.

1.3.3 *Entering Toponyms in Text Mode*

The designation of descriptors in graphics mode of geographical Objects is as simple to manage, thanks to GIS tools, as those by keywords or toponyms is complex. In the latter case, the user has to be steered towards his choice of keywords by an appropriate structuring of terms (the introduction of different links/ontological relationships between toponyms, mainly hierarchical, brings us to the formulation of the spatial ontology and again raises the question of links between thematic and spatial keywords posed in Section 2.1.3 and illustrated in the diagram in Fig. 1.3, (Barde et al., 2004b)). In fact, it is not practical to present all the toponyms from all the imported layers sequentially in a pull-down scrolling menu (more so when the number of available layers is large, which is certainly the case with ICZM).

Cartographic interfaces that are able to display a large number of layers of geographical information normally take recourse to a thematic structuring of available layers. This structure's most common form is that of a tree-structure (similar to file/folder explorers, structured by theme) which, as far as Integrated Management of a territory is concerned, should conform to a systemic representation of that space (Prélaz Droux, 1995). The major difficulty here is to offer an organised and consistent representation of this tree-structure for a geographical space or territory that is located at the land-sea interface. We have used this solution thanks to links mentioned in Fig. 1.3; the navigational logic in the thematic or spatial reference bases thus remains homogenous.

1.3.4 *Execution of Spatial Requests*

The idea is to let the user conduct metadata searches without having to type in any text but by using instead the cartographic interface to resolve questions like "Which are the resources listed in this space?".

1.3.4.1 The Three Types of Possible Requests

In addition to other key descriptors for formulating multicriteria information requests (such as temporal, thematic and resource-type aspects, see Fig. 1.4), we have insisted on the spatial criterion by offering three types of spatial requests based on the description modes (I, II and III) offered by ISO.

In general, modes I and II are the most powerful since they specify, in addition to the topological limits of the geographical Object, the associated toponym and thus the thematic nature of the space of the search (lagoons) or even the instance of the geographical Object of the search (lagoon of Thau). We can thus search the metadata associated with an identical geographical Object – the same toponym – (or of the same kind) in addition to searches based on topological criteria. The mode II, however, only allows spatial requests on topological criteria. The execution of

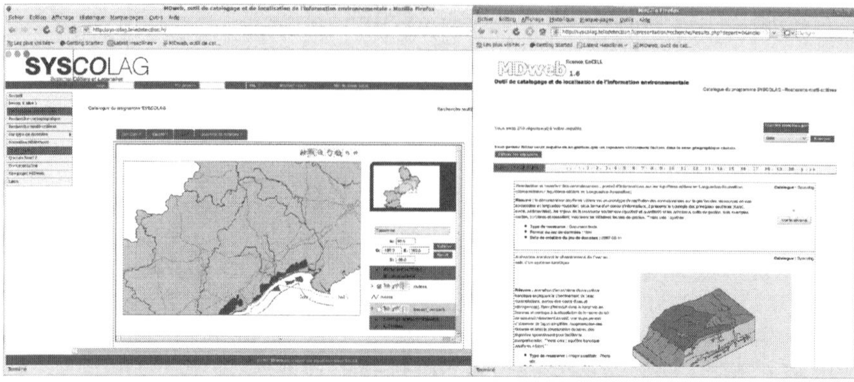

Fig. 1.5 Textual and cartographical responses to a multicriteria request (See also Plate 4 on Page 394 in Color Plate Section)

topological requests is done by spatial functions[9] of Postgis. Postgis offers a set of OGC compliant spatial functions that permit the execution of varied queries.

1.3.4.2 Compilation and Access to the Result Details with the Cartographic Interface

This involves displaying the different types of geographical objects that describe the metadata fulfilling the request made from the cartographic interface. In parallel, we summarise the answers in text form for the user (see illustration in Fig. 1.5). The manner of accessing the metadata details is then left to the user's preference (by a text or graphics hyperlink).

1.4 Outlook

The metadata service will be meaningful and effective only when the user finds pertinent geographical objects to describe, or search for, resources and knowledge that concern the ICZM in the LR region. This entails making, in advance, an inventory of these graphical objects that is as exhaustive as possible (on the basis of reference works on the subject or of the results of our project). This is akin to developing a GIS for the ICZM, which will surely prove to be very useful in its own right.

1.4.1 GI Inventory Tasks and Reference Concepts for the CZ

To the difficulty of collecting, in a single GIS, terrestrial and marine data produced and managed differently at the national level is added that of incorporating data

[9]Using the "within(geometry,geometry) and intersects(geometry, geometry)" Postgis functions.

originating from different disciplines. Problems of access restrictions, of distribution (example of geographical positions of artificial reefs) or of heterogeneity of already existing thematic and spatial reference bases only make the task more complex (example of the hydrographic zero and the choice of a reference base for the land-sea continuum). The ISO standard however permits, because of the richness of these metadata elements, to easily manage this diversity of usage modes.

There exists no current inventory of geographical objects or reference themes specific to the ICZM (Lehfeldt and Heidmann, 2002; Patrikalakis et al., 2000). However, initiatives such as the French littoral working group (Allain et al., 1999) cover broadly the themes involved in this domain.

The littoral working group has compiled an inventory of geographical information existing in the costal zone by broad thematic categories. However, this inventory is not exhaustive and relates to data more relevant at the national, rather than regional, level.

A detailed summary of reference projects is available (Barde et al., 2005).

1.4.2 Contribution of These Tools by the Taking of Local Particularities Into account

Depending on a project's regional characteristics, some geographical information layers may have special importance. For example, in the LR region, the lagoons which are distinctive to its coastal zone. In addition to the general geographical characteristics of a coastal zone, we often find some local geographical specificities or particularities which complement (or refine) the more general inventories made at the national level (generic or top-level approach). They contribute towards the compilation of an exhaustive inventory of geographical information relevant to ICZM problematics.

An interview survey was conducted to identify, structure and centralise the geographical concepts used by the different scientific disciplines in Syscolag (Maurel et al., 2005). We have also used the new toponyms supplied by the Syscolag user community to index the first metadata.

1.4.3 Contribution of Standards in Terms of Free Client Tools

The use of Postgis allows access to tools that use geographical information which are more powerful than our cartographic interface.

Standardised map servers (such as Mapserver, geoserver or geotools which implement the WMS or WFS standards of OGC and enhance online analysis capabilities) are rich in features and should allow the setting up of an atlas on ICZM in the LR region, accessible independently of the metadata service.

We can also benefit by using other tools such as GIS clients like Quantum GIS (see Fig. 1.6), UDIG or Jump, that allow a remote connection to our spatial data

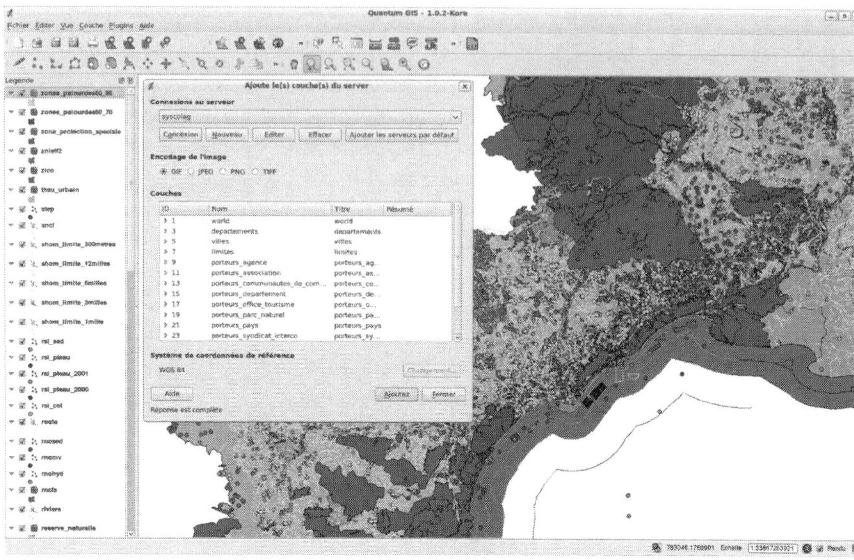

Fig. 1.6 Example of a connection to a Postgis database with the free software application *Quantum GIS* (See also Plate 5 on Page 395 in Color Plate Section)

server. It then becomes possible, from a GIS client, to run traditional spatial analysis functions on all the layers stored in the WMS server, indeed even to edit, modify or delete geographical information (by using the WFS specification). This is a major advantage for users in the many disciplines involved in the ICZM, users who are not conversant with GIS usage, mainly because of the high licensing fees required for proprietary software and reference geographical data.

1.5 Conclusions

The management of spatial information is becoming increasingly more important within information and knowledge systems devoted to sustainable development or to ICZM, i.e., systems in which space is a central concept.

By conforming to standards (ISO 19115 and OGC), we ensure the interoperability of our system with equivalent systems at the national or international level, and this without losing any of the adaptation to the characteristics of our particular application domain. The use of a spatial RDBMS associated with a cartographic interface allows relatively easy management of the description and, above all, the search of metadata using spatial criteria. Our design of the spatial interface benefits the user by overcoming, at least partially, the problem of semantic descriptions by providing an alternative to, or a complement of, the use of ontology.

The geographical information relevant to the domain that we have inventoried towards these ends constitutes a geographical reference base linked to the thematic knowledge of the domain.

The service we propose thus respects both an organisation's independence in managing its resources and the necessity of sharing common tools that are designed for interactions necessary for a proper centralisation of information and knowledge for resolving complex ICZM problematics.

This generic methodology can be adapted for different problems of management of environmental resources, as was the case in the common approach to this project, conducted in collaboration with the ROSELT[10] team.

Finally, the possibility of interoperability with other spatial-data servers (WMS OGC) for the coastal zone respects the concept of ICZM, the natural geography and the functioning of ecosystems, whose extents do not respect the boundaries of man-designated regulatory spaces.

References

Allain, S., Guillaumont B., Visage, C.L., Loubersac, L., and Populus, J. (1999). "Données géographiques de référence en domaine littoral", no. 25, Brest, pp. 67–79.

Barde, J., Libourel, T., and Maurel, P. (2004a). "Ontologies et métadonnées pour le partage d'information géographique", *Revue Internationale de Géomatique*, 14(2), pp. 199–217.

Barde, J., Libourel, T., and Maurel, P. (2004b). "Métadonnées et Information spatiale. Application à la gestion intégrée de la zone côtière", *Proc. Semaine du Document numérique, La Rochelle, France*, on CD-ROM.

Barde, J., Libourel, T., and Maurel, P. (2005). "A Metadata Service for Integrated Management of Knowledge Related to Coastal Areas", *Multimedia Tools and Applications*, 25, pp. 419–429.

Courtin, O. (2004). SIG orienté Web et standards ouverts. Implémentation de l'ISO 19128 avec PHP/Postgis et SVG/ECMAscript.

Craglia, M., and Evmorfopoulou, K. (2000). MADAME: Methods for Access to Data and Metadata in Europe.

Doody, J.P. (2003). "Information Required for Integrated Coastal Zone Management: Conclusions from the European Demonstration Programme", *Coastal Management*, 31, pp. 163–173.

Gruber, T. (1993). "A translation approach to portable ontology specifications. Knowledge Acquisition", *An International Journal of Knowledge Acquisition for Knowledge-Based Systems*, 5(2), pp. 199–220.

Henocque, Y., and Denis, J. (2001). "Des outils et des hommes pour une Gestion Intégrée des Zones Côtières", guide méthodologique Unesco, no. 42.

ISO (1985). Guidelines for the establishment and development of multilingual thesauri, ISO 5964. Geneva, Switzerland.

ISO (2003). International Organization for Standardization, Geographic Information – Metadata, ISO 19115, Geneva, Switzerland.

ISO (2004). Geographic Information – Web Map Server interface, ISO 19128, Geneva, Switzerland.

Kay, R., and Christie, P. (2001). "An Analysis of the Impact of the Internet on Coastal Management", *Coastal Management*, 29, pp. 157–181.

[10]ROSELT, http://www.roselt-oss.org/accueil.php.

Lehfeldt, R., and Heidmann, C. (2002). NOKIS: a metadata information system for German coastal regions of the North Sea and Baltic Sea.

Maurel, P., Libourel, T., and Barde, J. (2005). "Building knowledge infrastructures based on spatial cognition: an example for integrated coastal management", *Environment and Planning B*, submitted.

Patrikalakis, N. M., Abrams, S. L., Bellingham, J. G., Cho, W., Mihanetzis, K. P., Robinson, A. R., Schmidt, H., and Wariyapola, P. C. H. (2000). "The Digital Ocean", *Computer Graphics International*, pp. 45–54.

PNUE-PAM-PAP (1999). Cadre conceptuel et directives pour la gestion intégrée du littoral et des bassins fluviaux.

Prelaz Droux, R. (1995). Système d'Information et Gestion du Territoire. *Presses Polytechniques et Universitaires Romandes,* Collection META, EPFL, Lausanne.

Pullar, D., and Egenhofer, M. (1988). "Toward formal definitions of topological relations among spatial objects ", ed. Marble, D., *Proc. 3rd International Symposium on Spatial Data Handling, SDH'88, Sydney, Australia*, pp. 225–241.

Smits, P.C., Düren, U., Ostensen, O., Murre, L., Gould, M., Sandgren, U., Marinelli, M., Murray, K., Pross, E., Wirthmann, A., Salgé, F., and Konecny, M. (2002). Infrastructure for Spatial Information in Europe.

Chapter 2
The Development of an Australian Marine Spatial Information System (AMSIS) to Support Australian Government Ocean Policy and Multi-Use Marine Activities

Alister D. Nairn

Abstract The Australian Marine Jurisdiction is one of the largest in the world and if ratified by the United Nations Commission on the Limits of the Continental Shelf, just over half of Australia's land mass will be below the sea. Australia has national and international responsibility to manage and sustain the marine environment. This chapter describes an initiative being undertaken by Geoscience Australia (GA) to develop an Australian Marine Spatial Information System (AMSIS). The chapter outlines the progress made to date with this development. AMSIS will accurately locate marine features including interests and rights over the Australian Marine Jurisdiction. Arrangements are being made with other custodians of spatial boundary information so that AMSIS will become an accepted source of integrated information. In developing and implementing AMSIS, GA is working closely with many agencies to assist them produce or validate/correct spatial information and to determine the best way to capture and provide access to reliable information in a manner that is consistent with current principles for data management and spatial data infrastructure development. AMSIS is being developed as a decision support system to support regional marine planning, management of marine operations including regulation and enforcement of legislation, and industry development, especially in the identification of interests overlapping or adjacent to the annual offshore petroleum acreage releases. The project is being undertaken in close collaboration with key stakeholders including the Department of Industry, Tourism and Resources, The National Oceans Office and the Royal Australian Navy Hydrographic Service as well as other Australian Government departments that have interests in Australia's marine jurisdiction.

Keywords Marine GIS · Marine cadastre · Spatial data infrastructure · Web mapping

---

A.D. Nairn (✉)
Petroleum and Marine Division, Geoscience Australia, GPO Box 378, Canberra ACT 2601, Australia

D.R. Green (ed.), *Coastal and Marine Geospatial Technologies*, Coastal Systems and Continental Margins 13, DOI 10.1007/978-1-4020-9720-1_2,
© Springer Science+Business Media B.V. 2010

2.1 Introduction

2.1.1 Extent of Australia's Marine Jurisdiction

Australia's marine jurisdiction is large and diverse. Indeed much is still unknown about the physical environment and biodiversity. It has been said that we probably know more about the surface of Mars than we do about our own oceans. In fact some new submerged coral reefs have been discovered in the Gulf of Carpentaria as recently as April 2005 as part of work being undertaken by Geoscience Australia, the National Oceans Office and CSIRO to map bioregions of Australia. The extent of Australia's marine jurisdiction as submittd to the United Nations is shown in Fig. 2.1 below.

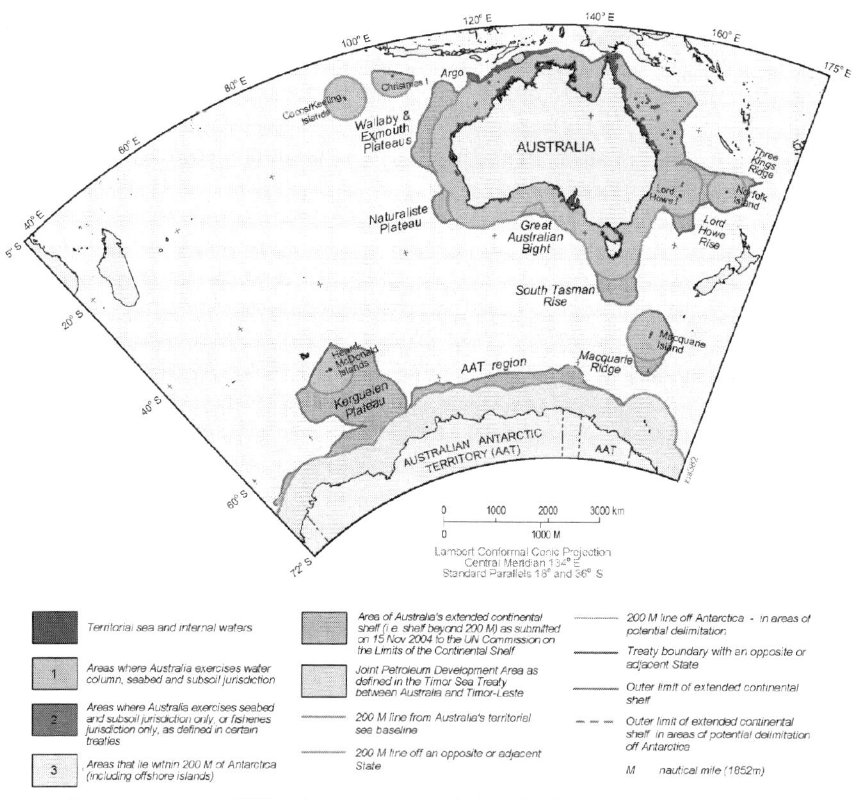

Fig. 2.1 Australia's marine jurisdiction (See also Plate 6 on Page 395 in Color Plate Section)

2.1.2 Project Drivers

There are a number of drivers that support the need for integrated spatial information. Under the United Nations Convention on the Law of the Sea (United Nations, 1983), countries are required to manage their marine jurisdictions and also to make information available. Domestically also there is a need to be able to map the existing interests that have been defined through legislation before embarking on activities such as regional marine planning and also the allocation of new rights such as marine reserves, acreage releases for petroleum exploration, defence training areas or fisheries. Also, when monitoring activities within the marine jurisdiction integrated accurate information on the location of various interests is critical. In some cases interception and prosecution of illegal activities is required, which also requires an accurate knowledge of boundaries and the ability to quickly locate and activity in relation to a particular boundary.

2.1.3 Marine Legislation

There is a range of legislation covering ocean use and ecosystem health. An audit of legislation covering the South East Marine Region of Australia by the National Oceans Office identified over one hundred pieces of Commonwealth legislation as well as a range of international treaties, conventions and customs (National Oceans Office, 2002).

Broadly classified, this legislation falls into a number of areas such as;

- Overarching marine regulation (eg UNCLOS, Seas and Submerged Lands Act, Offshore Constitutional Settlement)
- Shipping and related activities (eg Marine Pollution – MARPOL, Navigation Act)
- Indigenous interests (eg Native Title Act)
- Maritime security (Defence Act, Customs, Migration, Port Security)
- Environmental protection (Environmental Protection and Biodiversity Conservation Act, Sea Dumping Act)
- Living marine resources (Fisheries Acts)
- Seabed and subsoil activities (Petroleum and Submerged Lands Act, Offshore Minerals Act, Submarine Cables and Pipeline Protection Act, Historic Shipwrecks)

2.1.4 Project Aim

The AMSIS project aim initially is to:

- Support regional marine planning, industry development, policy development and operational requirements of the Australian Government by providing up to date, integrated marine spatial information & applications via Internet;

- Focus initially on Australian Government legal interests & infrastructure;
- Contribute towards the development of Marine Spatial Data Infrastructure (SDI) and link with other relevant initiatives.

2.2 Stakeholder Requirements and Applications

2.2.1 Identification of Stakeholders

A large part of the project initially was dedicated to meeting stakeholders and discussing their business needs for integrated spatial information. The selection of stakeholders was made by researching those that could contribute information (i.e. had some legislative responsibility) and agencies that could be potential users.

In Australia, the Australian government's responsibilities are defined in the Constitution and the government can legislate according to these powers. In general, State and territory governments' jurisdiction is limited can legislate out to the extent of Coastal waters, which extend three nautical miles from the Territorial Sea Baseline.

A decision was made to focus on the Australian government responsibilities for Phase one of AMSIS development. Integration with state systems would be considered as Phase two.

The agencies that were consulted with covered the following broad range of activities:

- Environmental protection and regional marine planning
- Fisheries management
- Industry development for offshore oil and gas exploration and offshore minerals
- Defence
- Customs and border protection
- Maritime safety
- Transport and port security
- Indigenous interests and native title
- Submarine cable and pipeline protection

2.2.2 Stakeholder Requirements and Applications

The analysis of the business requirements of stakeholders resulted in the identification of some common needs. These have been distilled down to:

- The need for up to date integrated spatial information available over the internet for use by management for policy formulation and decision making such

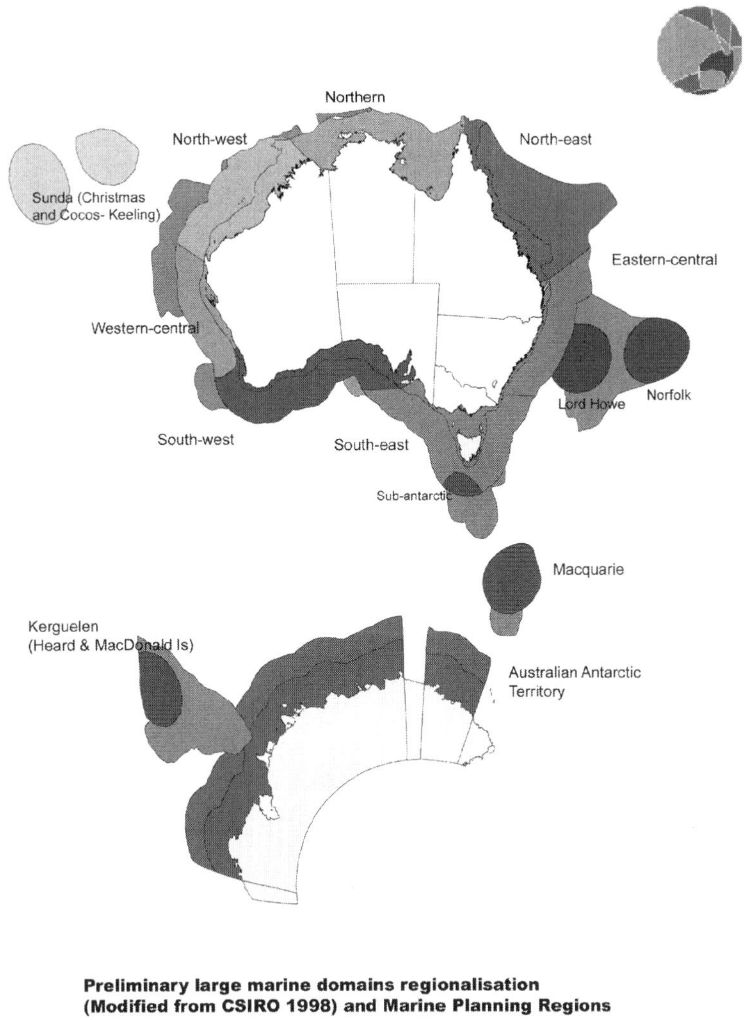

Fig. 2.2 Regional marine planning is being undertaken over a number of regions (See also Plate 7 on Page 396 in Color Plate Section)

as regional marine planning. Figure 2.2 shows the extents of marine planning regions where plans have been and are continuing to be developed (National Oceans Office, 2004)

- The need for access to quality spatial information for use in GIS systems;
- The ability to create maps for use in reports and for planning purposes;
- The need for specific applications. For example to determine the interests that exist within a defined area, and also and application that can calculate the distance from a nominated point to a nominated boundary. Figure 2.3 shows a typical diagram prepared to show the locations of a ship in relation to the Australian Fishing Zone.

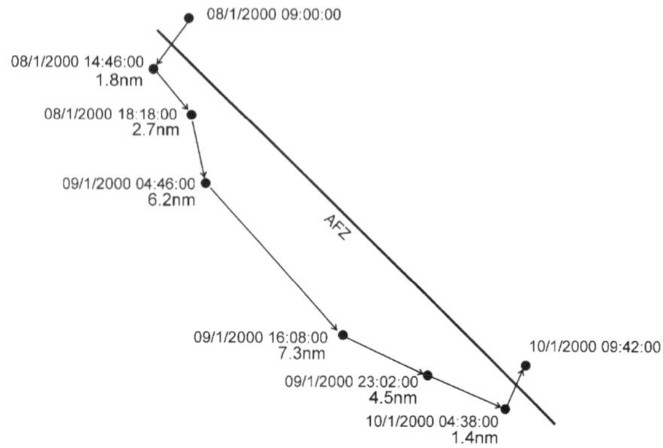

Fig. 2.3 A typical diagram showing the locations of a ship in relation to the Australian fishing zone

2.3 Data Identification

2.3.1 Priority Data for Inclusion in AMSIS

The priority data for AMSIS has been identified and is listed in Table 2.1 below. This data covers the majority of spatial boundaries defined in Commonwealth legislation.

Table 2.1 Priority data included in AMSIS

Administrative boundaries	Maritime limits (Coastal waters, territorial sea, contiguous zone, exclusive economic zone, extended continental shelf, coral sea limits). Territorial sea baselines and basepoints, treaty boundaries
	Petroleum and submerged land act boundary
	Offshore petroleum lease and offshore mineral lease boundaries
	Great barrier reef marine park boundary and planning zone boundaries
	World heritage areas. Commonwealth marine protected areas. Commonwealth national estate areas
	Commonwealth fisheries
	Indigeneous land use boundaries (agreements, native title determinations, native title applications)
	Australian search and rescue area boundary
	Defence firing practice and exercise areas
	Customs port limits, security port limits, immigration zone
Framework data	Shoreline, state borders, islands reefs, rocks, cays, shoals, seas
Bathymetry	Bathymetric image, isobaths
Coastal and offshore gazetteer	Cultural locations, land features, marine features
Anthropogenic features	Historic shipwrecks, ocean disposal sites
Transport	Ship reporting locations, derived shipping lanes, ferry routes
Infrastructure	Petroleum wells, platforms, pipelines, submarine cables, navigational aids
Geology	Seafloor features, sedimentary basins, tectonic elements
Environmental management	Bioregions. Marine planning regions

It is important that it will be continually revised as any new legislation is passed or boundaries are changed.

2.3.2 Data Audit

A data audit was undertaken to identify the availability and characteristics of spatial data that existed reflecting current legislation. This audit was conducted initially through online searching of directories such as the Australian Spatial Data Directory as well as the web sites of agencies with specific responsibility for the relevant legislation. Where insufficient information was available online, agencies were individually contacted.

The audit also obtained copies of datasets that we available for evaluation and these were used in an interim GIS.

The output from the audit was a report that consisted of metadata together with a list of issues for each dataset such as quality, format (analogue or digital), access constraints, and any other information thought necessary in terms of eventually publishing it on AMSIS.

2.3.3 Data Capture

Most of the data required has been obtained in digital GIS formats such as SHAPE or MID/MIF files and this was brought directly into AMSIS. In some instances transformation was required to bring all the data onto the WGS 84 datum that has been adopted for AMSIS. This datum was chosen to be compatible with the navigational charts produced by the Australian Hydrographic Service.

Geoscience Australia has worked with a number of agencies to define the spatial boundaries, to ensure they are in alignment with Australia's Marine Boundaries. These boundaries include Commonwealth fisheries, customs port limits, and security port limits. Figure 2.4 shows one of the 65 plans produced for the Australian Customs Service that defined these port limits.

Some spatial information was only available as coordinates in spreadsheets and in some cases the datum was undefined, so more research work was required to be confident that the information in the GIS was correct. Inconsistencies in boundary definitions have been found during the process of generating spatial data from legal descriptions and overlayed in the GIS and processes are now being developed to ensure that the definition of boundaries is more accurate. An assessment of the accuracy of the data was stored in feature level metadata in the database.

2.3.4 Development of the Data Model

A data model has been developed that identifies the attributes and relationships between data. As the organisation uses ESRI GIS environment together with Oracle, an ESRI personal geodatabase was initially created to test the model prior to loading into a corporate geodatabase model using SDE.

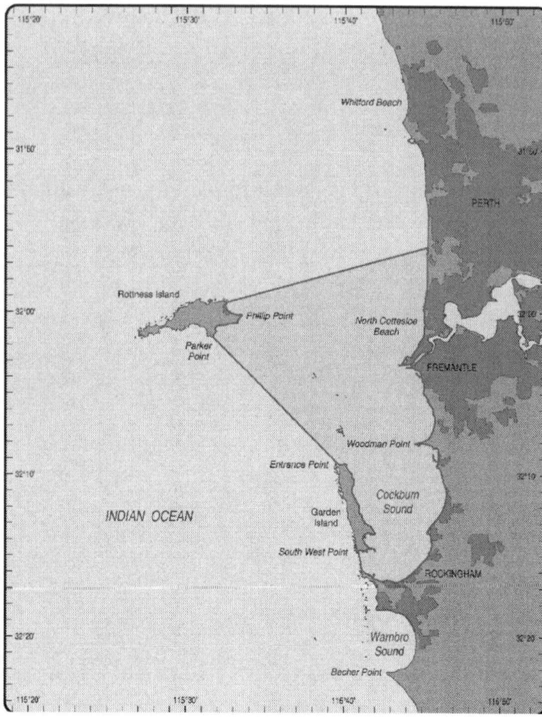

Fig. 2.4 Map produced to show port limits under the customs act – Fremantle, Western Australia (See also Plate 8 on Page 397 in Color Plate Section)

Some complexities within the data and the proposed application will affect the data model. For example Commonwealth Fisheries have the additional complexity that some areas within a fishery are subject to closures for particular times. If the end user of the system wants to know "active" fisheries easily, then queries will need to be able to search on a date field.

Whilst a more complex data model could have been considered to model many other types of marine data, the initial focus of this project is the marine boundaries element, and so a relatively straightforward model was adopted. The ArcGIS Marine Data Model (Wright et al., 2003) provides a good place to start for those considering modelling more complex marine data.

2.4 Application Development

An application was implemented for testing and demonstration purposes using ArcGIS. Figure 2.5 shows a screen from the prototype GIS showing overlapping interests and treaty boundaries in the Timor Sea. This prototype attracted significant interest from key stakeholders and provided something concrete on which to discuss requirements in more detail.

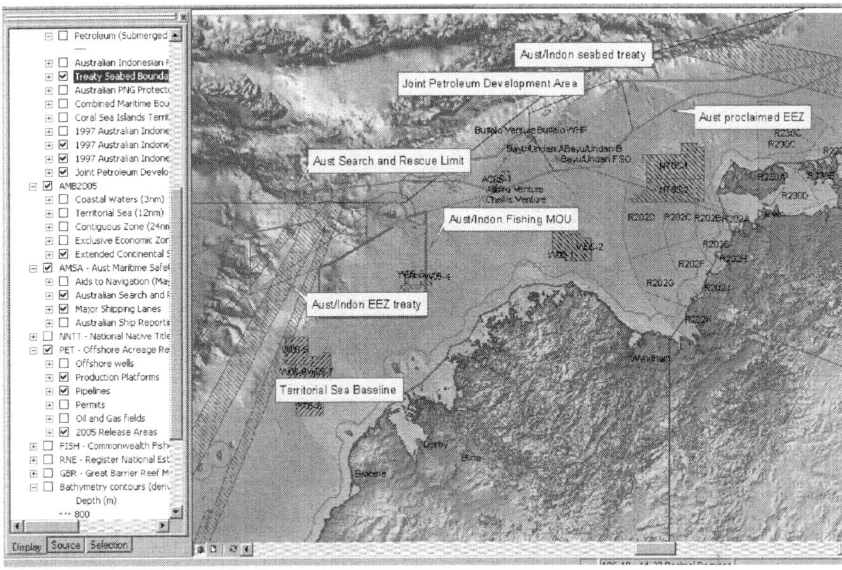

Fig. 2.5 Extent of overlapping interests and treaty boundaries in the Timor Sea (See also Plate 9 on Page 397 in Color Plate Section)

At the time of writing this chapter, the project is in the process of prototyping and evaluating the functionality of a number of web mapping systems against a list of requirements.

Experience to data has shown that as data is used in applications that require integrated information, problems and sometimes errors in the spatial data are encountered. A valuable outcome from this type of project is being able to understand these problems and work with the data custodians to reduce the uncertainties.

2.4.1 Functionality Requirement for the Web Application

The focus of the application will be to present high quality mapping functionality with functionality such as map composing – ability to select layers of information, window an area, zoom, pan, set a scale, measure and print a map (possibly via production of an A4 sized PDF). Also the ability to label features on the map based on attributes in the database is required. It is not initially necessary for the user to define symbology or linestyles etc. The ability to define "standard views" eg a standard view for petroleum industry is also seen as desirable, but not essential in first version of software.

The software system should support the OGC web map and at a later date, web feature functionality. The software architecture should ideally be able to access features from multiple corporate databases including SDE and SDO as the point of truth data is stored in both of these environments within our organisation (Fig. 2.6).

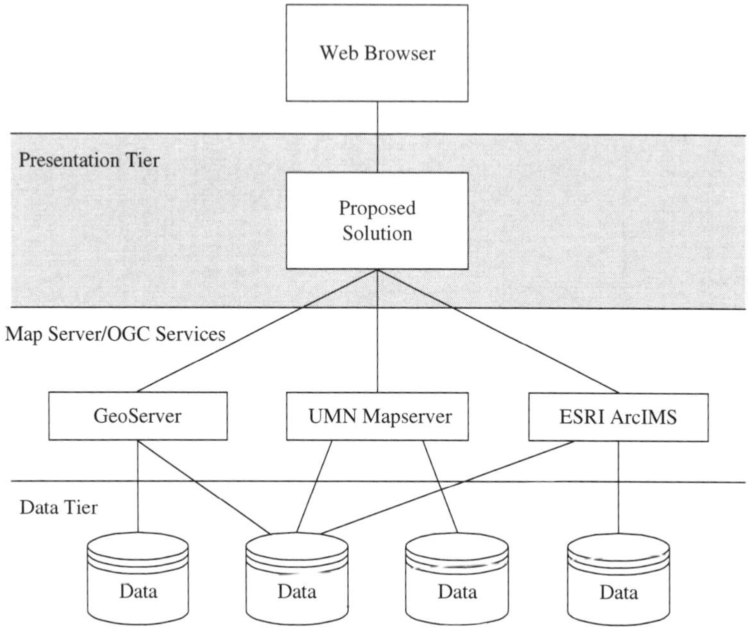

Fig. 2.6 Proposed web mapping architecture for AMSIS (See also Plate 10 on Page 398 in Color Plate Section)

In the longer term and subject to testing and being able to meet the functionality requirements, some data layers may be accessed directly from the custodian web sites using Open GIS standards.

The system will contain the ability for a user to register, and then make layers available for viewing based on the user profile. This is necessary as some more sensitive data (such as for example fish catch maps) is in some cases commercial in confidence.

Another requirement is for data download – the system should have a component for searching through the data layers, viewing metadata and linking to download sites, where the data licence allows this. In the case where the licence does not allow downloading, this should be conveyed to the user and a custodian contact provided. The link to download site or email address of contact will be stored in the data model. Geoscience Australia already has a data download and licence management facility and will offer use of this facility to other agencies.

A specific application that was of interest to many of the agencies consulted is the ability to determine the distance from a point to a nominated boundary. Also the ability to query the legal interests pertaining to a particular point or area is an application that will be developed.

The development team are also considering providing this application via a web service that can be accessed by another application run in another agency.

2.5 Other Project Linkages

There are a number of other information system initiatives underway within Australian Government agencies. The development of an "Oceans Portal" by the National Oceans Office has been commenced. This project aims to develop three distinct components, a web based portal, a marine catalogue, and a network of interoperable service and content providers. This development will be largely implementing Open Geospatial Consortium standards for the catalogue and the AMSIS project will contribute content and services.

Another project in the early stages of implementation is known as the Australian Oceanographic Data Centre Joint Facility that will also utilise the marine catalogue to improve access to a wide range of oceanographic data managed by a number of agencies.

2.6 Conclusion

The AMSIS project is developing data sets and services that will assist meeting the needs of Australian government agencies with responsibilities over the marine jurisdiction. By becoming an accepted source of integrated boundary information, AMSIS has the potential to improve spatial data definition and accessibility in an integrated online manner.

Care is being undertaken with the development to create linkages to other projects occurring and increasing use will be made of spatial data infrastructure principles as promoted by ANZLIC such as custodianship, data quality, metadata management, and web based interoperability standards such as catalogue services, web map and feature servers and gazetteers using OGC specifications.

References

ANZLIC web site http://www.anzlic.org.au/infrastructure_ASDI.html accessed June 2005.

National Oceans Office (2002). *Ocean Management, the legal framework. The South East Regional Marine Plan Assessment Reports*, The National Oceans Office, Hobart, Tasmania, Australia.

National Oceans Office (2004). *South-east Regional Marine Plan, Implementing Australia's Oceans Policy in the South-east Marine Region,* The National Oceans Office, Hobart, Tasmania, Australia.

United Nations (1983). *The Law of the Sea-United Nations Convention on the Law of the Sea,* United Nations.

Wright, D.J, Halpin, P.N., Blongewicz, M., Grise, S., and Breman, J. (2003). *ArcGIS Marine Data Model,* Source http://support.esri.com/index.cfm?fa=downloads.dataModels.filteredGateway &dmid=21

Chapter 3
Spatial Data Infrastructure in the Management of the Lagos Lagoon

P.C. Nwilo, D.A. Osanwuta, A.O. Onojeghuo, and O.O. Oni

Abstract The Lagos lagoon is the largest lagoon in West Africa and empties into the Atlantic Ocean through Lagos harbour. In recent years however, overpopulation, sand filling of the wetlands for property development and dumping of industrial and domestic wastes have adversely affected the health of the Lagos lagoon. This chapter discusses the need to establish SDI for the Lagos Lagoon stressing that the SDI will assist in planning and development in and around the lagoon. It emphasizes the need for standardization of datasets, development of a metadata bank and a clearinghouse.

Keywords Lagos lagoon · SDI

3.1 Introduction

The Lagos lagoon is one of the several lagoon systems in the West African sub region and the most extensive. It is part of the Barrier lagoon complex of Nigeria. The water is shallow and covers an area of about 208 km^2 (Ekundayo and Akpata, 1978). The average depth of the lagoon is generally about 1.5–3 m (Ibe, 1988), and made up of muddy and sandy bottom. The rivers of Ogun, Shasha, Oshun, Agboyi and Majidun, the Ogudu creeks and waters of Epe and Lekki lagoons feed the lagoon. The lagoon empties into the Atlantic Ocean through the entrance into the Lagos harbour. The southern margin of the Lagos lagoon is bounded by the Five Cowries Creek and on the eastern margin by the Palavar Island via Ikorodu. The lagoon is 50–60 km long and has two arms; one connects the Lekki Lagoon while the other leads northwards into the hinterland (Allen, 1965). The relief of the lagoon's bed is quite negligible.

P.C. Nwilo (✉)
Department of Surveying and Geoinformatics, University of Lagos, Lagos, Nigeria
e-mail: pcnwilo@yahoo.com

D.R. Green (ed.), *Coastal and Marine Geospatial Technologies*, Coastal Systems and Continental Margins 13, DOI 10.1007/978-1-4020-9720-1_3,

© Springer Science+Business Media B.V. 2010

The vegetation of the lagoon is characterised by mangrove, herbaceous, shrub plants and dotted by raffia palm, raffia sudancia, oil palm, Elaesis guineensis and coconut palms (Cocos nucifera). Floating plants like the water lettuce, pistia stratitotes and duckweed (Lemna sp) cover parts of the lagoon. After the rains, water hyacinth (Eichornia crassipes) invades the lagoon yearly between August and Octobers. Presently, a large portion of the vegetation has been destroyed as a result of sand filling, deforestation, agricultural activities and aquaculture.

The geomorphology of the lagoon is classified under the Barrier-Lagoon Complex, which extends for about 250 km from Nigeria/Benin Republic border to Aumo village. The complex consists of narrow ridges, which are aligned parallel to the coast. Adjourning the Lagos lagoon are six beaches namely Lagos, Lekki-Maiyegun, Kuramo, Badagry, Epe-Marina and Badagry-Marina beach (Dublin Green et al., 1997). The beach sediments vary from medium to coarse-grained sand.

With regards to the climatic condition of the area, it consists of rainy and dry seasons. The period of the rainy season is April to November, while that of the dry season is December to March. Between October and May, the lagoon surface temperatures ranges from 27 to 28°C while the range between June and October is 24–28°C.

The tide experienced in the Lagos Lagoon is semi-diurnal in nature. The lagoon is affected by the ocean surges, which periodically occur along the Lagos coastal areas. A great part of the Lagos Bar Beach (Kuramo) and Ahmadu Bello Way has at various times been cut off by the actions of the surges. The Federal Government of Nigeria has spent billions of naira in the beach nourishment and other beach preservative measures but the problem still persists. Tidal currents also transport the sediments brought into the Lagoon by the adjoining rivers.

The salinity of the lagoon is influenced by seasons, tides, floods, human and industrial pressures. The Lagos Lagoon is classified into fresh water, low, mid and high brackish waters with respect to salinity. A greater part of the Lagoon is tidal with salinity ranging from 10 to 24% during the dry season and zero during the rainy season due to the flooding from the rivers that drain into the lagoon and water arising directly from rainfall. The inflow of both sea and fresh water into the lagoon at different times determines the brackish environment of the lagoon and controls the ecology of the biota. With deforestation and farming activities in the hinterland, more sediments are being brought into the lagoon. Deposition occurs in five principal sub-environments and these are beach, channel and creek, lagoon delta and marginal swamp (Nummendal et al., 1987). A high percentage of these sediments are trapped on the bed of the lagoon, others enter into the Badagry Creeks or the nearby Lekki lagoon. Sediment mixing in the lagoon comes from the river floors and man-made water discharges into the lagoon. Figure 3.1 shows the Lagos lagoon, the creeks that empty into it and the adjoining islands.

3.2 Activities Within the Lagos Lagoon and Environs

Over the years, the population of Lagos has grown astronomically and this has had direct impact on the physical, chemical and biological characteristics of the lagoon.

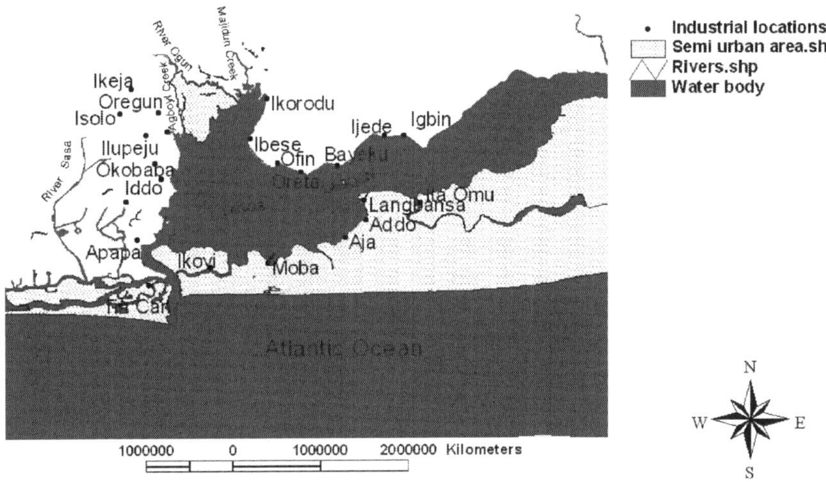

Fig. 3.1 Map of Lagos Lagoon with the rivers and creeks (See also Plate 11 on Page 398 in Color Plate Section)

The present state of the lagoon is determined largely by anthropogenic actions such as fishing, sand mining, pollution and transportation. Apart from these activities, the Lagoon is being used as a major dump for domestic and industrial effluents (Nwilo and Onuoha, 1993). It is important to note that the social life of the people of Lagos is tied to the lagoon.

3.2.1 Fishing

One of the most important contributions of the lagoon is source of protein through the fish production. Hence, fishing is the most popular activity on the Lagos lagoon. Most of the fishing is small scale and is mostly carried out by the tribe known as the "Ilajes". Several studies have been carried out on the fishing potentials of the Lagos lagoon. Kusemiju (1981) has documented the fisheries potentials of the lagoons in southwestern Nigeria. Fagade (1969) gave the marginal fishing area of this lagoon system to be 40–80 km^2. The Lagoon's productivity is of immense importance to the communities living along the lagoon. It serves as an important seaport and nursery ground for fisheries, a source of other natural shell fisheries and a water body used for recreational purposes. It provides a major source of fishes and economically important crustaceans and molluscs for inhabitants of Lagos. It has been reported to be the habitat for the early life history of the Banga – ethmalosa fimbriata and pink shrimp – panaeus diorarum (Ugwumba, 1984).

3.2.2 Sand Mining

Sand mining is another popular activity along the Lagos lagoon. Sand mining on the lagoon has destroyed the natural spawning grounds of many fishes, surface and ground waters and natural barriers against sea incursions. Due to the low-lying nature of Lagos, sea level rise often have adverse consequences on a large number of the estates around the lagoon. This has exacerbated the destruction of the wetlands. At present, there are serious flooding and erosion problems at the Victoria Island Beach. Sand mining, erosion and global rise in sea level may in future create easy connection between the lagoon and the Atlantic Ocean if unchecked (Uda et al., 1986).

3.2.3 Pollution

The Lagos lagoon has over the years grown to be a place for refuse disposal by the inhabitants of Lagos and its environs. Such effluences are from industries within Lagos state area and from sewage sites within the state. The lagoon also receives debris and waste oil carried by gutters, natural drainage and storm water. The pollution of the lagoon is by both private and government establishments. In 1993 alone, the Lagos City Council dumped 25.8 million litres of untreated sewage into the Lagoon. The Lagoon now serves as a major "sink" for various pollutants including lubricating oil and other petroleum products (Nwilo et al., 2000).

As a result of the above acts the aquatic life has been adversely affected hereby leading to the pollution of the nursery grounds of both fish and shrimps within the lagoon. There are other adverse effects on the aquatic fauna and flora as well as the bio-diversity within the Lagoon.

3.2.4 Storm and Ocean Surges

Storm and ocean surges have become a regular feature in the Lagos Lagoon. These storms actually result from the Atlantic Ocean of which are attributed to sea level rise and subsidence (Nwilo, 1995). Some of the resultant effects of storm surges include coastal erosion, flooding, inundation, destruction of roads and the beach, salinity intrusion into the lagoon and sources of water for domestic use (Nwilo, 1997).

3.3 Spatial Data Infrastructure

Spatial data infrastructure (SDI) provides a basis for spatial data discovery, evaluation and application for users and providers within all levels of government, the commercial sector, the non-profit sector, academia and citizens in general (SDI Cookbook, 2000). The SDI could also be described as an umbrella of policies, standards and procedures under which organizations and technologies interact to foster a more effective use, management and production of geospatial data (FGDC, 1997).

The SDI ensures the facilitation and coordination of spatial data exchange and sharing which when properly developed yields immense benefits. Amongst these benefits are: improved decision making by non-government, government and private individuals as it provides required data; adoption of an existing geospatial data standards; and availability of core data sets to the public through a Clearinghouse.

Having evaluated series of standards set up by different countries such as the US and Canada which have successfully embarked on SDI creation for managing coastal/marine environment one can then deduce that for an SDI to manage the Lagos lagoon the following components shall come to be, namely: data sources; standards; enabling technology and institutional policies.

3.3.1 Data Sources

One of the greatest challenges of setting up an SDI is the availability of data (i.e. geospatial information). The nature of the geospatial data in the marine/coastal region varies from that which is obtainable on land. Hydrography, which is the study of water as a whole shows that water, is dynamic in nature therefore making it really sensitive when it comes to issues of data acquisition. Operations such as bathymetric mapping, sounding operations to determine seabed depths, tide gauge determination, salinity determination are just a few of these operations. However the availability of such information is an issue of concern in Nigeria as there are different sources of data thereby showing the need for an SDI.

3.3.2 Accessibility of Geo-Information to the Public

In an SDI, the accessibility of relevant environmental/geospatial information is of great importance as it aids to improve the ecology of the lagoon by providing reliable data sets for use by decision makers in both the private and government sectors (Eagan et al., 1978). The ecology of the Lagos lagoon can however be improved by aggregating, evaluating, and increasing access to environmental/geospatial information as this is a major aim of an SDI that will manage its activities properly.

There are several reasons for the aggregation and dissemination of information pertaining to the Lagos lagoon:

- The pressure on the need to deliver public services has risen; hence there is now a need to deliver such information faster and economically.
- The need to engage public agencies and the entire public has proven to be beneficial, as they constitute part of the stakeholders in the SDI.
- There is a demand by public-interest groups for such environmental information (geoinformation) of the lagoon as it influences public policy development by the Government.

- The access to such environmental information about the lagoon to the public and private sectors will raise awareness and reduce the risk of pollution in the waters, as is the case of the Lagos lagoon.
- The availability of such information will give raise to better decision-making.

Presently the availability of geospatial information such as control points within the lagoon area, bathymetric and topographic data, salinity data and pollution sources/destinations are not readily available as there is no centralised data bank for the collection of such information. Alongside this challenge is the absence of standardised formats to be followed when acquisition of such data about the Lagos lagoon and its environs. There is therefore a need to take an inventory of all data types within the marine sector and sources of such data noted (this is the metadata). Also there is need to create a Clearinghouse. This will be discussed in the following sections of this chapter in more detail.

3.3.3 Metadata

The knowledge of data existence, contents, and fitness of an application are important aspects of data access and retrieval. Such knowledge is referred to as metadata, or data about information. Metadata describes the content, ancestry and source, quality, data base schema, and accuracy of data. Metadata support data sharing by providing information on many aspects of spatial data, each aspect having meaning in particular application contexts. Metadata that describe database contents include data dictionaries and definitions, attribute ranges, and data types. The origin or ancestry of data is critical for ascertaining the validity and suitability of data.

The first task to be undertaken in the establishment of an SDI is taking an inventory of the data available of which are the type and quality. It is normally recommended that a form be drawn to this effect that will aid in data collation of geospatial data sources and availability. The metadata could therefore be said to be a documentation of existing internal geospatial data resources within an organization. It permits structured search and composition of spatial data held by all parties involved in data acquisition. Consequently a metadata will provide all end-users adequate information that will be useful in whatever venture they desire to embark upon. The metadata will obviously consist of different geospatial information that will need to be properly harmonized for easy retrieval. It therefore provides framework for defining common data layers available in a geospatial database. An effective framework ought to feature representation of features from the highest to the least denominator. In managing the Lagos lagoon and its environs it is needful for the framework to involve stakeholders from key organizations. The framework should also support participants collecting, converting or associating information to common framework feature specification. The frameworks will therefore posses multiple representations of real world features at different scales and times by feature identifier and generalization. For the management of an SDI for Lagos lagoon for instance there will be a need to embark on continuous bathymetric data acquisition to determine the state of the sea bed at all times, alongside the salinity determination over a period of time for analysis.

In and around Lagos, there has been a problem of lack of homogeneity of the coordinate system. Coordinate transformation of all the controls will be of great importance as this will aid in data harmonization and standardization; analysis of topologic overlay services.

3.3.4 Creation of a Clearinghouse for SDI in Lagos Lagoon

The Clearinghouse is a means by which anyone could have access to digital spatial data through metadata. For instance, the United States FGDC has to its credit a sponsored programme called the Clearinghouse Activity (FGDC, 1997). The Clearinghouse activity is a decentralised system of servers located on the Internet that contain field-level descriptions of available digital spatial data. Clearinghouse uses readily available Web technology for the client side and uses the ANSI standard Z39.50 for the query, search, and presentation of search results to the Web client. On a broader perspective an SDI should have servers mounted at various clearinghouse node designation and a network for these servers established. The clearinghouse will then provide a uniform distributed search via the single user interface to all the servers in the other clearinghouse node locations.

Amongst the goals of the Clearinghouse will be a need to establish or develop procedures that will assist in the implementation of a discovery mechanism for digital geospatial data for the entire Lagos lagoon. The system when established will be such that the Clearinghouse has as one of its functionalities a catalogue service that has links to spatial data and graphics access through browsing the Internet. Such sites when established are encouraged to have hypertext linkages to spatial data and downloading provisions thereby allowing users interested in such information to have easy and fast access to such spatial information. However in cases where such spatial data, such as imageries, are too large for Internet availability there should be a linkage to an order form. Through the use of this order form the user can then send relevant personal information that will assist in delivery of the package to him at an agreed cost. The Clearinghouse metadata could therefore be said to be a low cost mode of advertisement for spatial data providers. The setting of a clearing house will minimise duplication in the collection of spatial data and create faster cooperation among users.

3.3.5 Standards

SDI comprises of underlying infrastructure that allows data to be shared between and within organisations, states, or countries. Such infrastructures that form the basis for SDI are different policies, standards and access networks. These policies and standards require the participation of stakeholders in policy creation and implementation, as they constitute the strength of the SDI. Standards specify regulations for data access, content, and exchange. This ensures the interoperability and integrateability of different datasets.

The International Standards Organization Technical Committee 211 (ISO TC/211) has recently developed a set of 40 Geographic Information related standards most of which are focused on terrestrial spatial data. In the marine environment the International Hydrographic Organization (IHO) in conjunction with the International Hydrographic Bureau have developed a transfer standard for digital hydrographic data (S-57) and are examining other standards for marine data (Williamson et al., 2004). Thee will be applied in the case of the Lagos lagoon.

3.3.6 Technology

This component of the SDI deals with the techniques behind the execution of an SDI. It comprises of the hardware, software, networks, database, and technical implementation plans of which are all embedded in Geographic Information System (GIS) technology.

Rhind (1989) defines the GIS as a computer system that can hold and use data describing places on the earth's surface. Burrough (1986) also defined GIS as a set of tools for collecting, storing, retrieving at will, transforming and displaying spatial data from the real world for a particular set of purposes. In general these definitions of GIS technology covers three main components of which are the hardware (i.e. the physical parts of the computer and associated peripherals – plotters and printers), software (the computer program and applications running on the computer), and appropriate procedures (or techniques and order of task implementation).

The means of spatial data acquisition is a vital issue in the SDI. Geomatics has brought to the "doorsteps" of professionals a faster, more accurate and reliable means of data acquisition. The most popular of these methods of data acquisition are Remote Sensing and Global Positioning System (GPS).

3.4 Acquisition and Sharing of Spatial Data

3.4.1 Data Acquisition

The management of the Lagos lagoon and its environs would require that a balance be reached upon in exploitation and conservation of its resources. This can be achieved by acquiring the right information, processing them for use by decision makers and making such geospatial data available. All these are embedded in the services rendered by an SDI when properly implemented and executed. The information needed to make decisions include bathymetry, topographic survey data, pollution, tidal variations, fish and fauna, mineral resources etc. It should also be observed that some of the data required are non-spatial thus going to illustrate their incorporation to the SDI.

GIS technology has as one of its major components data acquisition. Amongst the data sets to be processed for use in the GIS is a comprehensive base map of the entire Lagos lagoon. Also, maps of Lagos state environs will be useful for such a project. The bathymetric survey of the lagoon shall also be undertaken on a continuous basis as this will give researchers the status of the lagoon at all times. The Inland Waterways Authority could handle this operation as this falls under their jurisdiction and job description alongside with the Nigerian Navy. The Nigerian Ports Authority and the survey departments will handle some components of the data acquisition.

The use of the GPS devices will aid the acquisition of interested locations of certain activities such as saw milling, industrial waste disposal and other pollutions sites around the lagoon. Alongside the acquisition of these point locations, attributes of the locations are recorded, as the as the location data are being collected.

3.4.2 Sharing of Spatial Data

The main objective of a spatial data-sharing program is to increase benefits to society arising from the availability of spatial data. The benefits will accrue through the reduction of duplication of effort in collecting and maintaining of spatial data as well as through the increased use of this potentially valuable information. Due to the availability of spatially referenced data to the public there will definitely be improvements in data quality as professionals in the private sector could make suggestions resulting in value-added products and services. This will also foster a good environment for developing new business opportunities and enhancements of economic growth.

There are already in existence Data Sharing Programs in the United States of America and Canada. In the US, examples are found in the federal, state, and local levels (Department of Environment, 1987). A similar mode of data sharing should be encouraged here in Nigeria. The sharing of data could be in levels: the public and private sectors.

The data in question are of two types, the geographic base data and other spatial data. The base data are primary geographic spatial references that are produced to a recognised standard of accuracy and is subject to certified quality assurance programme. This data is produced by the Federal Government organization in this case being the Federal Ministry of Surveys, Nigeria. The Federal Surveys recently embarked on a nationwide project in which the entire base maps of Nigeria were digitised using AutoCAD and imported into a GIS environment, this is to serve as a data archive. The features represented in these maps included topography, vegetation cover, utilities, road network, etc. The other spatial data are available from a variety of producers whose standards for spatial accuracy may not meet certified quality. Although these data that are of less precision in locational control, they often contain valuable supplementary information.

3.4.3 Mechanism for Data Sharing Program Implementation

The data-sharing program being suggested for the managing of spatial data in the Lagos Lagoon will not have as its sole aim the dissemination of spatial data to users but awareness creation through the print, radio, and television media which shall be funded by the Government. But, its quality of import to the well being of Lagos lagoon will lie in the program being able to enhance participation of donors from the private sector, academia, and non-governmental organizations. The challenge of implementing this program will depend greatly on two factors:

i. Whether data sharing program will need a formal organizational structure?
ii. Is there sufficient fund to match the program?

Our response to the first question is that one of the parastatals of government such as the Inland Waterways Authority has to take the lead. As for funding, this is never enough but the lead agency has to provide some funding for the take off. The program should however be in line with the National Spatial Data Infrastructure Policy of the Federal Government of Nigeria. Guidelines or recommended regulations could also be drafted for how federally funded cost-shared projects and programs ought to make spatially referenced data available. Guidelines and legal conditions and liability limitations for organizations and agencies volunteering to deposit data under the program could also be documented.

The distribution policy for the spatial data sharing program would be defined for both the metadata base as well as the specific geographic data files. Costs associated with fulfilling the distribution of the data should be borne by the end user (consistent with current federal policy on data distribution). Guidelines for technical user support should also be specified under the distribution policy. The operation of the spatial data-sharing program will require some financial resources, but the bulk of the operational costs should be borne by the donors and recipients of data from the program.

The first few years of the implementation of such a program will undoubtedly expose many issues and difficulties, some of which may not be easy to resolve or reconcile. These difficulties should not be allowed to distract attention from the central theme of the initiative that a cooperative environment could greatly benefit the nation. Although some initial financial support from federal agencies may be required to initiate this program, it is believed that the benefits of this investment will greatly outweigh the initial costs.

3.5 Recommendations and Conclusion

3.5.1 Recommendations

In order to ensure a successful implementation of SDI for the Lagos lagoon, the following recommendations are made:

(i) There is a need a lead agency to coordinate the establishment of policies, strategies, and organizational structures that will assist in the collection, use and distribution of spatial data of the lagoon. The Inland Waterways Authority could play coordination role.
(ii) A Spatial Data Infrastructure policy for the Lagos Lagoon should be developed by a committee made up of various interest groups in the Lagos lagoon.

3.6 Conclusion

We have looked at the various issues pertaining to the Lagos lagoon including its bathymetry, activities on the lagoon and its strategic importance to the inhabitants of Lagos state and Nigeria as a whole. The lagoon provides the entrance into the sea ports of Lagos, which play very vital roles to the Nigerian economy.

Data/information are resident in different agencies such as the Inland Waterways Authority, the Lagos State Survey department, the Federal Surveys department, the Ports Authority, the Federal and State Ministries of Environment, the Nigerian Institute for Oceanography and Marine Research and the Private Sector. These data are not readily available to users and the standards vary from industry to industry and from provider to provider. Setting up of SDI will ensure that quality and accurate spatial data will be readily available for decision makers and users during planning and implementation of programs regarding the Lagos lagoon and its environs. It will also reduce cost and duplication of data collection.

References

Allen J.R. (1965). 'Late Quaternary of the Niger Delta, and Adjacent Areas: Sedimentary Environments and Lithofacies'. AAPG. V 49, PP 547–600.

Burrough, P.A. (1986). 'Principles of Geographic Information System for land Resource Assessment'. Clarendon Press, Oxford.

Department of Environment (1987). 'Handling Geographic Information'. Report of the Committee of Energy chaired by Lord Chorley, HMSO London.

Developing Spatial Data Infrastructures: The SDI Cookbook (2000). Ed by D. Nebert, Technical Working Group Chair, GSDI. Available on: http://www.gsdi.org/docs2004/Cookbook/cookbookV2.0.pdf

Dublin-Green, C.O., Awosika, and Ajao, E.A. (1997). 'Coastal Profile of Nigeria.' GOGLME Ecosystem Project.

Eagan, P.D., Weise, L.M., and Liebl, D.S. (2001). 'Public Access to Environmental Information': Available at http://www.books.nap.edu/catalog/6322.html.

Ekundayo and Akpata, 1978.

Fagade, S.O. (1969). Studies on the Biology of some Fishes and the Fisheries of Lagos Lagoon. Ph.D Thesis, University of Lagos. 385pp.

FGDC (1997). Geologic sub-Committee Progress Report. www.fgdc.com

Ibe, A.C. (1988). 'Coastline Erosion in Nigeria.' Ibadan University Press, Ibadan.

Kysemiju, K. (1981). The Hydrobiology and Fishes of Lekki Lagoon, Nigeria. Nigerian Journal of Natural Science 3 (1 & 2), Pages 135–196.

Nummendal, D., Pilkey, O.H., and Howard, J.D., (1987). Sea Level Fluctuations. Special Publication No. 41. Tulsa, Ok: Society of Economic Palaeontologist and Mineralogist, 267p.

Nwilo, P.C. (1995). 'Sea Level Variations and the Impacts along the Nigerian Coastal Areas'. PhD Thesis, Environmental Resources Unit, University of Salford, Salford, UK.

Nwilo, P.C., Peters K.O., and Badejo O.T. (2000). 'Development of a Lagos Lagoon Information System'. In: Environmental Review, Volume 3 Number 2, pp. 403–409.

Nwilo, P., and Onuoha, A. (1993). 'Environmental Impacts of Human Activities on the coastal areas of Nigeria.' In: Coastlines of Western Africa Ed. By L.F.Awosika & O. Mayoon. American Society of Engineers.

Nwilo, P.C. (1997). Managing the impacts of Storm Surges in Victoria Island. Destructive Water, Water – Caused Natural Disasters, their Abatement and Control. Proceedings of the Conference held at Anaheim, California. June 1996). IAHS Publication no. 239.

Rhind, D.W. (1989). Cartography Past, Present and Future. Elsevier Science and Technology.

Uda, T., Agemori, and Chijo, N. (1986). Beach Changes Caused by Offshore Dredging. Coastal Engineering in Japan, 29 215–226.

Ugwumba, A.O. (1984). The Biology of Ten Pounder Elops Lacerta (Val) in the Freshwater, Estuarine and Marine Environments. Ph.D Thesis University of Lagos, 297 pp.

Williamson I., Strain L., and Rajabifard, A. (2004). 'Spatial Data Infrastructure to Facilitate Coastal Zone Management.' In GIS for Coastal Management. Ed by Bartlet D. and Smith J., PP 1–15.

Chapter 4
Harmonising Marine Information Exchange in Ireland

Elizabeth O'Dea, Edward Dwyer, Valerie Cummins, Dídac Perales Í Giménez, and Declan Dunne

Abstract The Marine Irish Digital Atlas (MIDA) is a web-based resource which provides visualisation, information and data on Ireland's coastal and marine environment. The development from 2002 to 2005 highlighted a number of difficulties related to data identification, sourcing, acquisition and licensing costs. In Ireland, the MIDA is one of a number of web-based initiatives related to the public provision of marine related data and information, and as such, it follows a set of open standards in order to facilitate harmonisation between these systems. The experience gained is of relevance in the context of the development of the Irish Spatial Data Infrastructure.

Keywords Internet mapping · Data and information visualization · Coastal atlas · Spatial data infrastructures · Open standards

4.1 Introduction

The complexities of managing the diverse components of the coastal zone are reflected in the challenges of accessing coastal spatial data. Coastal zone management must incorporate input from a large quantity of assorted stakeholders, while integrating spatial data from a broad range of both terrestrial and marine organisations which hold data relevant to these various stakeholders. In Ireland these data owners include, but are not limited to, local and national government agencies, research institutions, industry, private companies and non-governmental organisations (Bartlett, 1999; Dwyer et al., 2003). The time-consuming process of data sourcing and acquisition from these data owners is complicated by added challenges such as the lack of data catalogues within organisations, nonexistent or poor-quality metadata and variations in data quality (Bartlett, 1999; McCormack, 2003; O'Dea

E. Dwyer (✉)
Coastal & Marine Resources Centre, University College Cork, Naval Base, Haulbowline, Cobh, County Cork, Ireland
e-mail: n.dwyer@ucc.ie

D.R. Green (ed.), *Coastal and Marine Geospatial Technologies*, Coastal Systems and Continental Margins 13, DOI 10.1007/978-1-4020-9720-1_4,
© Springer Science+Business Media B.V. 2010

et al., 2004a). Among recommendations from members of the coastal community is a stated need for tools to facilitate search and access to spatial data in order to better determine, evaluate and monitor what exists (Bartlett, 1999; Connolly et al., 2001; Dwyer et al., 2003).

The advances in web-based geographic information systems (GIS) over the last few years have revolutionised the way in which spatial data is accessed, in particular by making them available to a wider, non-specialist audience. It has also provided an opportunity for organisations to share their data holdings more easily with other professionals (Stachowicz, 2004). Some Irish government organisations, such as the Environmental Protection Agency (EPA) under the European Union's Water Framework Directive, are required to make data publicly available (Mills and O Riain, 2002). Other organisations, such as the Marine Institute and the Geological Survey of Ireland (GSI), share data through a web GIS in order to provide a public service or market their data. No matter what the purpose, web GIS developers and spatial data managers often run into similar challenges with various issues involved in displaying data on the web, such as variability in data quality, scale, data licencing and metadata (O'Dea et al., 2004a; Department of Environment, Food and Rural Affairs, 2002). When a web GIS displays spatial data from multiple data sources as opposed to from a single organisation, these issues are compounded by differences in data management practices, including the existence of or differences in data catalogues and standard specifications. Even if the technology is in place, the underpinning data issues must first be addressed in order for the technology to be useful (Bartlett et al., 2004).

In an effort to address the needs of the Irish coastal and marine professional community, the Coastal & Marine Resources Centre (CMRC) at University College Cork developed the Marine Irish Digital Atlas (MIDA). This interactive online tool, centred on a web GIS (http://mida.ucc.ie), was created in part to serve as a resource for professionals to view and source existing data from numerous data owners (Dwyer et al., 2003). In tandem with developing the technology behind the atlas, the process of collecting data for display in the MIDA provided the opportunity to gain a good understanding of the various data issues and needs specific to Ireland.

This chapter discusses the current situation in Ireland regarding the cataloguing and exchange of coastal and marine data in Ireland and summarises data and technology issues related to data sharing which the MIDA development team has encountered in its development. Finally Ireland's progress is discussed with regards to adopting standards as well as developing partnerships for creating distributed systems between key data holders.

4.2 Current Realities

The coast has always played an important role in Irish life historically, culturally and economically. With over 50% of the total population of 5.9 million (1.7 million in

Northern Ireland) living within 10 km of the coast (Northern Ireland Statistics and Research Agency and Central Statistics Office, 2008), the immense pressures on the coastal zone require adequate integrated coastal management in order to best sustain resources and the environment (Cummins et al.,2004; Devoy, 2000). Managers must have access to adequate data and information covering the diverse cross-section of coastal topics in order to make informed decisions.

In light of the historically poor accessibility of quality data a number of new approaches are being adopted to reshape how coastal and marine spatial data is shared. In the development of the MIDA a number of issues in regard to coastal and marine data sharing have been brought to light. These include the variability in data quality; the need for data cataloguing by data owners to improve data accessibility; and the adoption of standards to enable data sharing, including metadata and technology. There are no simple solutions to these issues; however raising awareness of their existence is an important step to making progress.

4.2.1 Data Access

While in some countries, such as the United States, base datasets are easy and free/inexpensive to acquire, in Ireland they are the first barriers one encounters when compiling data for a project. The Ordnance Survey of Ireland (OSI), the Ordnance Survey of Northern Ireland (OSNI) and the United Kingdom Hydrographic Office (UKHO) are responsible for Ireland's terrestrial and marine base data, and all charge for the licencing of their data as well as for datasets derived from their base maps. Base data such as coastline, bathymetry, digital terrain models, hydrographic chart data, roads, rivers and imagery can often be the most expensive acquisition cost in a project (O'Dea et al., 2004a). Access may also be difficult due to strategic and commercial factors that limit its availability (Bartlett, 1999). The MIDA encountered these when some owners were hesitant to share their data online due to the commercial nature of their data or the development of their own web GIS.

The vast number of data owners in Ireland is a significant factor which impinges on data accessibility. Administrative organisations responsible for Ireland's coastal zone include at least 45 local authorities such as county councils, urban district councils and port authorities (Bartlett, 1999). Numerous other organisations such as central government departments, state and semi-state organisations, non-governmental organisations and universities also work in the coastal zone, on the terrestrial and/or marine side of the coastal divide (Cummins et al., 2004; O'Dea et al., 2004a). The broad spectrum of coastal data owners across Ireland makes it challenging to learn who holds what. Data acquisition involves personal knowledge, word of mouth and time investment to contact the potential data sources to learn what exists since there are few resources in place to make data access easier (O'Dea et al., 2004a). However, some key data holders such as the Marine Institute and the Department of Communications, Energy and Natural Resources (DCENR) are aware that without appropriate data cataloguing systems, datasets get shelved and forgotten once a project is complete.

4.2.2 Data Quality

There is great variability in the quality of available coastal spatial data in Ireland (Bartlett, 1999; O'Dea et al., 2004a). Variations in scale are one contributing factor. The range of methods and base maps used for collecting data can lead to potential misinterpretation when data from different sources are displayed together. The cost of base data may affect the scale of base map used. The attributes within a dataset also impact quality. While some datasets have detailed tables with many fields, others contain very few, if any, attributes. Metadata quality is another significant issue. More often than not metadata does not exist for any given dataset (including base data), thus significantly reducing its value. If it does exist, quite often it is sparse and not in a standard format. Some Irish organisations do recognise the importance of metadata and efforts are under way to improve data documentation (O'Dea et al., 2004b).

The age of some base data, particularly some areas of UKHO Admiralty Charts, also affects the quality of data created from those base maps. Political issues are partly to blame. Since the Republic of Ireland gained its independence in 1922 the UKHO has only made efforts to update its Irish hydrographic charts on an ad-hoc basis, with ports getting the only regular attention. Because Ireland does not have its own hydrographic office, some sections of charts covering the coast are as old as when they were first mapped during the British Victorian period in 1840–1860 (Marine Institute 2009a). Added to the typical cartographic and scale factors that can affect data quality, the age issue is therefore significant if attempting to integrate coastal and terrestrial base maps for the entire Irish coastal zone. The current INFOMAR inshore marine survey, which is being carried out in partnership by the Marine Institute and the Geological Survey of Ireland, is addressing this shortcoming by surveying 26 bays and three coastal areas around the Republic (Geological Survey of Ireland and Marine Institute, 2008).

Identifying and sourcing other thematic coastal data are challenging and time-consuming activities, as previously discussed. When acquiring data for the MIDA, significant time was spent sourcing important national or island-wide datasets, in some cases taking months to locate. It was not uncommon to find that some desirable datasets expected to exist in GIS-ready format were not available, for example commercial, fishing and ferry ports. In these cases the CMRC undertook the process of generating the GIS datasets using all available sources of information (e.g. Internet, printed reports, personal knowledge). To acquire data for the MIDA once they were sourced required additional time due to issues such as the owner's data preparation time, the establishment of licence agreements and the determination of licencing costs (O'Dea et al., 2004a).

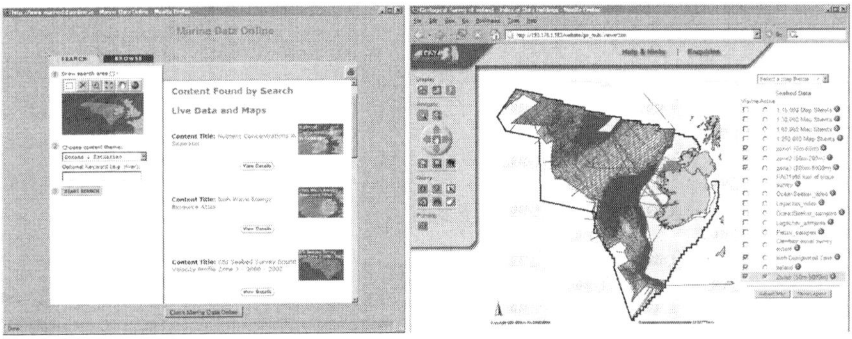

Fig. 4.1 The Marine Institute's marine data online portal (*left*), and the geological survey of Ireland's interactive web data delivery system (*right*)

4.2.3 Web Catalogue and Web GIS Efforts

Some organisations have developed or are developing web catalogue systems of their spatial data holdings which are either available via the Internet or an Intranet. The Marine Institute has been an early adopter of such systems and has developed Marine Data Online, a metadata database that enables users to search the Marine Institute's spatial data holdings (Fig. 4.1). In some cases the data can be displayed in a web GIS (Marine Institute, 2009b). The development of this system has provided a valuable opportunity for the Marine Institute to take stock of their holdings in a centralised system that benefits the organisation as well as the marine community as a whole. The EPA has created an online searchable metadata catalogue for data related to the Water Framework Directive, as well as the ENVision Internet site that provides public access to various EPA spatial data (EPA, 2009). The GSI have bypassed a catalogue and instead provide some of their terrestrial geological, Irish National Seabed and INFOMAR survey data in various web GIS (Fig. 4.1; Geological Survey of Ireland, 2009). The DCENR has also developed a couple of web GIS, one called the Geographic Exploration and Mining Services and another called the National Coastal Infrastructure Service, which is in partnership with a number of other organisations including the Marine Institute (Department of Communications, Energy and Natural Resources, 2009a and 2009b).

The MIDA was developed by the CMRC as a web portal to coastal and inshore marine spatial data and information in Ireland, with a web GIS at its core (Dwyer et al., 2003). While it is one of a number of efforts to bring Irish coastal and marine spatial data to the Internet, it is unique in that it displays data from a large number of data owners, ranging from government agencies to individuals. The atlas provides over 140 data layers and accompanying metadata from over thirty-five data owners (as of March 2009) as well as informational pages about various coastal topics and issues (Fig. 4.2).

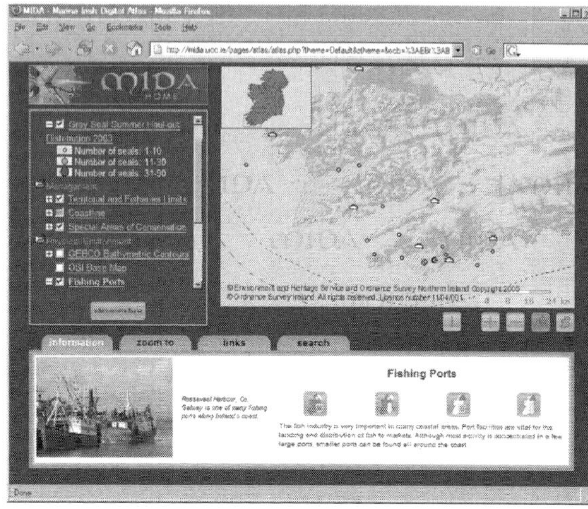

Fig. 4.2 The main atlas page of the Marine Irish Digital Atlas

Coastal and marine organisations in Ireland have chosen different products as their web mapping software to share their data online. Factors influencing the selection of software include the availability of financial and technological support within an organisation as well as the functionality and ease of use of the final product. A majority of organisations, including the Marine Institute, the GSI, DCENR and the EPA are currently using proprietary software for their web GIS efforts: ESRI's ArcIMS, which, while expensive, offers extensive functionality along with site creation and management tools. The MIDA uses open source software as its web mapping software: Minnesota MapServer, which is freely available and, while offering simple zoom and feature query functionality, is fully customisable with the appropriate technical resources.

4.2.4 Standards

There are currently no legal standards for data and metadata in Ireland. Some organisations, such as the Marine Institute, the EPA, the GSI and the CMRC have adopted various profiles of the ISO 19115 metadata standard, while other organisations are still in early development phases. In 2002, the Irish government began developing the Irish Spatial Data Infrastructure (ISDI; Department of the Environment, Heritage and Local Government, 2009). From the beginning, the ISDI has recognised the need to incorporate coastal and marine data (Bartlett et al., 2004). The development of the ISDI is closely aligned with implementation of the INSPIRE Directive for developing a European spatial data infrastructure. ISO 19115 is the core of the metadata standards of the ISDI and INSPIRE, therefore current coastal and marine metadata collection efforts are in line with these infrastructural developments (McCormack, 2004).

Technological standards have also been developed for web GIS. The popularity of web GIS from its inception has resulted in a wide selection of different software products on the market. All offer different levels of GIS functionality, web site design tools and capabilities. In response to this diversity of web GIS products, the Open Geospatial Consortium (OGC) developed the OGC Web Services as a standard for web mapping systems. This standard specifies the requirements to allow requests for geographic "resources" be implemented across the Web using platform-independent calls. The main OGC services include the Web Map Service (WMS), the Web Feature Service (WFS), the Web Coverage Service (WCS) and the Catalogue Service for the Web (CSW) (Open Geospatial Consortium, 2009). Despite the name it is not limited to open source software such as Minnesota MapServer. Some proprietary web GIS software developers have designed extensions that make their products OGC Web Service compliant, such as ESRI's ArcIMS OGC WMS Connector (Environmental Systems Research Institute, 2009). Thus current coastal and marine web GIS endeavours in Ireland have adopted technologies that meet these standards.

4.3 The Way Forward

While there have been many challenges in the past with coastal and marine data and their accessibility, some of the key data owners in Ireland are making efforts to improve the quality of their own data and metadata, adopt standards, catalogue their data holdings and make their data more accessible. Many of these activities are designed to tie in with known and predicted future developments in terms of standards and technology.

4.3.1 Metadata Standards

Both the ISDI and INSPIRE are works in progress and their development will be ongoing for the foreseeable future. Once the ISDI is in place it will provide, among other things, spatial data and metadata standards that government agencies will be required to implement. It is expected that the standards set out by the ISDI will also be adopted by outside organisations (McCormack, 2004). The establishment of the ISDI is part of the larger National Spatial Strategy for Ireland, established by the government, which also encompasses the development of an integrated coastal zone management strategy (Department of the Environment, Heritage and Local Government, 2002). Movements ahead in both of these areas on a national level are promising for the future of Irish integrated coastal zone management by improving data quality as well as data accessibility between diverse key data owners. Implementation of the ISDI will be a requirement of the INSPIRE Directive which has to be transposed into the national legislation of European Union member countries by May 15th, 2009.

The development of these standards by government and umbrella organisations is important, provided there are ample opportunities for input from members of the community. Empowering members of the community by considering their feedback in the final standards ensures a better chance of adoption once those standards are implemented. Both the ISDI and INSPIRE have made successful efforts in this regard. Key to the success of these initiatives is their adoption on the ground by those creating and collecting the data itself. The current trend among the various organisations in Ireland's coastal and marine community in adopting ISO 19115 as a metadata standard is encouraging.

4.3.2 Distributed Systems

Key coastal and marine data holders in Ireland who have developed web GIS are aware of the importance of using technology that is OGC Web Service compliant. While different coastal and marine organisations have implemented either proprietary or open source software as their web GIS for sharing their data, their software choice should not impact their ability to share data between each other across a distributed network. Having selected software that is OGC compliant, the sharing of metadata and data between these systems will be easier to implement.

The Irish Spatial Data Exchange (ISDE), funded by the Irish Government and developed by the Marine Institute, the DCENR, the EPA and the CMRC, is a distributed discovery service for spatial data residing in the online data catalogues of each participating organisation. The exchange architecture is distributed with a light central mediation service. The exchange is based around OGC and ISO standards, which allows it to be independent of any catalogue implementation technologies. Users can search the catalogues via a simple search interface (Marine Institute and partners, 2008). Other organisations holding spatial data of relevance to Ireland and implementing OGC and ISO standards are encouraged to participate. It is envisioned that future enhancements of the service will include the seamless display of partners' spatial data within each portal's web GIS. This will improve access to quality data by pointing web GIS to the data source instead of potentially outdated versions on individual servers.

4.4 Conclusions

Technology pertaining to web GIS and data catalogues is quickly evolving and there are pressures on organisations to keep up with these advances as well as meet the demands of the user community. It is important for organisations not to lose sight of the various data issues that must be addressed in tandem with focusing on the technology. Without quality spatial data and metadata to populate the tools developed, the technology is insufficient and will not meet the needs of the user community.

Experience from the development of the MIDA has shown that more work needs to be done within organisations themselves to address data and metadata issues. Continued efforts must be made to ensure that data and technological standards are properly adopted, data holdings are adequately catalogued and tools are in place to make data more accessible. The implementation of the ISDE, which provides seamless access to the metadata catalogues of a number of organisations, demonstrates awareness among key data owners in Ireland's coastal and marine community of how to improve data accessibility. It can act as a demonstration project for the implementation of a fully-fledged Irish Spatial Data Infrastructure.

References

Bartlett, D.J. (1999). Working on the Frontiers of Science: Applying GIS to the Coastal Zone, *Marine and Coastal Geographical Systems*, ed. Wright, D.J. and Bartlett, D.J., Taylor & Francis, London, pp. 11–24.

Bartlett, D., Longhorn, R., and Garriga, M.C. (2004). Marine and Coastal Data Infrastructures: a missing piece in the SDI puzzle?, *GSDI-7 Conference Proceedings*, Bangalore, India, February, viewed march 3, 2009, <http://gsdidocs.org/gsdiconf/GSDI-7/papers/FTmcg.pdf>.

Connolly, N., Buchanan, C., O'Connell, M., Cronin, M., O'Mahony, C., Sealy, H., Kay, D., and Buckley, S. (2001), *Assessment of Human Activity in the Coastal Zone,*Maritime Ireland/Wales INTERREG Report No. 9, Marine Institute, Dublin, viewed March 3, 2009 <http://www.marine.ie/home/publicationsdata/publications/Maritime+INTERREG+Series.htm>.

Cummins, V., O'Mahony, C., and Connolly, N. (2004). *Review of Integrated Coastal Zone Management and Principles of Best Practice*, The Heritage Council, viewed March 3, 2009 < http://cmrc.ucc.ie/publications/reports/Heritage_Council.pdf>.

Department of Communications, Energy and Natural Resources (2009a). *Geographic Exploration and Mining Services*, viewed March 3, 2009, <http://www.mineralsireland.ie/Online+Services/Interactive+Maps.htm >.

Department of Communications, Energy and Natural Resources (2009b). *National Coastal Infrastructure Service*, viewed March 3, 2009, <http://www.coastalinfo.ie/Home.asp>.

Department of the Environment, Heritage and Local Government (DEHLG) (2002). *National Spatial Strategy for Ireland: People, Places and Potential*, Stationery Office, Dublin, viewed March 3, 2009 <http://www.irishspatialstrategy.ie/>.

Department of Environment, Heritage and Local Government (DEHLG) (2009). *Irish Spatial Data Infrastructure Consultation Document,* viewed March 3, 2009, < http://www.irishspatialstrategy.ie/isdi/>.

Department of Environment, Food, and Rural Affairs (2002). *Delivering Integrated Marine Mapping for the UK: Report of DEFRA-funded workshop held at Church House*, London, 11 September 2002, compiled by F.L. Franklin, DEFRA, Burnham Laboratory.

Devoy, R.J.N. (2000). Climate warming and the links to the coast. *Inland Coastal Estuarine Waters* 3:4–5.

Dwyer, N, O'Dea, L, and Cummins, V (2003). The Marine Irish Digital Atlas: a Web Portal to Coastal and Marine Data in Ireland, *CoastGIS 2003 Conference Proceedings*, Genova, Italy, viewed March 3, 2009, <http://mida.ucc.ie/assets/documents/MIDA_CoastGIS2003.pdf>.

Environmental Protection Agency (2009). *ENVision, Maps at your fingertips*, viewed March 3, 2009, <http://maps.epa.ie/InternetMapViewer/mapviewer.aspx>.

Environmental Systems Research Institute (2009). *GIS Standards and IT Interoperability*, viewed march 3, 2009, <http://www.esri.com/software/standards/index.html>.

Geological Survey of Ireland (2009). *Interactive Web Data Delivery System*, viewed March 3, 2009, <http://www.gsiseabed.ie/>.

Geological Survey of Ireland and Marine Institute (2008). *INFOMAR, Integrated Mapping for the Sustainable Development of Ireland's Marine Resources*, viewed March 3, 2008, <http://www.infomar.ie/>.

Marine Institute (2009a). *Joint Irish Bathymetric Survey Project (JIBS)*, viewed March 3, 2009 <http://www.marine.ie/home/services/surveys/seabed/JIBS.htm>.

Marine Institute (2009b). *Marine Data Online*, viewed March 3, 2009 <http://www.marine-dataonline.ie/>.

Marine Institute and Partners (2008). *Irish Spatial Data Exchange*, viewed March 3, 2009 <http://www.isde.ie>.

McCormack, B. (2003). Irish Spatial Infrastructure. Proceedings from *GIS Ireland 2003*, Dublin, 14 October. Last Viewed July 3, 2009, <http://national.mobhaile.ie/Portals/45/ISDI-IRLOGI_2003.pdf>.

McCormack, B. (2004). Irish Spatial Data Infrastructure, *10th EC GI & GIS Workshop, ESDI State of the Art*, Warsaw, Poland, 23–25 June, viewed March 3, 2009, <http://www.ec-gis.org/Workshops/10ec-gis/papers/24june_mccormack.pdf>.

Mills, P., and O Riain, G. (2002). *Guidance on Information Management and Data Interchange between River Basin Management Systems and National Organisations*, Environmental Protection Agency.

Northern Ireland Statistics and Research Agency and Central Statisctics Office, Rebuplic of Ireland (2008). *Ireland north and South, A Statistical Profile*, NISRA, Regional Reporting and EU Programmes Branch, Belfast, CSO, Information Section Dublin, viewed March 3, 2009, <http://www.nisra.gov.uk/archive/uploads/publications/North_South_2008/North_South_2008.pdf>.

O'Dea, L., Cummins, V., and Dwyer, N (2004a). Developing an Informational Web Portal for Coastal Data in Ireland: Data Issues in the Marine Irish Digital Atlas. *Oceanology International London 2004 Conference Proceedings*, 18 March, London, England: pp.32–40, viewed March 3, 2009, <http://mida.ucc.ie/assets/documents/oceanology2004_mida.pdf>.

O'Dea, L., Dwyer, N., Cummins, V., Dunne, D., and Ó Tuama, É. (2004b). Improving Access to Coastal Information: Metadata in the Marine Irish Digital Atlas, *ECO-IMAGINE Conference Proceedings,* 13–15 May, Seville, Spain, viewed June 13, 2005, <http://mida.ucc.ie/assets/documents/ecoImagine_mida_paper.pdf>.

Open Geospatial Consortium (2009) *OpenGIS® Specifications,* viewed March 3, 2009, <http://www.opengeospatial.org/standards/is >.

Stachowicz, S. (2004). Geographical Data Sharing – Advantages of Web Based Technology to Local Government, *10th EC GI & GIS Workshop, ESDI State of the Art*, Warsaw, Poland, 23–25 June, viewed March 3, 2009, <http://www.ec-gis.org/Workshops/10ec-gis/papers/25june_stachowicz.pdf>.

Chapter 5
Creating Marine Spatial Data Infrastructure for the UK

Michael J. Osborne and John Pepper

Abstract The need for a Coastal Spatial Data Infrastructure (CSDI) as part of a national or regional framework has been recognized widely. However, as reported for example by Longhorn (2003), few nations have progressed significantly towards this goal. This chapter re-examines the implementation of CSDI and what it means to data custodians and end users. It provides a practical example of how the UK is creating a topographical baseline of data that includes inter-operable data from the Ordnance Survey (land) and the UK Hydrographic Office (sea) as part of achieving its own data infrastructure.

Keywords CSDI · MSDI · Spatial data infrastructure · SeaZone · Coastal zone

5.1 Introduction

The creation of true coastal spatial data infrastructure (CSDI) remains an elusive goal. This is not surprising, as many nations including the UK divide responsibility for land and marine mapping between different departments or agencies, with little reason historically for them to collaborate. The situation has been exacerbated by the mismatch in the purpose for which land and marine mapping is undertaken. On land, the aim is for accuracy and completeness, whilst in the marine environment, to support the creation of nautical charts, the aim is towards clarity of presentation and safety of navigation.

The paucity of marine data products for purposes other than for shipping and maritime leisure has left oceanographers and other marine scientists with no option but to use navigational charts as their only source of marine geographic information. Professional scientists, engineers and planners have been photocopying and digitising paper charts for many years. The lack of collaboration and source data,

Dr. M.J. Osborne (✉)
SeaZone Solutions Limited, Red Lion House, Bentley, Hampshire GU10 5HY, UK
e-mail: mike.osborne@SeaZone.com

D.R. Green (ed.), *Coastal and Marine Geospatial Technologies*, Coastal Systems and Continental Margins 13, DOI 10.1007/978-1-4020-9720-1_5,
© Springer Science+Business Media B.V. 2010

especially in open digital formats, has made the creation of a CSDI virtually impossible.

The corollary to this situation is that, for policy makers and regulators needing data for reasons such as shoreline protection and management, there has been little choice but for them create their own base maps, often with the financial support of central government departments that could have been more strategic in their thinking. The effort expended in undertaking basic survey work should have been spent to improve the mapping framework to which other datasets could be related. A combination of national agencies that have historically ignored the coastal zone and a lack of strategic thinking by government have left the UK with no coastal framework or spatial data infrastructure.

The need for baseline mapping to support applications such as engineering design, marine operations, situation awareness, marine and coastal zone planning, environmental modelling, flood protection and regulation has been around for a long time. As the marine environment comes under greater pressure and scrutiny, with the introduction of strategic environmental assessment and marine spatial planning, it is only recently that government is starting to properly address the call for better and consistent data. Through initiatives such as those described below, the UK may at last be starting the process of being able to deliver datasets that encompass both land and sea and support sustainable development and better management of resources.

5.2 ICZMap and SeaZone

The UK Government funded Integrated Coastal Zone Mapping data research project (www.ICZMap.com), which ended in July 2003, took data from Ordnance Survey (OS), the UK Hydrographic Office (UKHO) and British Geological Survey (BGS) to create an integrated dataset for three pilot areas. As described by Harrison et al. (2003), user feedback was overwhelmingly positive with a demand to roll out ICZMap nationally. However, it is now a private initiative by an end user that is doing so.

Around the time that the ICZMap pilot project was coming to a close, a commercial marine geographic data service called SeaZone was launched by Metoc plc and UKHO under a Licensed Partnering arrangement. The SeaZone service addresses the needs of professional scientists and engineers whose only source of data up to that point was (as described above) the navigational chart. SeaZone's vision is to create, improve and maintain true marine geographic information from source, thereby removing shortfalls, such as the shoal bias and generalisations, introduced to aid safety of navigation. As a consequence, SeaZone is producing data products supplied in open readable formats in a similar manner to that of land mapping agencies.

Of the follow up actions that came about as a result of the ICZMap pilot project two are of particular importance to the creation of CSDI for the UK. The first is the creation a common shoreline based on the OS MasterMap dataset. The production of the Mean High Water shoreline (Mean High Water Springs in Scotland) was completed in March 2005. The second was that the UKHO undertook to define the vertical reference framework needed to make the vertical adjustments. This project was completed in late 2006.

Picking up where the ICZMap pilot project finished, SeaZone is working closely with OS (and UKHO) to create and maintain digital marine datasets that are joined at the shore with land data. In this respect, the discrete integrated datasets, created as pilots and envisaged during the ICZMap pilot project, no longer seem like a good idea: why create two boundaries (5 km inland and 10 km offshore) when one (at the coast) will suffice? The key, therefore, to CSDI is to define a common boundary at the shore and to maintain a digital framework of topographically defined datasets for the land and sea that is harmonised across that boundary. With the shoreline and vertical reference issues being addressed, and with SeaZone's energy and commitment to creating marine geographic information, the UK is moving closer to achieving a national digital framework for both land and sea. Furthermore, SeaZone is looking to collaborate to ensure that this framework will join to similar frameworks that are being developed by the UK's European neighbours.

Of course, there are many challenges. One is to address the cartographic differences that exist at (and seaward) of the coastal boundary. Figure 5.1 demonstrates where the jetty shown is depicted differently by UKHO compared to that of OS (white). SeaZone has assumed (rightly) that the mapping of OS is more suitable inland. Bathymetry of course is the domain of hydrographic surveyors and SeaZone's new datasets (Fig. 5.2) are based on best available raw survey and other data, as it is recorded from vessels at sea, airborne survey and Earth observation, or captured from existing analogue survey sheets. Within the littoral zone, the situation is less clear. The fact that OS and UKHO define different shorelines (the UKHO uses Mean High Water Springs and Chart Datum, which approximates to Lowest Astronomical Tide) and differences in when and objects are captured, depicted and kept up to date means that there are large differences in the source datasets.

This is where government funded research, often to meet a specific sectoral goal, can come into play. This data, which – it could be argued – should have been collected by the mapping agencies in the first place, should now be made available to improve the UK's mapping of the coastal zone. With bringing together of these

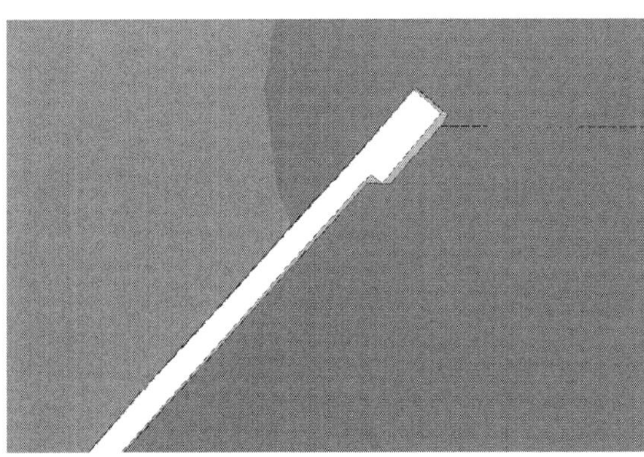

Fig. 5.1 Example of the cartographic differences being identified by SeaZone as follow up to the ICZMap pilot project (See also Plate 12 on Page 399 in Color Plate Section)

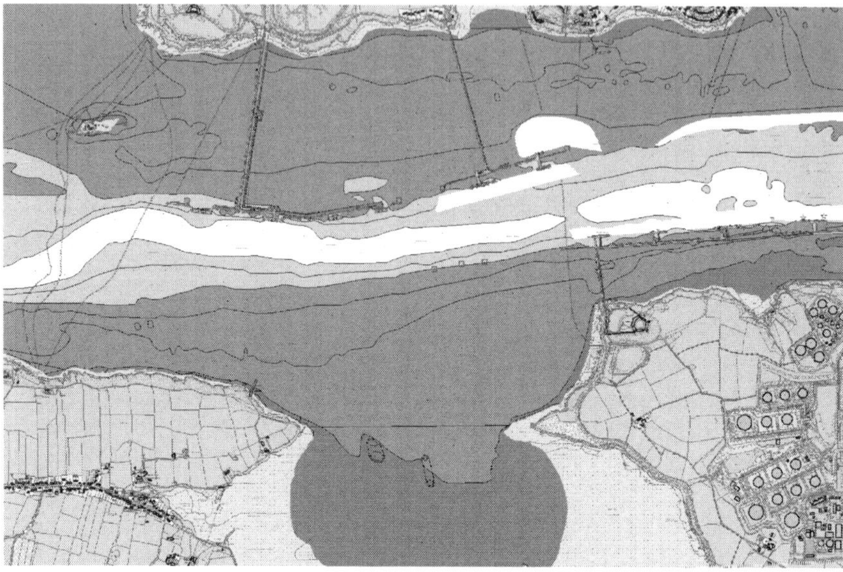

Fig. 5.2 Example of the joined marine geographic information created by SeaZone (land data courtesy of Ordnance Survey) (See also Plate 13 on Page 399 in Color Plate Section)

datasets in sound ways, the cartographic differences now seen can be investigated and a plan to allocate responsibility and improve these sources developed. The need for hydrographic and coastal data other than for navigation will mean surveying areas outside those used for shipping. This will require a major shift in emphasis, and major investment, tempered potentially by the use of lower cost technologies, such as LIDAR and Earth observation.

Through a better understanding of the vertical differences between Chart Datum and the local land datum, the marine geographic information is able to be adjusted, thereby creating a joined up 3-dimensional surface from land through to seabed. This has massive benefits for coastal zone flooding and shoreline management activities, as the work will allow extreme water levels, such as highest astronomical tide and surge levels, to be established. For the first time, the true effect of sea level changes through global climate change on our shorelines will be able to be determined.

5.3 Marine Data Information Partnership

The need to better define and manage marine data was recognised by the UK Government Department for the Environment and Rural Affairs (Defra) in its "Safeguarding our Seas" report published in 2002, as part of the drive to promote a more integrated ecosystem-based approach to marine management. With this as a starting point, the Inter-Agency Committee on Marine Science and Technology (IACMST) was asked to consider the case and to look at ways that marine environmental information could be better integrated.

In 2004, IACMST confirmed there would be benefit substantially from adopting a unified marine data management approach that encompasses the aspirations of e-Government and maintained compatibility with EU/international initiatives. IACMST produced a set of recommendations and a "Statement of Intent", which formed the basis and terms of reference for the UK's Marine Data and Information Partnership (MDIP) that was formally launched by Defra in March 2005.

MDIP has instigated a series of sub-groups that will work on specific aspects of marine data management and integration. One of these sub-groups is considering mapping and how real world objects are visualised. It is believed that this aspect of the work is vitally important, as it will define the geodetic and topographic framework to which other datasets can be related and referenced, and (if successful) the UK will have created a spatial data infrastructure that encompasses the coastal zone, not isolates it. MDIP was replaced by the Marine Environmental Data and Information Network (MEDIN) in Apr 2008 and the work that MDIP started continues.

5.4 Summary and Conclusions

The need for true marine mapping, interoperable with land mapping through a common agreed shoreline and forming the foundation of a CSDI, has never been greater, to support marine spatial planning, strategic environmental and sustainability assessment. Both the hydrographic and geographic communities have an opportunity to embrace this need, by helping to develop products that meet the aspirations of scientists and engineers, and to not promulgate the misuse of nautical charting products by continuing use them for a purpose for which they were never designed.

Agencies within the UK are starting the process of collaboration and are using the commercial sector to help in this process. Through initiatives such as SeaZone and the Marine Environmental Data and Information Network (MEDIN) the likelihood of the UK developing a spatial data infrastructure that encompasses the coastal zone is looking more certain.

References

Defra, (2002), "Safeguarding Our Seas – a Strategy for the Conservation and Sustainable Development of Our Marine Environment", Defra MWD, *Marine Stewardship Report*, 2002.

Harrison, M., James, C., Whitfield, M., Overton, D., Philpott, S., and Pepper, J. (2003), Integrated Mapping of the Coastal Zone of Britain: What Have We Learned and Where Do We Go From Here? *Fifth International Symposium on GIS and Computer Cartography for Coastal Zone Management*, 16th–18th October 2003 Palazzo Ducale, Genova, Italy.

Longhorn, R.A. (2003), Coastal Spatial Data Infrastructure as Part of National/regional SDI. *Fifth International Symposium on GIS and Computer Cartography for Coastal Zone Management*, 16th–18th October 2003 Palazzo Ducale, Genova, Italy.

Chapter 6
Unlocking the Marine Data Treasure Chest

John Pepper

Abstract The use of geospatial data in the marine environment is still a small "cottage" industry when compared to that on land, especially in the UK. This is partly due to a crucial lack of knowledge about what public or commercial sourced data exists, how to get hold of it, how much is digital and what it would cost. The current situation does not meet either Government expectations or that of data providers and users. External drivers such as the UK Government sponsored Marine Stewardship Report ("Safeguarding our Seas"), the EU Directives for ICZ Management and the Water Framework Directive are exerting increasing pressure to resolve the situation. Coupled with that is the growing need amongst data users for wider dissemination, digital outputs, data exchange agreements, licensing of data/information, interoperability and e-business. This presents an interesting challenge for organisations like the UKHO. This presentation explores how the UKHO and other data providers are unlocking the data "treasure chest" and the move towards the development of a marine geospatial data information framework (MSDI).

Keywords Geospatial data · Marine stewardship · UKHO · EU · Water framework directive

6.1 The Current Situation

A fast changing business environment means that demands placed upon organisations for information is growing. Changing user needs means the hunger for knowledge is placing a strain on traditional business models. Innovation is challenging the existing paradigm and moulding the future direction of the information business.

J. Pepper (✉)
Head of Commercial Development, The United Kingdom Hydrographic Office, Admiralty Way, Taunton TA1 2DN, Somerset, UK
e-mail: John.Pepper@ukho.gov.uk

D.R. Green (ed.), *Coastal and Marine Geospatial Technologies*, Coastal Systems and Continental Margins 13, DOI 10.1007/978-1-4020-9720-1_6,
© Crown Copyright and/or database rights. Reproduced by permission of the Controller of Her Majesty's Stationery Office and the UK Hydrographic Office (www.ukho.gov.uk) 2010

6.1.1 Where Are We Now?

Despite some progress to date, marine data is still held in a disparate manner, often incomplete and inconsistent in terms of capture, specification, storage and dissemination. In addition to this, awareness of what data exists is low compounded by a lack of how such data might assist in the decision making process amongst marine data users. Competitiveness between Government departments exists and a general lack of co-ordination of operations (no real joined-up approach) leads to a waste of resources and duplication of effort.

A further issue is the general lack of interoperability of data files. An example of this is data sets gathered with different horizontal and vertical datum (e.g. Ordnance Survey GB and UKHO). Indeed, until very recently, there was no agreed coastline for the UK.

6.1.2 Impetus from UK Government

Improved service delivery of public services via the e- Government initiative is beginning to place a need on data holders to make data more accessible. The review of HMSO Copyright, the creation of "fair trading" in 2000/2001 and the development of Public Private Partnerships (PPP) and Public Finance Initiatives are a few that spring to mind. In addition, the expectation of greater data exchange and sharing mechanisms, provision of metadata leading to greater Joined-up Government and Joined-up geography must not be overlooked. Other initiatives and Directives steering change includes the UK Government Marine Stewardship policy– "Safeguarding our Seas" which seeks to manage the whole marine environment on an ecosystem approach through long term stewardship and provides a two yearly review of progress through the "State of the Seas" report in the Department for Environment, Food and Rural Affairs (Defra). In addition the business models of some major public sector data providers is changing. This is evident at UKHO, the Environment Agency and the Natural Environmental Research Council (NERC).

6.1.3 Impetus from the European Union (EU)

The European Resolution for ICZ Management across the EU is promoting the need for each member state to have a demonstrable plan in place by 2006, which highlights how they will manage their littoral zone in future. The draft Directive for the creation of an Infrastructure for Spatial Information in Europe (INSPIRE) is also promoting greater interoperability across data themes (ranging from land mapping, hydrography, geology, oceanographic and meteorological data), metadata, network services and re use of public sector information. In addition, the EU Environmental Information Regulations (EIR), Freedom of Information (FOI) and the Public Sector Information (PSI) re-use Directive all impact. At the "working" level EU directives require Strategic Environmental Assessments for oil & gas and offshore wind farm developments; the Habitats and Birds Directive requires identification mapping and selection of Special Areas of Conservation (SACs) and Special Protection Areas (SPAs); the EU 6th Environmental Action Programme has a prime objective of halting the loss of biodiversity by 2010 and the Water Framework Directive (WFD) includes coastal areas and aims to achieve demanding water quality targets by 2015.

6.1.4 Changing Market Expectations

UK, in common with many other nations in the world are seeing a wide awakening of the value, marine geospatial data has the potential to play in decision making and the value and benefit attributed to that data provided it is provided in a format and to a specification that the user requires. This is amplified by the re purposing of UKHO "chart data" into a form that is beneficial for non-navigational use. Users are now seeking information superiority in a digital rather than analogue form and this has led to an upsurge in the licensed use of marine data/information over the past 2–3 years. Marine data is now at last being considered as part of the national geospatial infrastructure alongside terrestrial data. However, as we have seen in the provision of land mapping "solutions", bespoke outputs are required in order to meet users' needs.

6.1.5 The Challenges Ahead

There is a huge challenge globally in ensuring that as much data as possible is "unlocked" to allow the marine economy to develop and flourish and allow Governments to legislate with confidence rather than from a position of weakness. Building authoritative National Datasets is necessary but upon the paradigm of "Collect data once, use many times" rather than from ad hoc source material. Data holders need to create a framework allowing wider dissemination that does not hamper the users at the point of delivery whilst supporting the UK Government policy that "the user pays". Geospatial data providers must develop interoperability and/or harmonisation mechanisms between datasets (e.g ICZMap). Above all a significant culture change is necessary amongst the marine data communities to eliminate vested interests and sacred cows where data is concerned.

6.2 The UK Marine Spatial Data Infrastructure

6.2.1 What Should It Be?

An electronic discovery mechanism, knowledge base and transactional facility for marine geospatial information providing an underlying framework allowing data delivery via specific authoritative sources such as Central Government Departments and Agencies. In addition, links will be developed with Local Authorities and the commercial sector such as the oil and gas sector; architects, engineering and construction sector; the fishing industry.

It should allow people to make better decisions about using and protecting vital resources and allow extraction of data from diverse sources, blend it and come up with original perspectives and innovative solutions.

It should provide a metadata portal that will allow people to discover and access information about data and services easily.

6.2.2 Who Should Be Involved?

Professional, scientists, technologists across Government, commerce and Non Government Organisations (NGO's) involved the provision and use of data supporting:

- Habitat management
- ICZ management
- Renewable resource management
- Oil and Gas infrastructure
- Disaster management/emergency response
- Sovereignty and defence
- Estates and asset management
- Ocean and coastal monitoring
- Recreation and tourism
- Freshwater resource management
- Marine engineering works and services
- Education

...and many more

6.2.3 What Are the Benefits?

Such a capability connects private and public expertise and technology to stimulate development of new applications, providing interoperability at both organisational and data levels, provides metadata linkages, facilitates data access, exchange and trading (including the provision of network services) and in doing so meets the EU requirements for a more "joined up" approach. Such an infrastructure would support the Digital National Framework (DNF) for all geospatial information in the UK.

6.2.4 How to Make It Happen?

- Provide a framework for the UK marine data community with respect to data capture (includes ingestion), Interoperability, curation and dissemination.
- Develop, adopt and promote standards, specifications and procedures in support of such a framework.

- Contribute to the marine component of and support for the geospatial strategy for the UK.
- Harmonise stewardship and access to marine data and information to facilitate improved management of the seas around the UK.

6.3 Marine Data and Information Partnership

6.3.1 The Way Forward

Following the report commissioned by DEFRA "Marine Data and Information – Where to Now", an industry expert group produced a report entitled "Marine Data and Information Partnership: A Proposal". This document was presented to and adopted by the UK Government as a positive step forward and a programme of work has been subsequently developed. The Programme will be delivered in-line with the principles set out in the Partnership's "Statement of Intent" and the resulting framework will be developed so that it can be applied to Government departments, non-departmental public bodies, research institutes and the private sector. The overall output will be a national framework supported by a series of standards and procedures together with supporting guidance to help organisations and individuals adopt this more co-ordinated approach.

6.3.2 Programme Objectives

- To encourage a culture of co-ordinated data management and custodianship; clearly demonstrate the need for,and benefits of a co-ordinated data/information management framework
- Identify and deliver the key products/outputs needed to underpin such a framework
- Show through testing that these key products/outputs work
- Establish an implementation mechanism that enables government departments and other key data/information
- providers to adopt this approach

6.3.3 Programme Outputs

- Quantify the benefits of a national framework
- Identify key data/information providers
- Identify key data/information sets
- Identify organisations who can serve as national marine Data Archiving Centres (DACs)
- Develop a set of principles and standards for national marine metadata
- Develop a set of principles and standards for national marine DACs (including service level agreements)
- Develop a set of principles and standards for national marine mapping

- Develop a data/information management clause/term for use in contracts that involve the collection/provision/use of marine data & information
- Devise pilot/demonstration project(s)
- Develop a detailed business case and work-plan for pilot/product testing
- Devise & run a workshop to raise awareness, disseminate outputs & identify stakeholder issues
- Develop communications tools to inform the wider marine community and others about the Programme and its outputs
- Develop a comprehensive access point (Gateway) to UK marine data and information

6.3.4 The Role of Data Archiving Centres

- Database and bank with appropriate metadata standards
- Provide long term custodianship and updating of data sets
- Make datasets freely available wherever possible (not necessarily for low or zero cost)
- Raise awareness of the holdings
- Advise other organisations collecting similar types of data on procedures, and providing data-banking (warehousing) facilities for such similar data from other sources
- Promote the development of appropriate data products
- Exhibit evidence of expertise and a track record in the scientific area of the data
- Act as places of deposit for new marine data arising out of Government contracts.

6.3.5 The Benefits

- Centralised access to information.
- Increased efficiency at organisational and transactional levels.
- Data security.
- Improved communication.
- Reduced risk of obsolescence, isolation and support of multiple applications.
- Reduced costs (direct and indirect) associated with not sharing information at the individual, organisational, national and European level.
- Adherence to wider corporate, government or European initiatives.
- Public and commercial sector provision

6.3.6 The Barriers to Overcome

- Standards and Technology – much work has been undertaken in this area with the development of OGC compliant data formats and ISO 19XXX standards. However, the marine community needs to work to harmonise further to adopt generic standards wherever possible.

- Policy and Education – a lack of understanding restricts the achievement of wider interoperability. Organisational policy tends to be company rather than user centric which still results in:
- Domain silos giving rise to low levels of integration and quality.
- A lack of understanding of the real world needs for integrated and complex multidisciplinary solutions.
- A traditional approach to delivery rather than best practise.
- A lack of awareness and confidence data interoperability can bring in the medium and longer term.
- A lack of imaginative information management policies and strategies within organisations and across sectors.
- A perceived complexity of freedom of information and copyright laws and their application.
- Partnership and Coordination – there is a lack of understanding of the specific drivers for coordination. Differing commercial drivers in individual organisations leads to potential restrictions in any participation. Copyright and Intellectual Property Rights (IPR) legislation are interpreted in different ways by different organisations but above all there is a lack of an integrated approach to the needs of the users. However, existing examples of good partnership practise in the UK includes MAGIC, the Marine Biodiversity Network (MBN), Integrated Coastal Hydrography (ICH) and the ICZMap initiatives.

6.4 Conclusions

The creation of a Marine Data Infrastructure is a long-term initiative that will only succeed with the positive approach and good will of the stakeholders. An incremental approach to its creation is judged the best way forward in order that organisations can adapt to the obligations it places upon them. The first positive step along the road is the agreement of all relevant parties to act in partnership to deliver a mechanism to discover what data exists, where, its specification and terms of release but also links to those initiative specific sites that are already extant (e.g. UK DEAL). Only when this is in place can the partnership begin to consider the development of metadata catalogue services, creation of a geo portal data market and eventual conversion and downloading of information held at source by data owners. The result should be a framework under which customers can search for, assess the quality of, access and use data with confidence.

References

Cowling, M. (2005). Glasgow Marine Technology Centre, University of Glasgow, Marine Data and Information – *Where to Now*? January 2005.

DEFRA (2002). May 2002. Safeguarding Our Seas – *a Strategy for the Conservation and Sustainable Development of Our Marine Environment.*

DEFRA (2005). March 2005 'Charting Progress – *An Integrated Assessment of the State of Our Seas*'.

IACMST Expert Group (Cowling, M (IACMST); Greenaway, B (MWD Defra); Lawrence, B (BADC); Brown, J & Rickards, L (BODC); Guymer, T (IACMST); Morris, D & Harries, J (CEFAS); Rose, M (EA); Smallman, J & Millard, K (UKMIC); O'Carroll, K (DTI); Way, L (JNCC); Hancock, C (Office of e-Envoy); Pepper, J (UKHO); James, C (BGS); Maslen, J (Geowise); Easton, C (Scottish Executive), Spillard, R (MCA), Price, T (DIJE)) – January 2005. Marine Data and Information Partnership: *A Proposal, final draft report to DEFRA*.

Pepper, J. (2002). *Towards a Marine Geospatial Data Infrastructure v1.0*: October 2002.

Rose M, Gabriel M. and Jones N. (2004). Environment Agency: *Achieving Interoperability and Information Sharing*. March 2004.

Chapter 7
The European Marie Curie Project "ECO-IMAGINE"

Emanuele Roccatagliata and Giorgio Saio

Abstract ECO-IMAGINE is a group of eight conferences and training courses on ICM and GI/GIS. Four Working Groups has been created to address the main features in the field of ICM and GI: "Building Coastal Knowledge", "Coastal Governance, Planning and Design", "The Waterfront Management", "Geo-spatial Technologies". ECO-IMAGINE has two general conferences to introduce and summarise objectives and Working Groups, three thematic conferences and three Training Courses. An effective network building and generation of added value is ensured by the web-site that implements a Virtual Permanent Conference as discussion forum and reference thematic repository.

Keywords Training · Integrated coastal management · Network · Geo-information · Landscape

7.1 Introduction

Coastal areas play a vital role for most of the European countries, from the ecological, economical, social and cultural point of view. The high human pressure makes them very sensitive either for environmental or for social issues.

International and European Policies indicate Integrated Coastal Area Management (ICAM) as the main tool to reach the sustainable development of coastal areas.[1]

In so complex areas, a general information framework is essential for the elaboration of a management strategy. On the contrary, information is often available only according with the priority objectives, in particular in areas characterised by high

G. Saio (✉)
GISIG – GIS International Group, via Piacenza 54, 16138 Genova, Italy
e-mail: gisig@gisig.it

[1]According to the UNEP definition ICAM is a continuous, proactive and adaptive process of resource management for environmentally sustainable development in coastal areas and is not a substitute for sectoral planning, but focuses on the linkages between sectoral activities to achieve more comprehensive goals.

D.R. Green (ed.), *Coastal and Marine Geospatial Technologies*, Coastal Systems and Continental Margins 13, DOI 10.1007/978-1-4020-9720-1_7,
© Springer Science+Business Media B.V. 2010

human pressure or by profit-oriented objectives: data comparison and interoperability is in this way more difficult. Geo-Information is a key element of knowledge and GIS are actually useful tools to manage the Geo-Information.

The project aim is to enhance an effective forum for discussion among GI and Coastal Management scientists and operators, assuming a GIS based interdisciplinary approach for a structured knowledge of coastal landscape, able to effectively impact on coastal governance and planning.

7.2 The "ECO-IMAGINE" Project

ECO-IMAGINE – European Conferences and forum for Integrated coastal Management and Geo-INformation rEsearch is a EU funded project consisting in a group of eight events on the complementarity between Integrated Coastal Management (ICM) and Geo-information/GIS as support tool (Fig. 7.1) that took place between May 2004 and November 2007, within the 6th Framework Programme, as a Marie Curie Action. Its general objectives was fostering interdisciplinarity and building a structured knowledge in ICM through the use of GI technologies, under the paradigm of coastal landscape as unifying element. In this context, relevance has been given to planning and management of coastal landscape as the meeting point between natural and human features of the coast.

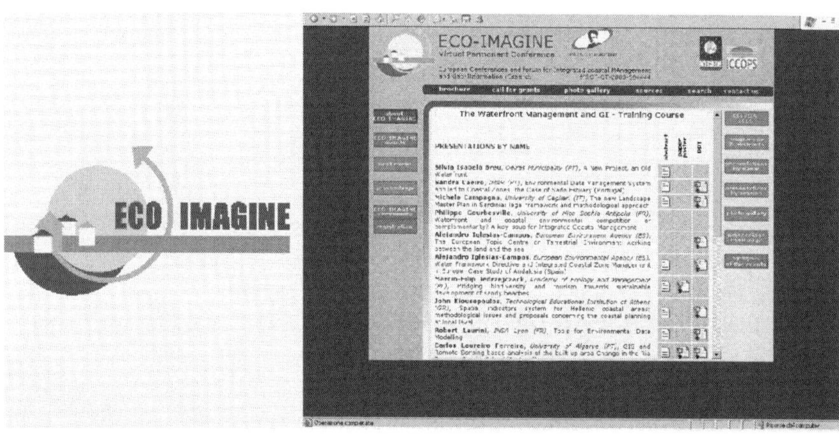

Fig. 7.1 The project logo (**a**) and a page of the virtual permanent conference (**b**) (See also Plate 14 on Page 399 in Color Plate Section)

7.3 Key Points

ECO-IMAGINE aims to integrate various thematic research activities related to ICM, under the paradigms of Geographical Information and coastal landscape, thus creating a comprehensive training environment. In particular, is to be stressed:

- the relevance of the coastal landscape, as unifying element since it is one of the most sensible evidence of the results of a management intervention
- the fostering of an interdisciplinary approach through the integration of the sectoral achievements and the possibility of sharing them
- the high scientific level ensured by the participation of established scientists
- the transfer of knowledge and principles from intergovernmental organisations and scientific environments to national and local realities
- the transfer of knowledge and experience from expert scientists to early stage researchers and to users, both in the public and in the private sector.

7.4 Objectives

ECO-IMAGINE targeted sharing experience in the fields of ICM and Information Technologies (IT), as well as the activation of new collaboration in the field. At an overall level, the initiative aimed at:

- collecting at the international level the expertise of scientists, enhancing the cross-sectoral opportunities supplied by GIS
- putting together academics, decision makers and experts, thus supporting a better decision making by authorities in charge for planning and control
- increasing the effective impact for thematic projects favouring investigations tuned to the decided policies
- disseminating the work of the participant structures and researcher by means of lectures, demonstrations and field work
- dealing with trans-boundaries issues
- creating a favourable environment for the establishment of an inter disciplinary forum for enhancing co-operation among different sectors
- training cross-cutting operational experts, able to include an interdisciplinary and science based approach in local management initiatives

7.5 Partners

The organizations that contributed to the project development are: GISIG – Geographical Information Systems International Group (IT), ICCOPS – Landscape Natural and Cultural Heritage Observatory (IT), University of Aberdeen – Centre for Marine and Coastal Zone Management (UK), University College of Cork – Coastal and Marine Resources Centre (IE), University of Nice – Sophia Antipolis (FR), University of Seville – Human Geography Department (ES), UNEP MAP PAP/RAC – Priority Actions Programme / Regional Activity Centre (HR).

7.6 Working Groups

Four Working Groups are foreseen, to address the main features in this field of ICM.

7.6.1 Building Coastal Knowledge and GI

A key issue for the practical implementation of ICM objectives is the availability and easy accessibility of appropriate data and information about coastal issues.

This implies the collection of data on a scientific base, the quick elaboration of raw data and the possibility to update them continuously; as well as the need to share the information and to make it suitable also for non expert final user.

In a landscape-based approach, information related with natural disciplines need to be integrated with data deriving from human sciences. Moreover an integrated approach to coastal area should deal with the stakeholders' needs. As a consequence, the perception of the area and of its problems, the hints for its management, historical and cultural issues contribute to create a comprehensive knowledge of the area.

7.6.2 Coastal Governance, Planning and Design and GI

The theme deals about the sustainable management of coastal resources, through the different phases of a ICM programme. On the governance level it includes the definition of the policies for the coastal area, the choices and the needs that lead to start an ICM programme, and its general objectives.

In the planning phase, strategies, specific objectives and the extension of the intervention are outlined, in an operational perspective. Integration among scientists, actors, planners and decision makers must be carefully set out to ensure the comprehensiveness of the following actions. Moreover, budget, timing, available skills, conflicting situation are key issues to discuss.

The strategies thus defined are implemented in the design phase through specific actions. The operational scale is often local and integration is required especially between the ICM promoter and local operators and communities.

GIS are a dynamic tool able to face complex and continuously evolving situations, such those characterising coasts, forecasting and showing the expected outputs and consequences of the different management options, and evaluating the effectiveness of the interventions and the impact of the management on the coastal area.

Finally, GIS are also a useful way to keep local communities in touch with operators and planners, by collecting and including local inputs, and supplying user friendly representations of the different options and the expected results of the management.

7.6.3 The Waterfront Management and GI

Waterfronts are often subject to high human pressure and support a multitude of uses. A lot of coastal area developed in a strict interdependence with the sea and their waterfronts are the evidence of their evolution, while other are trying now to ensure a harmonic development of this strip of land and sea. Natural features, economical, social, political, cultural and aesthetic issues are concentrated in these areas and are self evident in a very "dense" landscape.

Waterfronts are object of ICM programmes at different scales, from small architectonic projects to comprehensive planning interventions with expected impacts also in the organisation and in the socio-economic structure. They are areas subject to continuous changes due either to local or to global causes and their management should adapt to their development. GIS give the opportunity of managing the inputs provided by so many different actors, and are capable of adjustments of the programme and the creation of scenarios, according to the evolution of the situation.

7.6.4 Geo-Spatial Technologies

Technological features of GI are relevant for all the other themes and is instrumental to the implementation of ICM programmes. The ECO-IMAGINE initiative has been an opportunity, for Information Technologies experts, to tune up technologies and research on the actual needs of coastal management and to set up the bases for creating transferable models for shared information and data sets, and for ICM experts, to have a general view of the opportunities supplied by a specialised technology and to calibrate the future projects on these bases.

7.7 Events

Following there is the list of the convened events:

- Initial conference "GI and GIS for ICM", Seville (ES), 13th–15th May 2004,
- Thematic conference "Coastal Governance, Planning, Design and GI", Genova (IT), 16th–20th November 2004
- Training course "The Waterfront Management and GI", Lisbon (PT) 31st May–6th June 2005
- Training course: "Coastal Governance, Planning and Design and GI", Nice (FR) 21st–26th November 2005
- Thematic conference: "Building Coastal Knowledge and GI", Cork (IE) 13th–17th June 2006
- Thematic conference: "The Waterfront Management and GI", Genova (IT) 14th–18th November 2006
- Training course: "Geo-spatial Technologies", Aberdeen (UK), 23rd–28th July 2007
- Final conference: "Future Perspectives of GI for ICM", Genova (IT) 21st–23rd November 2007

The conferences, by means of background papers and presentations, introduced the general themes, focused on the issues and the state of the art relevant to each Working Group and synthesised the progress made and the eventual projects or collaborations arisen.

The training courses, instead, included tutorials, case studies, simulation exercises, and field trips in order to practically approach the themes discussed and to introduce to the use of GIS for ICM.

Within the Marie Curie financing scheme, some grants were awarded for each event to early stage and experienced researchers,[2] covering the travel and the accommodation costs. Grant holders, selected on the base of their curriculum, were asked to briefly present their work in the ICAM and GI field or to present a poster.

Moreover, from 2005 onward, in collaboration with the GIS software house ESRI, the ESRI Award was launched, a competition for the best GIS project proposed and for the best poster presented, offering the participants the possibility to win ESRI software and publications, and the occasion to present their own works and interests in specific dedicated sessions.

As a general rule, the choice of the speakers and the selection of the grant-holders were addressed to achieve the best integration and multidisciplinarity. Moreover, by involving representatives of managers, scientists, engineers and experts active in the field of coastal management and GI final users and researchers have been given the opportunity to interact with decision makers and experts.

7.8 Target Audience

The ECO-IMAGINE events have been addressed to:

- early stage researchers, that benefit from the knowledge experience of senior scientists, and directly establish contacts
- experienced researchers and scientists, as an opportunity to discuss their achievements in the perspective of an interdisciplinary approach.
- ICM operators and local administrators, as an opportunity of education, a series of best practice samples and the possibility of start collaborations
- representatives of NGOs and associations as an overview of the main coastal issues and an opportunity to interact with decision makers and scientists

7.9 Besides the ECO-IMAGINE Events

ECO-IMAGINE aimed also at creating a network of people and organisations working on Integrated Coastal Management (ICM) and Geo-information. Interested people have been invited to register to the ECO-IMAGINE Community, to be regularly informed about the ECO-IMAGINE events and ICAM relevant news and to have the opportunity to share experience with other research groups.

In order to strengthen the network, in 2005 the project Community was invited to join the SDIC GI-CLAN (Spatial Data Interest Community – Geo-Information Community in Coastal LANdscape), promoted to support the development and

---

[2] 226 grants were assigned to Early Stage researchers, more 80 grants to Experienced researchers.

implementation of INSPIRE through the voluntary involvement of stakeholders and operators in the field of GI for ICAM and coastal landscape.

At the end of 2007 about 480 people were registered to the ECO_IMAGINE Community, from 56 different countries.[3]

An effective tool to advertise the project and disseminate its outcomes has been the ECO-IMAGINE web site (www.gisig.it/eco-imagine), maintained also after the end of the project itself, above all to support the liveliness of the ECO-IMAGINE Community. Besides providing some general information about the project, it hosted and made available other relevant material, such as the conference announcements and the forms for the registration to the conference and ask for a grant. In particular, the proceedings of each event have been published in a dedicated section of the web site, called Virtual Permanent Conference (Fig. 7.1). Currently, the proceedings stored into the ECO-IMAGINE web site are 740 coming from all the 8 organised events[4]

---

[3]Albania, Algeria, Argentina, Belgium, Bulgaria, Brazil, Canada, Chile, China, Colombia, Croatia, Czech Republic, Denmark, Ecuador, Egypt, Estonia, Fiji, Finland, France, Germany, Ghana, Greece, Hungary, Ireland, India, Indonesia, Iran, Israel, Italy, Kazakhstan, Kenya, Lithuania, Malta, Mexico, Monaco, Morocco, Netherlands, Nigeria, Pakistan, Poland, Portugal, Russia, Senegal, Sierra Leone, Slovenia, Spain, Sweden, Syria, Taiwan, Tanzania, Tunisia, Turkey, Ukraine, United Kingdom, United States, Uruguay.

[4]343 abstracts, 227 ppt presentations, 110 posters, 60 project descriptions from the ESRI questionnaire.

Chapter 8
Coastal Ocean Information Network (ATLANTIC): From Concept to Reality: A Status Report

Andrew G. Sherin, Michael J.A. Butler, Claudette Leblanc, Randy Gillespie, and Norval Collins

Abstract The Atlantic Coastal Zone Information Steering Committee (ACZISC) has embarked on the first phase in the development of a geospatial data infrastructure in Atlantic Canada termed COINAtlantic. Many workshops have been held to date to refine the concept and to test it for acceptance with both potential users and data providers. A consensus was achieved at the third workshop to the effect that COINAtlantic would be an important component of an Integrated Coastal Management (ICM) Community of Practice in Atlantic Canada. The chapter provides a brief history of COIN, a summary of the results of the first three workshops and presents evidence of the technical and organizational readiness of the stakeholder community for COINAtlantic. This status report has been updated to document progress in the development of COINAtlantic to the beginning of 2009.

Keywords Integrated coastal management · Spatial data infrastructure · Community of practice · Coinatlantic · Results-based management

8.1 Introduction

The Atlantic Coastal Zone Information Steering Committee (ACZISC) is exploring the need and feasibility to establish a geospatial data infrastructure in Atlantic Canada termed COINAtlantic. In concept COINAtlantic will support the integrated management of coastal and ocean resources and increase the competitiveness of regional enterprises in both the private and public sector through more effective, efficient and quicker access to a wide range of quality data and information needed

---

A.G. Sherin (✉)
Natural Resources Canada, Earth Science Sector, Dartmouth, Nova Scotia, Canada
e-mail: sherin@nrcan.gc.ca

D.R. Green (ed.), *Coastal and Marine Geospatial Technologies*, Coastal Systems and Continental Margins 13, DOI 10.1007/978-1-4020-9720-1_8,
© Her Majesty the Queen in Right of Canada 2010

for Integrated Coastal Management (ICM). The ACZISC is an intergovernmental and intersectoral committee that provides a forum for collaboration in Atlantic Canada for the sharing of information in support of integrated coastal management. The ACZISC has been active since 1992 and is composed of representatives from both the federal and provincial levels of government in Atlantic Canada, and representatives from regional industry associations and academe. Their regular meetings are open to the wider coastal community and often involve community and non-government organizations. COINAtlantic is the most recent initiative of the ACZISC to realize its vision of "regional cooperation in Atlantic Canada with regards to coastal mapping, geomatics and integrated coastal management (ICM)."

8.2 A Brief History of COIN

The Coastal and Ocean Information Network (COIN) concept was formally documented by the Maritime Resource Management Service for the Canadian Hydrographic Service of the Canadian federal Department of Fisheries and Oceans (Butler et al., 1988). Between December 1988 and May 1989 workshops were held across Canada to discuss the COIN concept culminating in a national forum in Fredericton in June 1989. The initiative, renamed ICOIN (Inland Waters, Coastal and Ocean Information Network), was proposed to: (1) Provide rapid access to integrated information from government, industry and the academic sector for strategic and operational decision-making; (2) Implement standards that facilitate data sharing and exchange; and (3) Promote the development of a new industry based on information services, systems and value added products.

A 1991 article (Forrest 1991) reported on ICOIN and optimistically suggested that a prototype ICOIN would be in place within five years. Since then there has been substantial progress but the goal of overarching co-ordination and integration has been elusive.

Recognising that the needs identified in the originally conceived ICOIN concept have substantially increased and that the current technical and funding environment has made the concept more feasible, a concept chapter for COINAtlantic was submitted by CCMC, a member organisation of the ACZISC, to the Committee at their 41st meeting in Charlottetown P.E.I. in January 2004. The ACZISC established a Task Group to review the COINAtlantic proposal and make recommendations to the ACZISC on whether/how to move forward with the proposal. In June 2004 the Task Group recommended that the ACZISC pursue the development of the initiative. The Task Group subsequently organised two COINAtlantic workshops: the 1st held in September 2004 and the 2nd in January 2005. The results of these workshops are presented below. At a 3rd workshop held in May 2005 to initiate the development of a Community of Practice for Integrated Coastal Management, COINAtlantic was introduced as a catalyst. The results of the May 2005 workshop are also presented below.

8.3 COINAtlantic and the Canadian Geospatial Data Infrastructure (CGDI)

Since 2000, Natural Resources Canada, a department of the Canadian federal government, has led the GeoConnections program in the development of a Canadian Geospatial Data Infrastructure (CGDI) (http://www.geoconnections.org), in partnership with other federal agencies and levels of government, academe and industry. The GeoConnections program has five components: access, framework data, innovation, architecture, and supportive policy environments and partnerships. The Canadian federal government invested $60 million over five years for the first phase (2000–2005) which has resulted in a substantial overall investment ($150 million) and significant progress in and commitment to the development of the CGDI. In the February 2005 federal budget, the Government of Canada committed to invest a further $60 million over the next five years.

Gillespie et al. (2000) originally described the Marine Geospatial Data Infrastructure (MGDI), a marine-based initiative linked to the CGDI and built on the ICOIN concepts. The MGDI was endorsed by the GeoConnections Marine Advisory Committee, one of the advisory committees within the structure of the GeoConnections Program Advisory Network.

In the latter part of the 2000–2005 funding period for GeoConnections, the program changed its emphasis from building the CGDI to encouraging its use. It adopted the concept of Community of Practice to focus its efforts. The GeoConnections Program defines a Community of Practice as a community that shares common interests, has common requirements of the CGDI and will develop applications to meet the needs of the community. The ACZISC, and in particular the COINAtlantic initiative, is being developed to be consistent with the GeoConnections concept of Community of Practice. In addition COINAtlantic will adopt all standards endorsed by the CGDI and be compliant with the CGDI architecture.

8.4 Progress in the Establishment of COINAtlantic

The ACZISC has held two workshops to refine the COINAtlantic concept and to test it for acceptance with both potential users and data providers. The organisers used the Open Space Technology meeting facilitation techniques originated by Harrison Owen (1997).

The first workshop held in September 2004 defined COINAtlantic as a structure to access and share data and information and to provide open access to regional data and information within the CGDI. In addition, COINAtlantic would provide access to databases and metadata maintained by data owners and providers, provide access to web-based tools, and facilitate discussion. The workshop identified the users and their data and information needs. Particularly challenging were the needs identified for data and information on coastal and ocean usage, historical data and time series. In addition to the thematic data and information needs were performance requirements such as a reduction in time and complexity to find information, reduced access restrictions and the need to provide tools for the evaluation, integration and minimisation of inappropriate use of data and information.

Most of the challenges to sharing were organisational or attitudinal in nature and not technical. These included restrictive policies on use, licensing and pricing; on data owner/provider attitudes related to misuse or misinterpretation of data and information; and on loss of competitive advantage, both in an entrepreneurial and academic context. The cost of new technology and the need for training were identified as impediments for both the data providers and users. Legislation (i.e. intellectual property, privacy, and access to information) was also identified as a potential impediment.

The workshop identified a number of concerns with regards to quality assurance and liability. Of particular concern: would COINAtlantic deal with the responsibilities for quality assurance and any liability attached to determinations of data validity or data certification? These issues will clearly have to be studied in greater depth. There is a potential conflict between the safe "buyer beware" or "as is where is" approach for data discovered or accessed and the higher risk approach of independently developed information and tools for data quality or appropriate assessment which may in fact provide greater benefit to COINAtlantic users.

The second workshop was held in January 2005 to build on the work of the first workshop and to identify specific actions. The actions defined at this workshop focused upon the development of a vision and principles, the development of a business model and prototype, and the identification of the geographic scope and relationship with other initiatives. User requirements and technical and architecture issues were also discussed. The vision and principles suggested from these deliberations were: (1) COINAtlantic must be more then a portal, (2) COINAtlantic must address the different user communities with appropriate levels of access and security, (3) COINAtlantic must be "profitable" to all users, and (4) COINAtlantic must advocate increased access to all data for all people.

Several immediate actions were also identified:

1. Examine existing business models for similar initiatives and prepare recommendations;
2. Identify other groups with similar objectives and define a method of co-ordination with COINAtlantic;
3. Identify the potential data providers and their technical readiness to participate in COINAtlantic;
4. Organise a workshop to establish an Integrated Coastal Management (ICM) Community of Practice to provide "user pull" for COINAtlantic; and
5. Conduct an inventory of ICM initiatives in Atlantic Canada.

8.5 COINAtlantic Logic Model

At the forefront of all the discussions on COINAtlantic have been the benefits COINAtlantic would bring to users. Without significant benefits to users,

COINAtlantic will not attract the interest, investment or commitment to develop and sustain it. Several benefits were outlined at the first COINAtlantic workshop:

1. Data collected once and used many times;
2. Increased usage of data and information, therefore increased return on investment;
3. A structured and transparent distribution system for data and information;
4. Greater effectiveness in locating data;
5. A vehicle for contributing data and information;
6. Data and information validation, i.e. through metadata and documented user experience;
7. Increased competitive advantage for the private sector;
8. Availability of information for risk management and for "smart regulations";
9. Transparency of governance;
10. Sharing of ideas, best practices and lessons learned; and
11. Regional consistency in data and information used for decision-making.

How are these benefits to be realised? How do you ensure the right investments are made, the right activities are executed and the right outputs are delivered to better ensure success?

A technique in wide use within the business community and becoming more common in the public sector is the development of a logic model or a results-chain for a particular project or program. Table 8.1 outlines a draft logic model or results-chain for COINAtlantic, developed by the ACZISC's COINAtlantic Task Group after the second COINAtlantic workshop. A full logic model includes activities and resource inputs as well as performance indicators. These additional components have not yet been developed. The ACZISC is using this logic model to discuss the benefits and build consensus within the COINAtlantic stakeholder community.

8.6 COINAtlantic and an ICM Community of Practice

A strong and active Community of Practice is a logical and necessary mechanism to enable the societal benefits of Integrated Coastal Management (ICM). The ACZISC, although not using the term, has been working to build such a community for over 10 years. It has brought together two key sets of actors for ICM, the informatics / geomatics specialists and their issues, and the policy and management specialists and their issues. A Spatial Data Infrastructure (SDI) underpins the community as described in the logic model above. COINAtlantic could be one component of such an SDI. An important centre of growth for a Community of Practice is a set of communal resources that represents the knowledge capital of the community. Databases and related documentation represent the repository of this shared capital, although intangible aspects such as procedures, policies, rituals and specific idioms are also included (Wenger, 1998). For effective communication to take place certain

Table 8.1 Draft logic model for COINAtlantic

Final outcome	Intermediate outcome	Immediate outcome	Outputs
The long term expected results describes a social / economic benefit or change in behaviour	*The medium term expected results, showing changes in attitude or acceptance outside the direct control of COINAtlantic*	*The short term expected results under the control of COINAtlantic*	*The tangible products from COINAtlantic*
Increased confidence of coastal stakeholders in coastal management decisions through comprehensive accessible information	Coastal stakeholders acknowledge COINAtlantic as a valuable and definitive source of information	Coastal stakeholders are aware of the benefits COINAtlantic can provide	User requirement documentation User benefit package Prototype project
Increased competitiveness of Atlantic Canadian coastal-related enterprises through efficient and certain management regimes supported by comprehensive accessible information	Atlantic coastal related enterprises acknowledge the COINAtlantic contribution to increasing the efficiency and certainty of management regimes	Coastal managers in Atlantic Canada are committed to use a comprehensive definitive source of information to support their decision-making	User agreements Service level agreements Web-based services Co-ordination services
Increased national and international competitiveness of Atlantic Canadian marine geomatics enterprises	Atlantic marine geomatics enterprises increase their capacity through contributions to COINAtlantic development and maintenance	Atlantic Canadian marine geomatics enterprises are committed to invest in the development and maintenance of COINAtlantic	Business case Web based services Web-based tools Value-added services Project proposals

information must be shared; this information is called "common ground" (Clark and Brenna, 1991). The SDI is an approach to identify this "common ground" for a diverse and virtual Community of Practice, such as proposed for ICM.

In May 2005, the ACZISC conducted a workshop with the objective of explicitly establishing a Community of Practice for Integrated Coastal Management in Atlantic Canada. The ACZISC wanted to explore the interest in assembling ICM practitioners and interested individuals into a cohesive and mutually supportive ICM Community of Practice where they can share their knowledge and experience. The specific objectives of the community were stated as (1) sharing lessons learnt more rapidly with members; and (2) increased capacity to replicate best practices and lessons. The issues discussed included: (1) *Trust*: Trust is the "glue" that binds communities of practice but it takes time to develop; (2) *Leadership*: A Community of Practice can be implemented through the development of clearly defined plans and objectives, a "culture" of mutual support and by the building of links to other Communities of Practice; (3) *Communication*: Face-to-face communication is essential, especially in the early stages of building trust between members of the community. Other forms of communication could include: newsletters (e.g., the ACZISC Coastal Update); annual workshop(s); electronic discussion groups; and roundtable sharing of experiences (e.g., ACZISC meetings); (4) *Resources and funding*: Financial, technical and other resources may be necessary; and (5) *Technology*: Technology must meet user requirements, (e.g., the development of a database of application and simulation tools.)

In the context of infrastructure and technology, COINAtlantic was presented to the participants of the ACZISC workshop on a Community of Practice for ICM in Atlantic Canada. There was a consensus achieved at the workshop to the effect that COINAtlantic would be a key catalyst for the implementation of an ICM Community of Practice in Atlantic Canada.

8.7 Potential Technical Architecture for COINAtlantic

COINAtlantic, as already mentioned, will be consistent with the architecture and standards established by GeoConnections for the CGDI. It will not replace existing initiatives but fill in gaps and pull initiatives together. It will build upon the accomplishments and experience of organisations which currently provide leadership in Atlantic Canada. These include: (1) GeoNova, the Province of Nova Scotia's initiative to share and distribute data, a partner in the CGDI (http://www.geonova.ca); (2) the DFO GeoPortal and other data management, metadata and web publishing initiatives of the Canadian Federal Department of Fisheries and Oceans (also a partner of the CGDI); (3) the Canadian Geoscience Knowledge Network initiative of the Canadian National Geological Surveys Committee; (4) the Gulf of Maine Management Initiative (GOMMI) and (5) the SmartBay Demonstration project in Placentia Bay, Newfoundland (http://www.smartbay.ca).

The technical architecture will be consistent with that published by Gillespie et al. (2000) for the MGDI as shown in Fig. 8.1.

8.8 Technical Readiness for COINAtlantic

One measure of the technical readiness for COINAtlantic is the maturity of the private and public sector use of the technologies adopted by the CGDI. An approximation of this measure can be determined from the number, quality and size of Open Geospatial Consortium (OGC) compliant web mapping services accessible on the

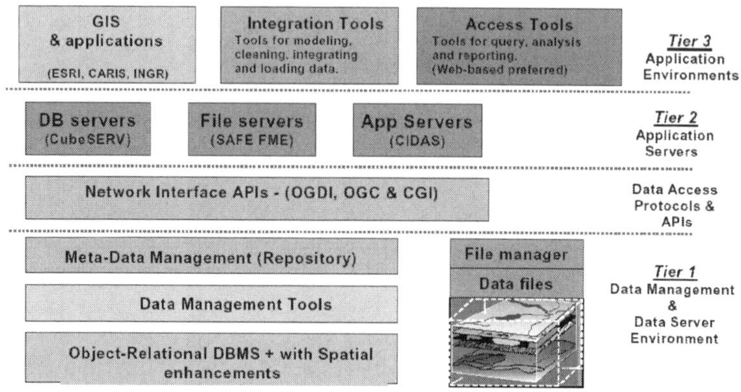

Fig. 8.1 Technical architecture of the marine geospatial data infrastructure (See also Plate 15 on Page 400 in Color Plate Section)

Table 8.2 Selected OGC compliant web mapping services relevant to COINAtlantic

OGC compliant server	URL
COINAtlantic utility and web mapping application	www.coinatlantic.ca
Province of Nova Scotia GeoNova Portal	www.gov.ns.ca/geonova/home/default.asp
DFO Web Map Service (GeoPortal)	public.geoportal-geoportail.gc.ca/index_en.html
Geoscience Data Repository	gdr.nrcan.gc.ca/index_e.php
Gulf of Maine Mapping Portal	www.gommap.org
Canadian Geoscience Knowledge Network	cgkn.net
Atlas of Canada	atlas.gc.ca
Natural Resources Canada Web Mapping Services (Topographic Information, Remote Sensing, Geographical Names)	www.ctis.nrcan.gc.ca/ www.ccrs.nrcan.gc.ca/ cgns.nrcan.gc.ca
NFIS – Forest Resources Web Map Service	nfis.org
Pockwock-Bowater Watershed Study (Environment Canada) – WMS	map.ns.ec.gc.ca/forest
Sackville High School NS, First Lake Ecology Project WMS	142.227.220.230/lakeone/
Climate Change (Atlantic Region) – WMS – Environment Canada	map.ns.ec.gc.ca/clchange
Eastern Prince Edward Island Clean River Quest WMS – Southeastern Environmental Association	www.seapei.ca/modules/articles/index.php?cat_id=1
ENVIRODAT – Newfoundland and Labrador Water Quality Agreement Database (WMS) – Environment Canada	http://map.ns.ec.gc.ca/envdat/map.aspx?service=WMS&version=1.1.1&request=GetCapabilities
Environmental Health Mapping Portal WMS – New Brunswick Lung Association	www.nb.lung.ca/mapping/
Storm Tracking – WMS – Environment Canada	www.atl.ec.gc.ca/msc/index_e.html
Wildlife Habitat Canada Citizen Science WMS	www.eman-rese.ca/

World Wide Web. Refraction Research prepared an inventory of discoverable Web Mapping Services (WMS) and Web Feature Services (WFS) (Anon. (2004)). They found 230 OGC compliant web servers world-wide on November 30, 2004. This has risen to 290 web servers when the inventory was updated on March 8, 2005. Of the servers discovered in November, 65 (25%) were from Canada. Several of these are publishing OGC compliant map information for themes of direct interest to the potential COINAtlantic. In Table 8.2 the servers of particular relevance to COINAtlantic are listed with their URLs. The DFO GeoPortal is of particular relevance since it publishes to the World Wide Web several hundred map layers both from the federal Department of Fisheries and Oceans as well as map layers from other organisations and servers. An image of a map produced from the COINAtlantic Web Mapping Utility is shown in Fig. 8.2. In addition, several servers identified in the survey are maintained by Canadian geomatics companies indicating an existing private sector capacity to contribute to the development of COINAtlantic.

The use of OGC compliant web mapping technologies by schools and non-governmental organisations, as shown by their inclusion in Table 8.2, demonstrates a widening regional acceptance of the technologies to meet organisational objectives of information dissemination.

In addition to the availability of web mapping services, another measure of technical readiness is the availability of metadata. The development of metadata relevant to the ICM community has been a focus of the work of the ACZISC since its inception in 1992. Initially the metadata was distributed as text files and paper inventories. This task was subsequently assumed by GeoConnections to make best use of its mandate to develop and maintain the developing standards and on-line technology. The metadata work of the GeoConnections Discovery Portal, Canada's national metadata is key to the success of the ACZISC COINAtlantic initiative.

The measures of availability of web services and metadata for Atlantic Canada would suggest that there is a technical readiness for COINAtlantic.

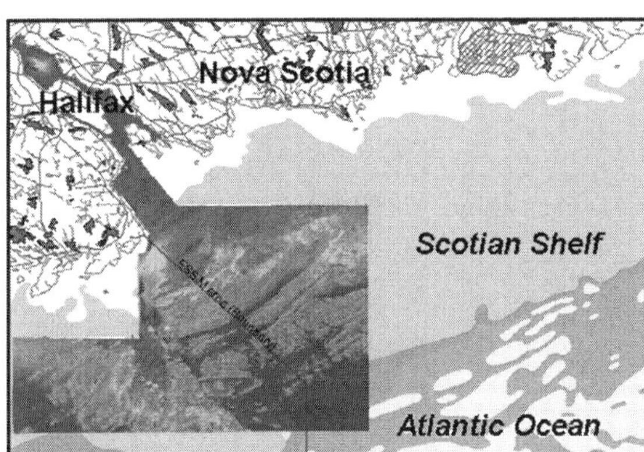

Fig. 8.2 Map generated by the COINAtlantic web mapping application (www.coinatlantic.ca) (See also Plate 16 on Page 400 in Color Plate Section)

Table 8.3 Geospatial collaborative success factors (after Johnson et al., 2001) and COINAtlantic status (updated to January 2009)

Success factor	Status	Comments
Broad support for vision and expectations	Significant progress	COINAtlantic development Phase 1 is underway with broad spectrum of collaborating organizations
Champion individuals/community support	Some progress	Multiple workshops have recognised the value of COINAtlantic but a Champion has not stepped forward
Knowledgeable, respected participants	Significant progress	ACZISC has endorsed, respected technical teams from several collaborators on side
Frequent contact with national (higher order) organisations	Significant progress	Commitment to national CGDI and Ocean Action Plan objectives; GeoConnections funding secured for Phase 1, advisory committee with national and North American members engaged
Proactive, open, and inclusive process/procedures to enable maximum participation/diverse perspectives	Significant progress	Open, inclusive consultations conducted, user support tools under development
Improved understanding/outreach	Undeveloped	Commitment in the vision and principles

8.9 COINAtlantic: Organisational Readiness

Johnson et al. (2001) published a study on six successful geospatial data collaboratives in the United States. All of these collaborations shared six characteristics that are shown in Table 8.3. The table also shows COINAtlantic's status with regards to these characteristics. The measurement of COINAtlantic against these attributes can be considered a measure of the organisational readiness for COINAtlantic.

8.10 Epilogue: COINAtlantic Progress up to January 2009

We have described the history of the COIN concept in Canada, the results of workshops to define COINAtlantic, the relationship between COINAtlantic and the Integrated Coastal and Ocean Management Community of Practice (COP) and the technical and organizational readiness of this COP for a joint enterprise like COINAtlantic.

In January 2008, the ACZISC entered into an agreement with GeoConnections for investment to develop Phase 1 of COINAtlantic with significant in-kind support of a number of major collaborators and has identified specific information services which will completed by June 2009. Through using modern information management and web-based search tools the ACZISC has established a COINAtlantic web mapping application that can be accessed from www.coinatlantic.ca

Discussion with regard to sustaining COINAtlantic was presented in Sherin (2007) and included the development of a sustainability plan, work with data providers to ensure the long term availability and reliability of web published data layers and the support and expansion of the COINAtlantic user community.

8.10.1 Governance to Build Regional and National Coherence

The governance structure for COINAtlantic is shown in Fig. 8.3. The legal accountability for COINAtlantic flows along the solid lines from the legal entity for the ACZISC, the ACZISC Association. The ACZISC Association is a not-for-profit association incorporated under the Nova Scotia Societies Act. The Association employs the Secretariat, one member of which chairs the COINAtlantic Management Committee that gives direction to the Project Manager.

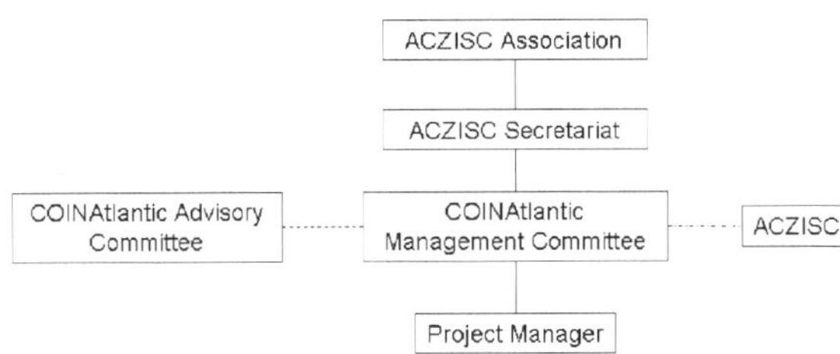

Fig. 8.3 COINAtlantic governance structure

Strong oversight is given by the full ACZISC Committee that represents key data providers and users with a mandate for coastal and ocean mandate. This Committee has been an effective organization bringing coherence to coastal and ocean information initiatives in Atlantic Canada for over seventeen years.

The development of further regional coherence is the objective of establishing the COINAtlantic Advisory Committee. In Atlantic Canada several similar coastal and ocean information infrastructures have been initiated and are at different stages of development. The COINAtlantic Advisory Committee is designed so that COINAtlantic development would benefit from the experience of these existing initiatives and encourage the sharing of information, tools and approaches.

On a national basis, proponents of the COINPacific initiative participate in the Advisory Committee. Canada's east and west coasts have substantial differences not only from a biophysical perspective but also from a socio-economic and political perspective. These differences make it less feasible for a single COIN solution, but participation in COINAtlantic development by west coast proponents will better ensure national coherence and provide a foundation for a national information system that supports coastal and ocean management.

The Government of British Columbia on behalf of COINPacific has provided all of the code for their web mapping application to COINAtlantic. This is a concrete demonstration of national collaboration. The west coast experience has been invaluable in the development of the COINAtlantic web mapping application which in January 2009 is available as a Beta version.

8.10.2 User Centred Design

User centered design (UCD) shifts the focus from the designer of the application/system/website to the user of that application, system, or website. The goal of UCD is to concurrently improve the usability of applications and the usefulness and utility of applications. This is achieved by making the purpose, scope and target audience clear to the user by integrating the user into the design and development process.

The ACZISC COINAtlantic Task Group held several workshops to solicit user and data provider requirements. A clear consensus identified the need for an effective search and access utility and not the development of "canned" applications. To continue this user driven focus, key stakeholders will be tasked with the user validation function for COINAtlantic. Members of the ACZISC will also play an important role in validating the usefulness of COINAtlantic. The step-wise details of user driven design that will be followed in the development of COINAtlantic are documented in GeoConnections (2006). COINAtlantic progress is discussed at thrice yearly ACZISC meeting. User support tools are scheduled to be rolled out in early 2009 along with a major next steps workshop.

8.10.3 Web Mapping Applications

As of January 2009 the first COINAtlantic web mapping application is available on-line from http://COINAtlantic.ca. Figure 8.2 shows the following web mapping services (WMS):

1. Geological Survey of Canada Canadian Marine Multibeam Bathymetric Data Web Map Service displaying layer: Halifax Harbour and Inner Shelf 1990–2002, Scotian Shelf, offshore eastern Canada;
2. Geological Survey of Canada Marine Geoscience Map Vectors Web Map Service displaying Regional Surficial Geology of the Scotian Shelf, Atlantic Ocean, Canada;
3. Canadian Department of Fisheries and Oceans Web Map Service displaying Eastern Scotian Shelf Integrated Management Boundary Line;
4. Bird Studies Canada and the Canadian Nature Federation Web Map Service displaying Canadian Important Bird Areas; and
5. The Province of Nova Scotia 1:500,000 Planimetric Web Map Service displaying coastline, roads, lakes and rivers.

8.10.4 The Future

A planned Next Steps workshop in early 2009, organised by the ACZISC, will provide user input and guidance to the management committee with regard to the future development of COINAtlantic. Within the context of the Canadian Geospatial Data Infrastructure (CGDI) and an anticipated national strategy for the collection, management and use of coastal and ocean data and information, the COINAtlantic utility will play an important role.

References

Anon. (2004). A Survey of OGC Deployment, *DigitalEarth.org collaborative geotech weblog*, http://digitalearth.org/story/2004/12/1/15658/1000 (May 30, 2005).

Butler, M.J.A., LeBlanc, C., and Stanley, J.M. (1988). *Coastal and Ocean Information Network (COIN)*. Report submitted to the Canadian Hydrographic Service, Department of Fisheries and Oceans by MRMS Inc., 85 p.

Clark, H., and Brennan, S E., (1991). "Grounding in Communication", In L. B. Resnick, J. M. Levine, and S. D. Teasley (eds.), *Perspectives on Socially Shared Cognition*. Washington, DC: American Psychological Association. pp. 127–149.

Forrest, D., (1991). New National GIS Network Gaining Support, *GIS World – GIS Canada*, http://www.innovationwatch.com/gisworldscan.1991.10.00.htm (May 29, 2005).

Gillespie, R., Butler, M.J.A, Anderson, N., Kucera, H., and LeBlanc, C. (2000). "MGDI: An Information Infrastructure to Support Integrated Coastal Management in Canada", *GeoCoast*, vol. 1, no. 1, pp. 15–24.

Johnson, R., Nedovic-Budic, Z., and Covert, K., (2001). *Lessons from practice A guidebook to organizing and sustaining geodata collaboratives*, Geodata Alliance, Reston, Virginia.

Owen, H., (1997). *Open Space Technology: A User's Guide, 2nd Edition*, Berrett-Koehler Publishers : San Francisco.

Sherin, A.G. (2007) COINAtlantic: Planning for a Sustainable Future. In: *Proceedings: CoastGIS 2007* (Santander : October 2007) http://www.gioc.unican.es/CoastGIS07/files/Sherin_CoastGIS07_paper.doc (January, 11, 2009).

Wenger E., (1998). *Communities of Practice: Learning, Meaning and Identity*. Cambridge University Press.

Chapter 9
A National Database on Coastal Dunes: Emilia-Romagna and Southern Veneto Littorals (Italy)

Umberto Simeoni, Edi Valpreda, and Corinne Corbau

Abstract This chapter describes the role of the coastal dunes in the context of coastal erosion and presents the possibility of making complex analyses on coastal dunes presence and state using a geodatabase, developed within a national research program. Data and parameters, such as littoral uses, defenses and coastal dune features, have been organized in the same geodatabase structure. This national homogeneous approach makes this project unique for Southern Mediterranean coasts. Applying this approach to data gathered from the Emilia-Romagna and Southern Veneto regions highlighted the complex analyses possibilities and their implication for coastal management and monitoring.

Keywords Foredunes · Beach erosion · Northern adriatic · GIS · Metadata

9.1 Introduction

This chapter analyses the role and significance of the coastal dunes within the coastal evolution analysis and monitoring. It also describes the results obtained by a national research financed by the MIUR (the Italian National Ministry for Education, University and Research) on the coastal dunes inside the beach-dunes coastal systems dynamic.

The role of the foredunes inside sandy beach environment is well known. In sedimentological and morphological coastal system, the coastal dunes are found above the high water level marks of the sandy beaches. The foredune environments play a dynamic role in coastal evolution: coastal sediments, eroded by the sea and transported alongshore, are accumulated as dunes, and may supply during the time shoreface deposits. The constant exchange of material between beach and dune is very important in the complex coastal morphodynamic interaction, which

U. Simeoni (✉)
Dipartimento di Scienze della Terra, Università di Ferrara, 44100 Ferrara, Italy
e-mail: g23@unife.it

D.R. Green (ed.), *Coastal and Marine Geospatial Technologies*, Coastal Systems and Continental Margins 13, DOI 10.1007/978-1-4020-9720-1_9,
© Springer Science+Business Media B.V. 2010

is not already completely understood. In Italy, there has been no long-standing tradition of geomorphologic studies on coastal dunes related to shoreline morphodynamic at national scale. Moreover environmental significances and interests to study these morphologies still growth with the awareness of their potential role in coastal management and natural risk mitigation (Psuty, 1989; Arens and Wiersma, 1994). Around 1800, the first experiences to re-shape foredunes in order to impede coastal flooding and to safeguard a forest occurred in France. In the last decades, in many European countries, researches on coastal dunes are focused on shore protection, recognizing their importance as a natural sea defense against floodings. At the same time, land planners have been interested on the dunescapes, because these provided an ideal location for recreational activities. As a consequence, in many countries around the world (Spain, Mexico, USA or Australia) coastal dunes were exploited for tourist initiatives often without any consideration for their natural or ecological value (Carter, 1990). Also in Italy, human impact affects coastal dunes due to land use demands: a broad lack of the interest on coastal dunes prevails, with rare exception of local awareness for their ecological values, but, even in these rare cases, there are no strategies to meet competing land use requirements.

This project provided a consistent national geodatabase especially developed for the presence and state of the coastal dunes. This geodatabase was planned and elaborated taking in consideration discussions and debates with all seven Operative Units involved in the project (belonging to eleven Italian research institutions).

The geodatabase represents a first homogeneous and noteworthy case of representation, classification and assessment of coastal dune morphologies at a national scale. Although many background data for Italian coastline have been published, no national plan exists to present those in a stochastic knowledge of the coastal system. Beaches Atlas, achieved during the eighties, is the main and unique former experience for Italy of a national homogeneous research on coastal state. In particular (and differently from what happens in other European countries) no detailed national geographic database exists on coastal topics. Many scientists involved in the Beaches Atlas project also took part in this opportunity to initiate an analogous initiative on coastal dunes and beaches erosion themes.

Keeping in mind the work on Italian coasts from La Monica and Landini (1983), our main objective was to provide a homogeneous awareness on the occurrence and the status of the foredunes in Italy where more than 45% of beaches are unprotected. This aim was achieved using a Geographical Information System (GIS) application. This application is an innovative and useful tool to describe and analyse the coastal dunes vulnerability in relation to coastal natural hazard.

The chapter describes the national geodatabase structure and contents furnishing, a systematic knowledge of coastal dunes occurrence and state. Moreover, this chapter presents the results obtained by the application of this newly created geodatabase to describe the status of the dunes in a coastal tract of the North Eastern Italy.

The studied littoral extends approximately 190 kilometers along the beaches of Emilia-Romagna and Souther Veneto regions where the continuous sandy beaches are greatly affected by human pressure and sea erosion.

9.2 A National Geodatabase for Coastal Dunes

Recently the evolution in Europe, territorial strategies mainly for coastal areas, highlighted the importance and the value of geodatabase availability as a very useful management tool. In turn, new attention on database, standard and metadata was drawn (INSPIRE, 2002). In the framework of the national research project, a vector geodatabase was planned to organise the coastal dune knowledge, accordingly to European standard for Geographic Information Sharing applying condition, datum and projection as defined for geographic "core data" as coastline.

The database design was built to be implemented inside ARCVIEW8.2 using previous studies from the seven Operative Units, taking in consideration existing data, data improvement methodologies and realistic expectations of the different partners.

All collected information correspond to historical data, field surveys, and image analyses. Georeferred and updated images (homogenous for the entire country), coming from the Italian Ministry of Environment, were used through the Cartographic web-Portal.

Having in mind the goals of the project as well as data source limitations, a composite core data was concerted through all project partners to describe main coastal dune features. In addition, to keep useful data, which could help to assess coastal erosion state and human impact, other information on beaches characterised by the presence of coastal dunes were gathered (Table 9.1). As understanding of aeolian processes is very important for dune evolution assessment, the positions of the main anemology stations as well as the data acquisition times were also included in the system.

Dunes features are described by their dimensions, their role inside the beach-dune interaction (active and re-active/fixed dune type); their vegetative cover (prevalent kind of vegetation, among woods, shrubs, grass); the presence or absence of urban areas (or scattered houses) to assess their naturalness degree; the main

DUNES	**DUNE ACTIVITY DEGREE-** (2 classes): active /fixed dunes **(AT/NAT)** **DUNE VEGETATION COVER** (6 classes): woods /shrubs/grasses/ woods and shrubs/shrubs& woods /no vegetation **(WOO/SHR/GRAS/WO-SH/SH-WOO/NO-V)** **DISTANCE TO SEA (real distance in metres)**
	DUNE HEIGHT (real values)
	WEAKENINGS (3 classes): tracks/streets toward sea/ tracks and streets on the ridges
	NATURALNESS DEGREE (3 classes): (urbanization/scattered houses/natural dune)
BEACHES	**BEACH WIDTH** (3 classes): **0-20/20-60/>60 m (1, 2, 3)**
	BEACH TREND 3 classes: prograding/retreating/stable seaside resorts
	DEFENCES (4 classes): beach nourishment/ seawalls/ nourishment and seawalls

Table 9.1 Dunes and beaches core data. *Light grey* background highlights data directly related to dune-polygon features; dark grey background shows data linked to polyline-features drawn on the present beaches; white background shows data linked to point features placed inside dune polygons. The items short form are indicated inside the brackets

direction of the dunes ridges and the main ridges elevation. Finally the presence of tracks (or streets) in the coastal dune systems, when these reach the sea or are placed on dunes ridges, are considered as indicators of dune lateral continuity.

Dunes are classified as active (or reactivated) or as established dune. Active dune class includes foredunes as well as old dunes having new wave cut scarps, depending on their present interaction with beach sediments.

There is a known relationship between wind energy, sea distance, and vegetation cover with a sort of spatial geobotanic classification. The coastal vegetation types and association change moving toward inner beach with a successive sequence, depending on the bio-geographic zone and on the physic and chemical circumstances. Coastal species association classes, proposed in the present work, partly refer to Short and Hesp morphodynamic foredune classification (1982). Since in most dune system there is a superimposition of old and new dune elements related to coastline changes, dune vegetation covers, and more specially association of typical coastal plants, describe these coastline evolution. Six different vegetation classes could be identified trough image analysis with the aim to highlight species mix that depends from dissipative continuum beach nearshore processes.

Data on beaches have been collected only when adjacent to foredunes with the aim to identify the possible interaction between beach features and dunes state. Some features have been gathered as geoindicators of dune-beaches system evolution: beaches width, beaches trend, beaches uses and defences. Information on beach width (measured on images) corresponds to foredune distance to the sea and is gathered as geoindicators of sediment supply for foredunes development. Beaches trend is not an absolute value but determines an overall tendency coming from many years data surveys and direct awareness from local academic field works. This option was used as a result of the impossibility to find a common data set able to support at national level comparable assessment.

Sandy beaches have great relevance in bathing seasonal activities: tourist initiatives affect beach and dune that, in our country, do not normally have any protection against pedestrian pressures. Conscious of this condition, the "beach use" was inserted inside the project legend to allow possible comparisons between dunes presence and state with fixed/seasonal seaside establishments. Their presence is considered as an index of seasonal human pressure on the beach-dune system. Beach defences are considered only when they potentially interact with beach-dune system: only seawalls and/or beach nourishments are gathered.

This geodatabase structure allows a possible integrated spatial analysis using all the information.

To accomplish this integration, the conceptual data model joins information on dunes and information of in front beaches (see Table 9.1). Beach width and beach trend are linked as information to dunes. Each dune feature has unique conditions and there is one-to-one correspondence between a geographic feature and a row in the data table with the attributes. Instead dunes ridges direction, tracks streets on dunes, dunes elevations, beach defences are organised in separate data class.

In this way, database matrix lets to have a composite data set on coastal dunes allowing, at the same time, complex spatial analyses among coastal dune "state" and beach "state".

The model used considers three geometric feature kinds: polygons, polylines and points. Dunes are drawn as polygons; dunes ridges, beach defences and uses are drawn as polylines; tracks streets on dunes are drawn as points. Four homogeneous ambits (Feature Data Classes) compose the GIS structure: 1-Dunes (Polygons); 3 Beach defences (Polylines); 4-Beach uses (Polylines); 5- Dune tracks (Points).

GIS schema consists of seven "geographic ambits" (*Features Data Set*) corresponding to each Project operative units. Metadata were produced in this phase according to European warning for geographic data production. Many routines have been made to simplify stored attribute information. Firstly, one whole geodatabase outline was created to be implemented by partners. Then a final re-compose of whole geodata structure is considered. This guarantees the geodatabase congruence and, at the same time, gives to every O.U. (Operative Unit) the necessary operative autonomy keeping their right property and responsibility on produced data.

9.3 Using Geodatabase to Depict Dunes and Littoral State in Northern Adriatic

Using the geodatabase built for the study of the Emilia Romagna and the Po Delta littorals, coastal dunes state and their relationship with beach uses and shoreline trend can be clearly defined.

In this coastal tract (Fig. 9.1a), the presence of low sandy beaches is characterized by a large surf zone corresponding to the dissipative beach type described by "Wright and Short" (1983). Sediments come from various rivers: Adige from North- Eastern Italy, Po from northern Italy Padana Plain, Reno and Bevano from Emilia Romagna Apennines chain. Since the middle of the past century, sediments supplied from these rivers have been drastically reduced (over 50%). The coastal evolution in the latest decades was strongly conditioned both by environmental and human interacting factors (Simeoni and Bondesan, 1997). Natural and mainly man induced subsidence (related both to methane and water pumping due to industrial or near coast reclaimed land management) greatly affects these coastal areas: having a maximum rate during 1984–1987 (from 5 to 50 mm/year), coastal lowering recently reduced to 2–36 mm/year (1987–1993) (Idroser, 1996). Large parts of former marsh lands reclaimed from the sixties are nowadays many metres below present sea level. Contemporary major building development occurred close to beaches. This induced in these low beaches not only coastline retreats but also shallow water slope increases with a consequent sandy beach volume reduction. These environmental morphologies greatly amplify the coastal dunes role as a way to protect inner flat and under-sea level territories against main storms and sea flooding.

Out of the 192 km of considered coasts (140 km in the Emilia-Romagna region and around 50 km in the Veneto region) only 38% are still bordered by dunes (43 km in the Emilia-Romagna region and 30 km in the Veneto region. An analysis

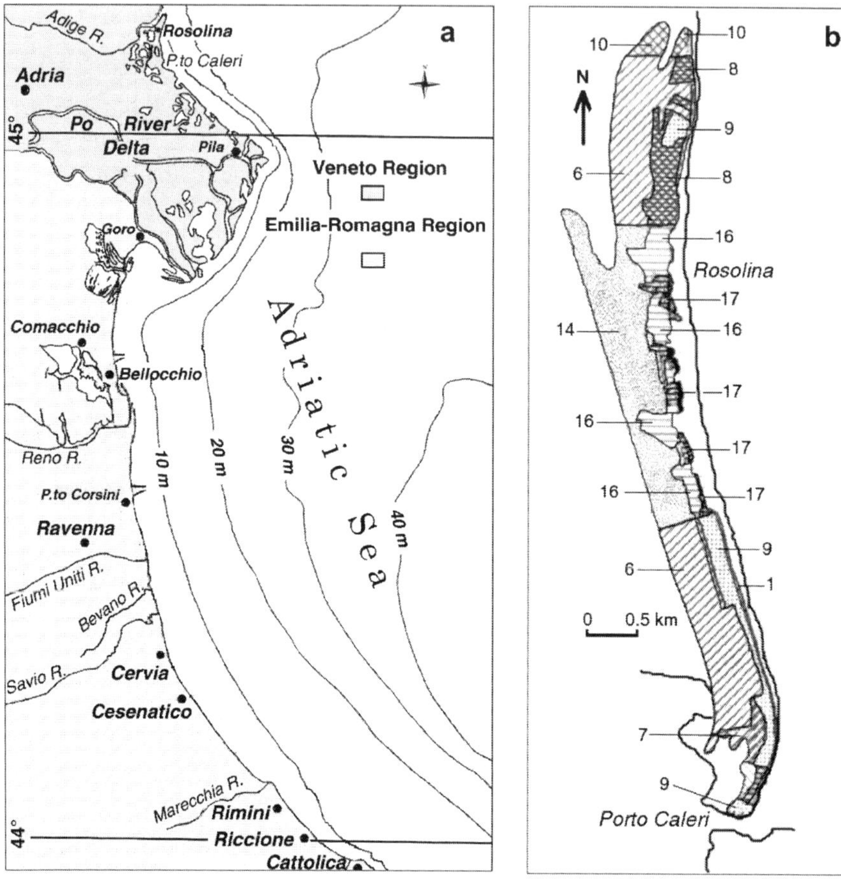

Fig. 9.1 (**a**) Northern Adriatic area sketch which refers coastal dunes data discussion (**b**) Details from Adige River to Porto Caleri. Example of complex spatial analysis on dunes activity referred to beach width and dune vegetation type. Simplified keys: *1*=AT/2/NO-V; *2*=AT/2/GRS; *3*=AT/1/SHR; *4*=AT/1/GRS; *5*=AT/3/NO-V; *6*=NA/3/WOO; *7*=NA/2/SHR; *8*=NA/2/NO-V; *9*=NA/2/GRS; *10*=NA/1/WOO; *11*=NA/1/SHR; *12*=NA/1/NO-V; *13*=NA/1/GRS; *14*=NA/3/WOO; *15*=NA/3/SHR; *16*= NA/3/NO-V; *17*=NA/3/GRS. Items short forms are explained in Table 9.1

of dune development in the past century clearly indicates that coastal dunes bodies are currently largely obliterated to enlarge shore for recreational facilities and are affected by sea wave erosion. In addition, near shore building development contributes to this devastation: study results highlight that about 10% of coastal dunes (roughly 2.5 km^2) are actually covered by urban developments. Houses and coastal infrastructures are mainly present over internal dunes ridges.

In the Northern part of the area, between Adige mouth and Porto Caleri site (approximately 8 km which, around 6.5 bordered by dunes, Fig. 9.1b), beaches are greatly retreating and their width is very small, having an average of roughly

Fig. 9.2 Example of coastal conflict between dunes and bathing establishments: **a** Rosolina Mare beach; **b,c,d** Emilia Romagna beaches

32 m. In this area, foredunes have almost disappeared and when present they present marked sea scarps. As a result of excavation performed to enlarge the area of tourist facilities a scarce continuity of the coastal dunes is observed (Fig. 9.2). Inside large part of the beaches (about 5.5 km) rough dunes elevation, ranging between 2.5 and 10 m, can be observed, due to seasonal resort activities. Many small blowouts are present along the dominant wind direction (NE). These blowouts are principally observed in the inferior sector, while in the upper coastal tract, near the Adige mouth and inside Rosolina Mare protected areas (Natural Park), dunes are well developed with greater lateral continuity and only few located breaks, due to pedestrian frequentation, are observed. Here beaches are wider (40–60 m): foredunes elevation ranges between 2 and 5.8 m, while internal dune altitude reaches about 6 m. Moreover, in all this Northern tract very high hazard condition for dune erosion are present locally as indicated by the presence and distribution of the vegetation: the present occurrence of shrubs on the "active dunes" is a clear index of persisting landward movements of the coastline. In fact, in this region, dunes vegetation profile, characterising the sea side slopes of well developed, established foredunes, indicates that the pioneer and dominant vegetation species is *Ammophila*. The two next dominant species are Cakileto and Agropireto respectively, followed by various shrubs species like *Echinophoro spinosae-Ammophiletum arundinacea* or *Bromo tectorum-Phleetum arenarii*. On fixed dunes *Pinus pinea* and *Pinus pinaster*, planted during past reforestation prevail over other vegetation species.

Between Porto Caleri and Po Delta (60 km with approximately 26 km of dunes, Fig. 9.3a), human pressure is minor and coastal lagoons are still preserved. Here

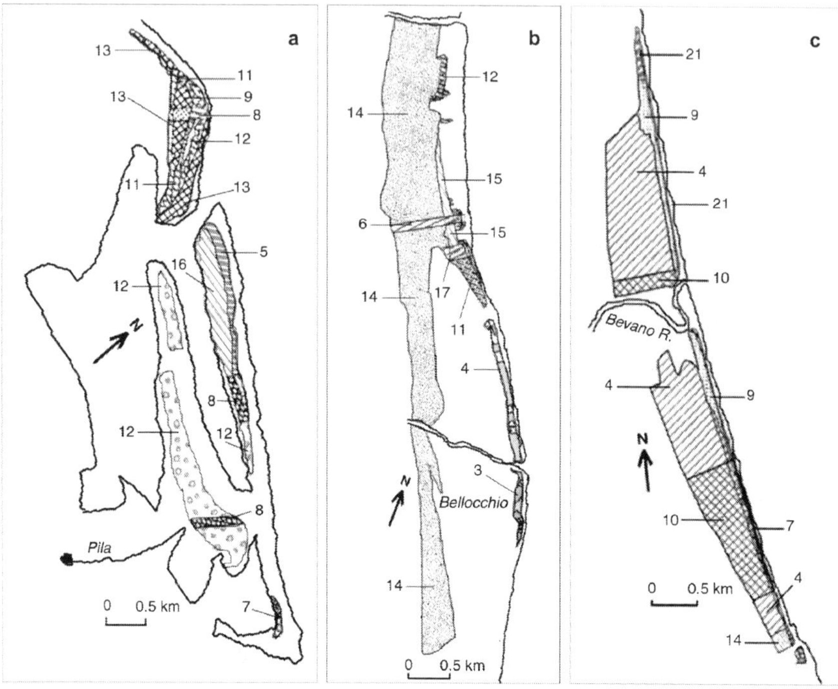

Fig. 9.3 Dunes classification related to dunes activity, beach width and dune vegetation type (item explained in Fig. 9.1). Details refer to littorals: **a** between Porto Caleri and Po Delta area; **b** Bellocchio area, near Reno River mouth; **c** near Bevano River mouth littoral

there is a well developed barrier-lagoon with a complex beach-dune dynamic. Despite the absence of human pressure, important beach erosion, initiated in the middle of the 19th century, affects this dune system. Most of the beaches are characterised by a width inferior to 20 m and dunes altitude ranging from 2 to 3.4 m. The coastal dune system is however complete: foredunes and inner ridges are both present while shrubs prevail on internal dunes (Fig. 9.3a). In Southern coastal tract (130 km) belonging to the Emilia-Romagna region, about 43 km are bordered by dunes that are in general very deteriorated: residual edges are incorporated inside buildings and infrastructures were constructed near the beaches. Very large seaside resort infrastructures are found on the beaches. The main use of beaches for seasonal tourism induced levelling use that reaches dunes bases. This results in anomalous slopes and in sand body loss that sometime affect the dune face when no protections are present. Pedestrian passages are common in the beaches and induce damages in vegetation, which is already compromised by slope steepness and instability. Here, it recently becomes usual to protect beach buildings from winter sand accumulation by hard defence placed in the middle of the beach. This obviously affected the natural beach equilibrium, not restricted to dunes.

Only inside the Bellocchio natural protected area (around 6.6 km length near to Reno river mouth) and in the coast close to Bevano river mouth (around 6 km), dunes are still well preserved (Fig. 9.3b,c). In Bellocchio area, dunes present altitudes ranging from 1 to 7 m and width ranging from 100 to 600 m. Here, few years ago, dunes extension was much wider, reaching 3.3 km in length where, nowadays, only 2.3 km persist. Near Bevano mouth, the difficult human frequentation allowed the best dune ridge preservation found in the Emilia- Romagna region with 6 km long and up to 60 m width (covering about 20 hectares). On the inner side of the foredune, a coastal pinewood grows on fixed dune ridges. The foredunes present elevation from 3 to 4 m to few centimetres nearest to shore, due to wave destruction. In this tract, beaches are very narrow, with width inferior to 20 m (20% superior of Bevano littoral area).

The considered coastal area presents various types of defences, built from sixties. The research analysed the connection between dunes and defences that have potential impact on beach-dunes sedimentary balance. While many seawalls are present near the dunes, more than 7 km of beaches, in front of the dunes, have been renourished during the last years.

9.4 Conclusions

The present study points out that no favourable condition for dune development are observed in the Emilia-Romagna and Po delta coastal zones as evidenced by the width of the beaches, which is lower than 20 m. The system is very helpful to assess the coastal state: it offers the possibility to shows not only "if" there is a dune but compare, in the same location, more different conditions happening. The possibility to use the information geographic component when comparing collected data, improves the analyses potential. A quantitative approach, that allows overtaking the "problem presence knowledge" with a "problem dimension evaluation", becomes also possible.

Moreover, more complex analyses become feasible: high hazard state for beaches erosion can be deduced from the kind of vegetation on the dunes related to their activity. In fact, the presence of shrubs of woods on foredunes acts as geoindicator of dissipative coastal environment where sea reaches the inner side dunes inducing their retreat (in a natural coastal environment) and/or their destruction.

Results highlights that this dissipative coastal state is very common along the Adriatic coast and indicates that, besides a great diffusion of defences, and shorelines mainly advancing, the coastal environments are conditioned by the former retreating trend inducing the lost of many foredune ridges not already counterbalanced by beaches renourishment or steadiness.

Moreover this analysis highlights a great stress conditions of the Northern Adriatic littoral. From geodata analysis it results that, except in the Bevano area, bathing establishments are located in front of larger still preserved dune systems. Presently, only the Po river mouth keeps a whole natural condition.

The realisation of a similar tool for large part of the Italian country may really support a national comparison. Its application also better highlight the importance to consider such coastal environments inside coastal planning not only as natural heritages but also as integral part of coastal risk mitigation strategies. In fact, coastal dunes and specially their environmental peculiarities have to be considered as natural barrier (where dunes are still well preserved) against flooding and land inundation especially where the territories are below the present sea level. This condition is common for Italian littorals, as in many other littorals around the world. In 1997, coastal dune has been reconstructed in order to act as Civil Defence against inundation in some test areas of the Emilia Romagna Region. This solution is not significantly diffused in the coastal planning practices in Italy as well as in other country. That highlights the importance of a coastal dunes pan European good practice development inside Coastal Zone Integrated Management (ICZM) general approach.

Acknowledgments We thank Ing. G. Onorato and Dr. S. Costabile from Italian Ministry of Environment and Territory for supplied the Project. Prof. G. Fierro, Project national coordinator, sustaining our work. Dr. A. Zamariolo, C. Schiavi and N. Bonora for field surveys. All colleagues from other Project Operative Units collaborating to data model content setting up.

Bibliography

Arens, S.M., and Wiersma, J. (1994). "The Dutch Foredunes: Inventory and Classification", Journal of Coastal Research, vol. 10, pp. 189–202

Carter, R.W.G. (1990). The geomorphology of coastal dunes in Ireland, in: Bakker, Th.W.M., Jungerius, P.D.and Klijn, J.A. (eds.), *Dune of the European Coasts,* Catena Supplement, vol. 18, pp. 31–39

Idroser (1996). *Progetto di Piano per la Difesa dal Mare e la Riqualificazione Ambientale del Litorale della Regione Emilia-Romagna, Relazione Generale*, Regione Emilia-Romagna, Bologna

La Monica, G.B., and Landini, B. (1983). "Tendenze evolutive delle coste basse della penisola italiana", *Atti XXIII Cong. Geogr. It.*, Catania, pp. 209–217

Piccoli, F., and Gerdol, R. (1990). La vegetazione attuale, in: Parco del Delta del Po, Spazio Libri (eds),. *Studi e immagini*, Ferrara, pp. 87–122

Psuty, N.P. (1989). "An application of science to the management of coastal dunes along the Atlantic coast of the U.S.A", *Proc. of the Royal Society of Edinburgh,* Edinburgh, vol. 96(B), pp. 289–307

RDM Working Group, (2002). *INSPIRE Reference Data and Metadata Position Paper*, Rase, D., Björnsson, A., Probert M., and Haupt, M.F., (eds), Eurostat, 45 pp

Short, A.D., and Hesp, P.A. (1982). "Wave, beach and dune interactions in southeastern Australia", *Marine. Geology*, vol. 48, pp. 259–284

Simeoni, U., and Bondesan, M. (1997). The role and responsibility of man in the evolution of the Adriatic alluvial coasts of Italy, in: Briand, F. and Maldonado, A. (eds), *Transformations and evolution of the Mediterranean coastline*, Commission Internationale pour l'Exploration Scientifique de la Mer Méditerranée (CIESM), Science Series, vol.18, no 3, pp.111–132

Chapter 10
Establishment of Marine and Coastal Spatial Data Infrastructure in Indonesia

Suwahyuono Wahyu, Gatot H. Pramono, and Bebas Purnawan

Abstract As the largest archipelagic nation with thousands of islands, the establishment of infrastructure for marine and coastal spatial data is indispensable. Six technical working groups have been formed as a component of spatial data information management in the Marine and Coastal Resources Management Project (MCRMP). The groups' responsibilities are to determine the standard of spatial database, the data clearinghouse and custodianship among stakeholders and improvement of human resource. After the discussion with stakeholders, some standards have been established and some agreements have been reached to be implemented in the further stage of national spatial data infrastructure.

Keywords Spatial data infrastructure · Data standard · Data exchange · Indonesian marine and coast

10.1 Introduction

Indonesia is an archipelagic state, a country with 17,508 islands and 81,791 km of coastal length. The wide variety of land, coastal and marine resources are available. This natural potential is managed by the government of Indonesia to improve economy and social well being of the people.

The direction of national development of Indonesian government has been changed since the 1990 s. At that time, the government decided to explore the marine and coastal resources in addition to the exhausted land resources. The exploration was meant to overcome the booming of population in Indonesia.

The exploration the marine and coastal resources nevertheless needs a comprehensive planning based on accurate geospatial data. Then the development of

S. Wahyu (✉)
BAKOSURTANAL (National Coordinating Agency for Surveys and Mapping), Indonesia
e-mail: wahyulaut@yahoo.com

D.R. Green (ed.), *Coastal and Marine Geospatial Technologies*, Coastal Systems and Continental Margins 13, DOI 10.1007/978-1-4020-9720-1_10,
© Springer Science+Business Media B.V. 2010

geospatial data particularly marine and coastal data becomes indispensable to the development of marine and coastal area.

This chapter describes the latest development of marine and coastal spatial data infrastructure in Indonesia. The focus is on the role of technical working group (TWG) under Marine and Coastal Resources Management Project (MCRMP) in establishing the standards of spatial data infrastructure. The tasks, members and results of TWG are discussed.

10.2 Indonesian Spatial Data Infrastructure

Geospatial data has an important role in the development of national economy, social and physics (Craglia, 2008). In Indonesia, the need of spatial data becomes so obvious after the deliberation of the Law number 32 in 2004 addressing Local Government issues. It states that in a planning of development, local government is required to utilize geospatial data. Moreover, the local government should develop a local spatial information system which will be integrated to a national data infrastructure.

The availability of integrated geospatial data becomes a high priority for both national and local governments according the above Law. These governments, especially geospatial related government institutions, have to make nationally integrated geospatial data available and accessible to the community users.

However, there are several problems with the availability of the integrated geospatial data in Indonesia. First, existing data managed by different institutions is saved under various formats, standards, naming conventions and database systems. Second, there is data duplication among the institutions. Similar data are produced by more than one stakeholder. Third, the existing data is not yet well managed so that it is difficult to access. Finally, the character of each institution is to keep their own data and to allow limited access to the data.

To cope with these problems, the National Coordinating Agency for Surveys for Mapping (BAKOSURTANAL) took an initiative by introducing the Indonesian Spatial Data Infrastructure (ISDI) in 2000. The ISDI envisions to make integrated geospatial data within Indonesian territory available and accessible for community users. The ISDI has five components or aspects namely institutional, legal, fundamental datasets, science and technology and human resource aspects (Bakosurtanal, 2004). These components are implemented into the following programs:

1. standardization
2. custodianship
3. national clearinghouse
4. partnership
5. human resource

Standardization program consists of developing fundamental dataset and its metadata standards according to the adopted standards.

The custodianship program deals with the development of custodianship guidelines, indication of data custodians, and the implementation of these guidelines to the indicated custodians.

The national clearinghouse forms an access infrastructure to Indonesian geospatial metadata which is stored in custodians. This clearinghouse consists of a network connecting all custodian servers, clearinghouse gateway and users through internet.

The partnerships program has to do with institutional coordination among geospatial data stakeholders including central, provincial, district and municipal government institutions, private sector, universities, and professional associations.

The human resources program attempts to support national efforts in capacity building including support and facilitate of certification and accreditation of survey and mapping professionals.

Since its establishment in 2000, the ISDI initiative has gained strong support from many institutions. In the last ISDI meeting held in Bali in May 2005, the development of ISDI is concentrated on finalising of national data and metadata standard, custodianship guidelines, operating of national clearinghouse and a government regulation for facilitating of partnership and human resources development.

10.3 Marine and Coastal Resources Management Project

In 2001, a national programme called the Marine and Coastal Resources Management Project (MCRMP) was initiated by the government of Indonesia (Asian Development Bank, 2001). This project is aimed to manage marine and coastal area of over 15 provinces and 40 districts in Indonesia to utilize and manage the natural resources of the area. The project is financed by the loan from the Asian Development Bank.

One of the target of the project is the development of the national spatial data infrastructure of Indonesia focussing on marine and coastal datasets. For this purpose, the project is aimed to develop of metadata standard, marine and coastal data standard, the custodianship guidelines, the clearinghouse guidelines, and related training guidelines.

The development activities are implemented by the Technical Advisory Group (TAG), an inter-institution group, established for guiding the development of the Indonesian Spatial Data Infrastructure within the marine and coastal data institutions. To implement its task, the TAG establishes expert groups called the Technical Working Group (TWG).

10.4 Technical Working Groups

The technical working groups (TWG) are formed to help the TAG in preparing all guidelines, as requested by the project that will be used by the TAG to guide the institutions to establish the documents for developing marine spatial data infrastructure (Bakosurtanal, 2005). There are six technical working groups such as TWG 1 (Communication and information), TWG 2 (Clearinghouse), TWG 3 (Database standard), TWG 4 (Thematic map), TWG 5 (Custodianship) and TWG 6 (Human resources). The tasks of each technical working group are shown in Table 10.1. It can be seen that each TWG has specific task.

Table 10.1 Tasks of technical working groups

TWG	Tasks
1 (Communication and information)	1. Preparing and creating documents for socialization such as presentation materials, scientific writings and demonstration regarding marine spatial data infrastructure 2. Defining activities in the national and local levels which are related to support marine spatial data infrastructure 3. Performing presentations, scientific talks, lectures or demonstration
2 (Clearinghouse)	1. Performing research, development and application of technology in spatial data exchange 2. Defining technical specifications for marine spatial data 3. Defining technical specifications for making metadata 4. Developing clearinghouse of data spatial for spatial data exchange
3 (Database standard)	1. Identifying spatial data which are necessary for fundamental data set in coastal zone management 2. Defining the database standard for management of fundamental data set referring to the existing standards 3. Defining specification for developing and quality control for thematic data set
4 (Thematic map)	1. Identifying spatial data which are necessary for thematic data set in coastal zone management 2. Defining the database standard for management of thematic data set referring to the existing standards 3. Defining specification for developing and quality control for thematic data set
5 (Custodianship)	1. Proposing the custodianship between national, local and other stake holders in the management of spatial data for the coastal area 2. Reducing the problems with regulation and administration
6 (Human resources)	1. Identifying specification of human resource capability to develop spatial data infrastructure 2. Developing curriculum and training modules 3. Monitoring and evaluating the training

TWG 1 is responsible for publicizing the results of other working groups. The clearinghouse group is aimed to propose the concept of data exchange among stakeholders. TWG 3 and 4 are specifically designed for database of fundamental data sets and thematic data respectively. Fundamental data set is data which is used as basic information for a map such as administrative boundary, road network, coastline and bathymetry. Custodianship deals with how the relation between stakeholders is formed. The last working group deals with the enhancement of human resources to support spatial data infrastructure.

Members of each working group are between 7 and 12 people from various backgrounds. Five ministries which participate in TWG are from ministry of ocean-fishery, mining, environment, forestry and agriculture. Eleven government agencies are involved such as agencies for surveys/mapping, development planning, public facilities, meteorology/geophysics, geology, science, cadastre, statistics, flight/space, hydrology and military mapping. Two universities and association of survey and mapping users are also members of TWG. The variety of members are intended to achieve comprehensive approach for building national spatial data infrastructure.

In 2004, nineteen meetings of TWG have been conducted. TWG 5 dealing with custodianship has met five times. The meeting is not inclusive for only certain working group. People from other groups are sometimes invited to get a common perspective. For example, TWG 3 (database standard) meeting was attended by member of TWG 4 (thematic) and TWG 6 (human resources). The MCRMP consultant also attended several meetings to provide some suggestions.

The results of each working group are documents and/or proposals for developing national spatial data infrastructure. Table 10.2 shows the results of each TWG. Several action plans to publicize the results of other TWG were defined by TWG 1. The

Table 10.2 Results of technical working groups

TWG	Results
1 (Communication and information)	1. Plan of publication in 2005 and 2006
	2. Materials for publication: clearinghouse, metadata and custodianship
2 (Clearinghouse)	1. Document of building clearinghouse for national data infrastructure
	2. Document of building metadata
3 (Database standard)	1. Draft of data dictionary for spatial database consisting of data structure, labelling, themes, relational tables and features of database
4 (Thematic map)	1. Classification of thematic data on the basis of its usage
	2. Identification of data for coastal management
	3. Standard of database for thematic data on the basis of the existing standard
	4. Defining the specification of developing and control quality of dataset
5 (Custodianship)	1. Document of managing custodianship, users and coordinating institution
	2. Proposal of main data and custodianship
	3. Proposal of custodianship for the national and local governments
6 (Human resources)	1. Specifications of human resource needed to build spatial data infrastructure
	2. Types of training for developing spatial data infrastructure
	3. Syllabus for training of Integrated Coastal Zone Management (ICZM)
	4. Syllabus for training of rapid integrated survey

Table 10.3 Problems faced by technical working groups

TWG	Problems
1 (Communication and information)	Results from other groups are finished by the end of 2004. As a result, materials for publication are compiled late
2 (Clearinghouse)	None
3 (Database standard)	None
4 (Thematic map)	1. Variability of data types and data classification 2. Variability of the thematic map production and its naming conventions 3. The dynamicity of the spatial data in the coastal area which is often updated
5 (Custodianship)	1. The concept of custodianship is not well understood by the members 2. The individual interest of each institution is to keep their own data 4. Overlap of mapping activities among institutions
6 (Human resources)	The lack of coordination among stakeholders of spatial data for marine and coastal area

plans will be carried out in 2005 and 2006. The spatial data and metadata exchange are proposed by TWG 2. The standard for fundamental dataset and thematic map has been set by TWG 3 and 4 respectively. TWG 5 has proposed the custodianship for national and local governments. Specifications for upgrading human resources and trainings have been proposed by TWG 6. These results then are delivered to technical advisory for further step in developing ISDI.

10.5 Evaluation

Besides results of working group activities, problems associated to each group are identified. These problems are evaluated for better results in next stage of ISDI. Table 10.3 shows the problems encountered by several working groups. TWG 2 and 3 do not find any significant problems. TWG 1 works late because the materials for publication are the results of other groups. TWG 4 finds the problem in variability and dynamicity of spatial data. Custodianship and human resource groups concern with the relation among stakeholders of spatial data.

10.6 Conclusion and Recommendations

A conclusion can be derived from the meetings of technical working groups. TWG meeting is an effective approach to help the development of national data infrastructure for marine and coastal areas. Several documents, specifications of the data format, structure, exchange and usage have been suggested. These results will be used by technical advisory group to perform further stage.

Several points can be recommended for the further step in building national spatial data infrastructure:

1. The working group activities should be continued to finish several points which are not yet completely finished.
2. The members of thematic map (TWG 4) should be strengthened to accommodate variety of spatial data.
3. The concept of custodianship should be clearly explained so that all stakeholders understand the benefits of defining custodianship.
4. The coordination among stakeholders should be strengthened by performing joint meetings, projects or researches.

References

Asian Development Bank. 2001. Marine and Coastal Resource Management Project (MCRMP) Loan Agreement.

Asian Development Bank. 2001. Marine and Coastal Resource Management Project (MCRMP) Loan Agreement and Project Administration Memorandum (PAM).

Bakosurtanal. 2004. Guideline for Establishment of Naional Spatial Data Infrastructure.

Bakosurtanal. 2005. Report of Technical Working Group (TWG) Marine and Coastal Resources Management Project (MCRMP).

Craglia, Massimo. 2008. The Socio-Economic Impact of the Spatial Data Infrastructure of Catalonia. JRC Scientific and Technical Reports.

Part II
GIS Tools

Chapter 11
Prestige Marine Pollution: A GIS Tool

Francisco Carreño, Inmaculada Rodríguez, Isabel Montoya, and Maria J. Sánchez

Abstract The accident and sink of the tank ship Prestige, that occurred offshore of the Galicia coast (northwest Spain) on November 2002, caused a large oil spill with important ecological and economic consequences for the Galician and Cantabrian littoral. Different data sets obtained through monitoring of the oil spill by over flight observation, satellite data, measurements from buoys and modelling of oceanic parameters (e.g. temperature, salinity, density, pH, fluorescence, dissolved oxygen) have been integrated into a Geographic Information System (GIS) for their access and study. The use of this GIS will allow the civil authorities and the scientific community to manage and to generate thematic cartography and to perform spatial and geostatistical analysis of the data in order to better analyze these marine pollution scenarios.

Keywords GIS · Remote sensing · Marine pollution · Spatial analysis · Geostatistical analysis · Prestige

11.1 Introduction

The accident and sink of the tank ship Prestige, that occurred offshore of the Galician coast (northwest Spain) on November 2002, spill into the sea more than 50000 tons of heavy hydrocarbon residue. The subsequent oil slicks that reached the coast resulted in severe ecological and economic consequences for the Galician coast and the Bay of Biscay.

The first oil slick of about 10000 tons was detected November 13, 2002, when the ship was towed northwards away from the coast. On November 15, the Prestige is located 60 nautical miles from the coast and is towed to the south with a change

F. Carreño (✉)
Departamento de Biología y Geología, Universidad Rey Juan Carlos, Grupo de Geología, C/ Tulipán s/n, 28933 Móstoles. Madrid, Spain
e-mail: francisco.carreno@urjc.es

D.R. Green (ed.), *Coastal and Marine Geospatial Technologies*, Coastal Systems and Continental Margins 13, DOI 10.1007/978-1-4020-9720-1_11,
© Springer Science+Business Media B.V. 2010

on course to the southwest on November 16. On November, 19, the Prestige split in two, and sank completely about 130 nautical miles from the coast. The ship broke in two and its sinking generates a second large oil slick of 20000 tons that in the following days reached the coast driven by the main currents and wind regime.

The wreckage, located more than 3500 meters deep in the southwest limit of the Galicia Bank, continued leaking oil with a decreasing rate; 80 tons/day on January 8 to 2 tons/day on January 29.

The seriousness of the Prestige's oil spill result in an important operative to follow and control the possible effects of the spill. This operative is coordinated by SASEMAR (Sociedad de Salvamento y Seguridad Maritima), the Spanish organization responsible for maritime rescue operations. Besides SASEMAR different research groups from Spanish universities and research institutes, together with institutions from other countries of the European Union, started complimentary research projects in order to analyze the evolution of the oil spill and the impact on the environment that could be derived of it. As result of these different studies a great amount of data and information with different source and formats is generated. These data sets that must be treated and stored for future analysis. This situation have put on relieve the necessity for a system that allow to homogenise and integrate all the available information into a geodatabase in order to facilitate the planning and response operations to fight against marine pollution to help the decision-making process and management of oil spill crisis in the future (Trigueros, 2002).

The cartographic document is a key management element, as more information needs to be used for a correct management of a hazard situation. To achieve a correct management of these situations it is necessary to use and update the cartographic documents with all the available information. The necessity for a versatile cartography together with the possibility of the using databases with different elements and processes implied in the study make necessary the use of GIS systems as a support tool for management purposes (Li et al., 2000).

This work shows the use of a GIS developed in the *Universidad Rey Juan Carlos*together with *SASEMAR*that takes advantage of different technologies in the field of geoinformation for the design and development of a geodatabase. In this work we treat the design, structure and integration of the different data sets that are necessary to study the pollution related to the Prestige oil spill.

11.2 Software

The software used for the project is the ESRI ArcGIS®. This system stores data in a geodatabase (GDB) that represent a significant advantage with respect to the traditional file structure used by other systems (ESRI, 2001). The GDB model manages geographic information such as coverages, shapefiles, grids, images, and TINs and stores vector data, raster data, nets, topology, notes, tables, etc, using standard relational database technology. Thus, the geodatabase provides integrity and consistency of the geographic information.

Different elements are inside the geodatabase: feature class, topology, geometric network, tables, relationship, raster dataset, metadata document, that we will develop.

We developed a relational database (DBMS) using Microsoft Access®, designed to facilitate the data integration and for the elaboration of tables that are permanently connected with the GIS. This DBMS was a logical choice because of its broad distribution and its easy use.

For image treatment of digital raster and satellite images we have used the program ENVI® 4.0.

11.3 Methodology

The procedure followed to achieve the goal proposed for this work started with the compilation of the different data sets. Alphanumeric data are arranged and structured into a relational database, meanwhile the graphic information (e.g. maps, satellite images, etc) are directly integrated inside the GIS that have been previously design and developed. The DBMS is a tool for the consult and management of the alphanumerical information that is also integrated into the GIS. Information with geographical references is stored and represented in the GIS as a cartographic element in thematic layers or as an attribute of existing geographic elements. The result is a versatile system because the access and use of the data information can be realized through the DBMS with SQL sentences, and the advantages of the ArcGIS system (e.g. editing and data automation, visualization, mapping and map-based task, spatial consult, spatial analysis, geostatistical analysis, etc).

11.3.1 Database Design

A correct design is the first and more important of the different phases on database development. A good design is based in an abstraction of the reality that result in a data model integrated in tables that related in order to represent elements and processes.

Geoinformation is stored in the GIS to allow treatment, consult, analysis and interpretation so a proper design is crucial for a correct system operation. Database design was made together with *SASEMAR*, whose personnel posses a great experience in the oil spills management. Besides, the database design benefits from the experience of numerous other researchers from different institutions with experience in marine pollution that helped to assure a correct function and optimum performance of the GIS tool (Carreño and Rodríguez, 2003). The first step of the database design was to establish the general goals for the system to comply with; a prototype tool to help to manage oil spills by *SASEMAR*, but also a compilation of all the information about the Prestige's oil spill in order to carry out a research on the evolution of the oil slicks and the factors that control their behaviour. In both cases it was necessary to establish the type of spatial analysis, the necessary cartographic products and their characteristics and the output formats for the products generated by the GIS. It is was also important to assure flexibility in order to add new features in the future or modified existing ones in case of new necessities.

Once the goals for the system were established we determined the data sets necessary to achieve them. An inventory and compilation of all the different data sets obtained through the operations of surveillance and control of the Prestige's oil slicks managed by *SASEMAR*was made, together with the technical reports of different institutions, bibliographical documentation, existing cartography and other relevant ancillary data (e.g. bathymetry, meteorological and oceanographic constraints, hydrography, satellite data, etc.). The next step in database design was to determine the constituent elements of the model; tables related to each of the elements, relation between them and the different types of geometry employed to represent them correctly. It was also important to establish the data model to be supported by the GIS (i.e. vector or raster), the projection used for mapping, coordinates, scale and the topological relations between the elements of the model.

Finally, we establish the final structure of the GIS: data and elements to represent, associated tables and their relations, measurements units, consults, analysis and to cartographic documents, statistics and reports that will be generated by the GIS.

11.3.2 Data Integration

Different data sets were used for the GIS. These data sets came from different sources and show a great heterogeneity regarding origin and format. Previous to its incorporation into the GIS, these data sets were homogenised and structured for the use in the geodatabase in digital format. The different criteria used are result of our previous experience in GIS development.

Graphic and cartographic information is structured in different levels of information. Alphanumeric information database are associated to each of elements of the GIS. In some cases cartography has been generated form the alphanumeric information but in other cases we have to generate or complete the different attributes of the exiting cartographical elements.

The cartographic base used was extracted from the nautical charts edited by the Instituto Hidrográfico de la Marina, usually nautical chart useful for *SASEMAR*. Other information levels have their origin in other sources that have been digitalized and georeferenced for integration into GIS.

Resulting cartography is in Mercator projection (conformal true direction and rhumb lines), whose latitude of true scale is defined in 44.00'. Reference system used is WGS-84 with geographical coordinates to optimize compatibility with GPS measures. Minimum scale used is 1:400.000 to cover the great geographical extent of oil spills (IPIECA, 1994) (see Fig. 11.1).

Database are the other component that form this tool. In these databases is where information is related to geographic elements by the use of tables. Design and the relations between them guaranteed coherence and format of the data, and allow integration with the cartographic base. Forms are used as an way to enter

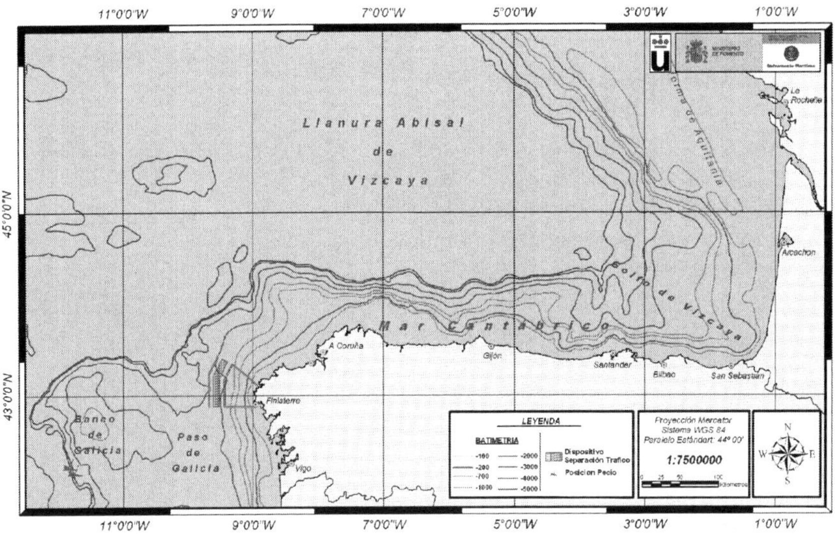

Fig. 11.1 Cartography elements and base map (See also Plate 17 on Page 400 in Color Plate Section)

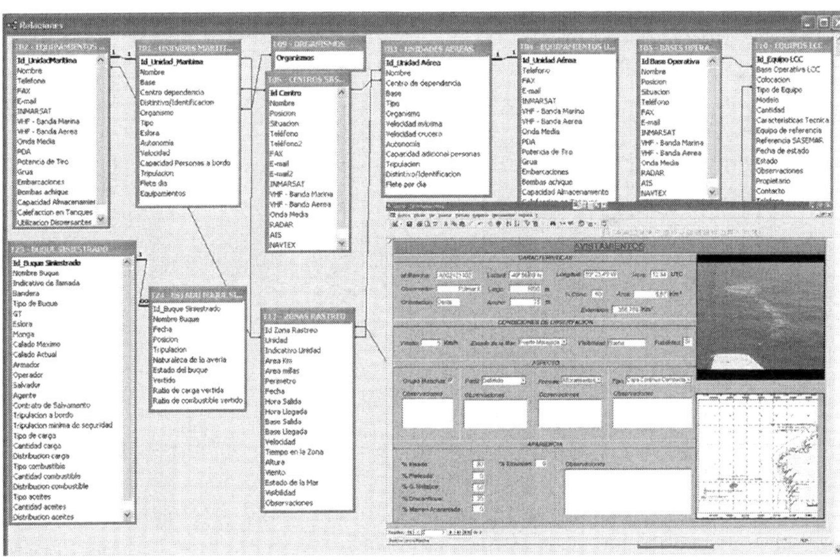

Fig. 11.2 Database design with table relationships and data access form of forms oil slicks sights

data into a database table (see Fig. 11.2). These forms detect and filter errors and incoherencies during the process of data introduction. We define two different groups of tables. One group is related to the operative part of the tool where resources for surveillance, control and fight against pollution are registered. The other group of tables is related to the oil spill and its evolution.

In the first group of tables an inventory of the resources is made, assigning a unique identification number (ID) to each of the resource, its geographic localization and their characteristics. The tables used are:

- Aerial surveillance units
- SASEMAR maritime units
- Marine units that belong to other institutions
- Rescue ships and resources for pollution response
- Other maritime units
- Fishing boats
- Damaged Ship
- Operative bases
- Rescue and emergency centers

The second group of tables include:

- Oil slicks sights: position, time, shape, profile, size, colour, observation conditions, etc.
- Buoys: position, speed, direction.

The first table consist of 4324 registers that correspond with sighing of oil slicks from November 14, 2002 to February 28, 2003. A great effort was made to assure coherence and homogeneity between formats taking into account their different source (fax, maps, schemes, reports, etc). Each registry represents a cartographic element that is represented by its dimension, shape and look. These data represent an important source of information to the study of the evolution of the oil spill and its precedence.

The buoy operative allowed obtaining real-time information about their position. This makes possible to measure distances between measurements and to calculate velocity and direction of the marine currents. This table consist of more than 50000 registers with the successive positions of the buoys, showing the track of the water masses and giving information regarding surface currents that are very important to know the spatial evolution of the oil slicks.

To investigate other processes that could have condition the behaviour of the oil slicks other information from other institutions has been included in the databases. These parameters are temperature, salinity at different depths, amount of fuel in the Galician and Cantabrian platforms and amount of hydrocarbons in the water column in both platforms. These datasets were obtained by the different survey carried out by the Instituto Español de Oceanografía (IEO). The digital model of the ocean bottom was obtained through the NOAA's Laboratory for Satellite Altimetry (Miller et al., 2003).

Satellite data was also used because can offer complementary data in a wide area and variable temporal resolutions. The processing of this information allowed to obtain other information like sea temperature, presence of terrigenous material from the continent and biological phenomena like Algae bloom. This information

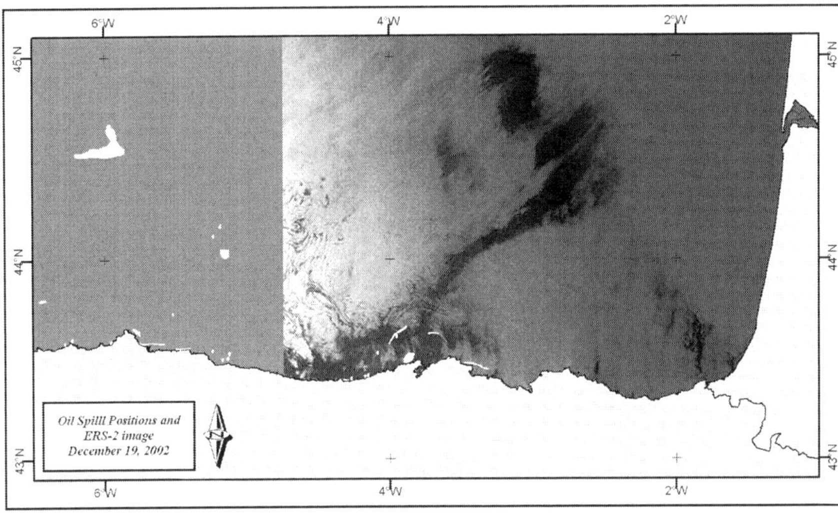

Fig. 11.3 Chlorophyll_a concentration on March 1, 2003, derived from SeaWifs satellite

is very important to determine the oceanographic conditions during the oil spill and its correct interpretation.

We used processed radar images from the European satellites Envisat (ASAR), ERS-1 and ERS-2 that were obtained trough the European Space Agency (ESA). Other data, obtained trough the Instituto Nacional de Técnica Aeroespacial (INTA) and its CREPAD program: This data sets include products derived from the AVHRR sensors (all individual bands, radiances, reflectances, brightness temperature, and sea superface temperature – SST), SEAWIFS (all individual bands, Chlorophyll_a) y MOS B/C (all individual bands), of the Iberian Peninsula zone and Bay of Biscay area. Between November, ten, 2002 and July, thirty, 2003. (see Fig. 11.3)

11.3.3 Spatial and Geostatistical Analysis

A high advantage of using GIS is the ability to perform spatial analysis of the stored information. The software can make automatically different calculations between the operating variables: spatial consults (e.g. length, surface and volume calculations, location, etc), correlation, superpositions, distances, proximity analysis (buffer), cost distance, contour, etc.. This calculation can be represented to generate new levels of information to be used. We have used the extension Spatial analysis of ArcGIS to carry out spatial analysis.

Another extension, Geostatistical analysis, allowed to perform this type of analysis and allow physical and mathematical modelling of spatially distributed phenomena. Spatial distribution of oceanic parameters (temperature, salinity, density, pH, fluorescence, dissolved oxygen) can be interpolated by different means, creating continuous surface from a set of disperse measurements. Results of this analysis are stored in the system as new layer of information that can be managed by the GIS.

11.4 Results

The result of all the described process is a system in which all the information regarding the Prestige's oil spill has been integrated. The system also has a great flexibility to respond to new problems and necessities. Each level of information has been represented with a simbology that allow describing location, geometry, spatial relation with other elements, together with other qualitative and quantitative characteristics. The stored information is continuously updated in the GDB for easy access and use.

The goal of the thematic cartography is to represent qualitative and quantitative geographic phenomena in a cartographic base that allow to study their spatial and temporal distribution.

The cartographic products that can be obtained, are:

- Positions of the Prestige.
- Location and characteristics of the marine pollution sights (see Fig. 11.4.)
- Zones of maritime and aerial survey.
- Maps with the position lagrangian buoys.
- Spatial distribution of different oceanographic parameters.
- Interpretation of remote sensing data.

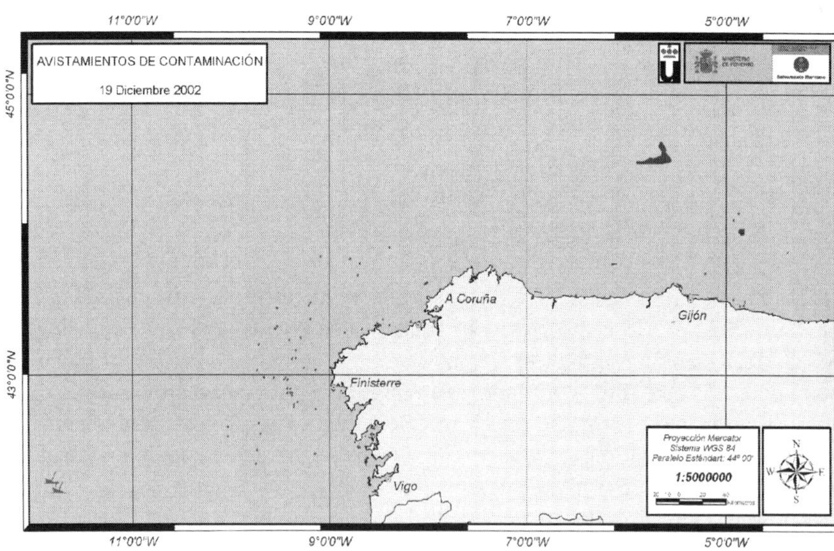

Fig. 11.4 Example of daily map with the oil spill distribution (See also Plate 18 on Page 401 in Color Plate Section)

The information shown by the GIS is the information that exist in the fields that have been defined in the GDB. The queries select records from one or more tables in a database so they can be viewed, analyzed, and sorted on a provided common datasheet. The user can used criteria or expressions for the query with wildcard symbols and arithmetic operators by common SQL sentences.

Spatial consults regarding position, distances and areal measurements can be performed directly on the cartography. When these consults are made the elements that match the consult parameters are selected in the tables and in the graphical output.

The structure of the GDB also allows obtaining reports, graphics and statistics. The system is flexible and allows to be updated and to obtain new documents when new situations require it.

The study of the oil spill caused by the tank ship Prestige is an important part of this research work. The main goal was to show the location, extent and other characteristics of the different oil slicks in useful cartographic documents (see Fig. 11.4.) To carry out this goal was necessary to generate the different cartographic elements with the help of ArcGIS's edition tools.

Regarding the operative part of the GIS, the main goal was the characterization and trajectories of the oil spill, together with that of the different available resources to fight the pollution caused. The GIS allowed to make easy consults of the location, distances and area necessary to manage these hazard situations. As information is structured in days we cab generate diary maps on the location of the oil slicks, survey area for the different units, etc.

The reconstruction of what happened during the Prestige crisis through the use of maps and reports is necessary to analyze and review the different actions carried out by the people in charge of the surveillance operative that tried to fight the oils spill. Mapping of trajectories buoys show the circulation patterns ob the sea and allow to make predictions regarding the evolutions of the oil slicks. Maps with the air and maritime surveillance areas allowed to evaluate the effectiveness of the operations and to carry out the planning for new missions that optimize the available resources.

The design of the database forms and the standardization of the variables offer a systematic and coherent integration (Rodríguez, 1997) This, automates and improves the new information in order to reduce the time employed in data management. (Ministerio de Fomento, 2001). This automation allow to generate specific cartographies (e.g. projection, scale, symbols, etc) together with reports and tables that make easier the decision-making process.

The environmental information of the GIS: hydrography (e.g. temperature, salinity, Chlorophyll, etc.), meteorology, currents, bathymetry, together with the extracted through remote sensing, allowed to determine the conditions of the sea and the process that can be related to the trajectory and behaviour of oil slicks.

11.5 Conclusions

The protection of marine environment and the oil pollution management from accidents and illegal operational requires taking into account a great number of parameters to control the operating processes and to plan response situations. GIS tools are very useful to treat information regarding the environment, allow to manipulate large databases, help to generate cartographic products, and to generate documents that help the decision-making process and the optimizitation of the available resources.

The design of the GIS is a key process to assure operability . It is very important to involve the final users in the design to better determine the necessities and capabilities of the system.

Our GIS is pioneer because of high amount of data treated and the implication that the main users throughout the project.

It is necessary to review the different actuations carried out by the different institutions in the case of the Prestige's oil spill in order to evaluate its efficiency and to analyze mistakes and success to improve future actuations.

Study and analysis of the information is helping to understand the patterns on the evolution of the oil spill and the important of the environments conditions and natural processes on the evolution of this type of marine pollution.

Systematic and structured storage of the information in the GDB represent an archive of the information regarding the Prestige catastrophe for a future use in scientific studies and operational management of these hazardous situations.

References

Carreño, F. and Rodríguez, I. (2003). "Aportaciones de los Sistemas de Información Geográfica (SIG) a la gestión de vertidos contaminantes en el mar." *II Congreso Internacional de Ciencia y Tecnología Marina. "Océanos III Milenio"*. Alicante.

ESRI. Environmental Systems Research Institute. (2001). *What is ArcGIS?*. Environmental Systems Research Institute, Inc., 71.

IMO/IPIECA. International Maritime Organization, International Petroleum Industry Environmental Conservation Association (1994). *Sensitivity mapping for oil spill response*. Ipieca Report Series. Vol 1.

Li, Y., Brimicombe, A.J., and Ralphs, M.P. (2000). "Spatial data quality and sensitivity analysis in GIS and environmental modelling: the case of coastal oil spill." *Computers, Environment and Urban System*, 24, 95–108.

Miller, M., Smith, W., Kuhn, J., and Sandwell, D. (2003). An Interactive Global Map of Sea Floor Topography Based on Satellite Altimetry & Ship Depth Soundings. <http://ibis.grdl.noaa.gov/cgi-bin/bathy/bathD.pl>.

Ministerio de Fomento, (2001). Orden de 23 de febrero de 2001 por la que se aprueba el Plan Nacional de Contingencias por Contaminación Marina.

Rodríguez, I. (1997). "Aplicación de los Sistemas de Información Geográfica en estudios de evolución costera. Ejemplo la desembocadura del Río Ebro." *Ingeniería Civil*, No 105: 25–31.

Trigueros, J. (2002). "El desafío de la gestión de las zonas costeras en Europa." *1er Foro Europeo sobre Estrategias Comunitarias para la Gestión Integrada de las Zonas Costeras*. Alicante. Spain.

Chapter 12
Aquaculture in Sicily: Ecological and Economic Aspects

Paolina Curro and Vincenzo Maccarrone

Abstract In Sicily, the main Italian island, both intensive and extensive systems of culture of euryhaline fish species have been developed. In Sicily in 1998, the production was 200 t, 20% of the national produce. In order for aquaculture to be successful, activity must not cause unacceptable modifications in water quality or interfere with other uses, such as tourism or industry of coastal areas. The aspects are extensively evaluated in studies on impacts of fish farms along the coastal zones. In this chapter we wish to delineate structure and problems of aquaculture in Sicily, emphasizing developments, especially of offshore aquaculture and its socioeconomic role.

Keywords Aquaculture · SSD · GIS · ICZM · Planning

12.1 Introduction

The development of aquaculture in Italy, and more generally in industrialized countries, has revealed a series of new problems arising from the need to identify models of sustainable development able to reorganize economic and ecological aspects (Folke and Kautsky, 1992).Aquaculture is an important alternative to fishing, but has very important environmental implications (Fernandes et al., 2001; Munday et al., 1992). It operates on two highly sensitive components of the ecosystem- water and the fish population. The quality of the water inevitably deteriorates and the fish are exposed to situations which interfere directly or indirectly with their behaviour and genetic structure (EEA, 2003).

Any more or less relevant repercussions and implications on the dynamics of ecosystems, which represent, goods of collective interest, have given another

P. Curro (✉)
University of Messina-Dipartimento di Studi su Risorse, Impresa, Ambiente e Metodologie Quantitative-Piazza Pugliatti 1- 98124 Messina, Italy
e-mail: pcurro@unime.it

D.R. Green (ed.), *Coastal and Marine Geospatial Technologies*, Coastal Systems and Continental Margins 13, DOI 10.1007/978-1-4020-9720-1_12,
© Springer Science+Business Media B.V. 2010

dimension to the positive aspect of aquaculture in areas which are more sensitive to environmental problems.

Therefore opinions contrary to a "cowboy" aquaculture (Edaw and Chapman/Hill, 1986; SWCL, 1988) have stimulated a more careful and stricter research of models with good environmental and social compatibility . In order to promote a sustainable management of coastal areas, the Territorial and Environmental Dept. of the Regione Siciliana has recently proposed some rules to mitigate the impact of aquaculture on the ecosystem and on activities occurring on the same territory, which must be observed by people authorized to carry out aquaculture.

In this chapter we consider compatibility between existing and planned plants in relation to the model proposed by the Regione Siciliana as far as the following important parameters are concerned: the distance from urban areas; agricultural and industrial types of activity; recruiting areas of fish species. These distances must be respected before the maritime authorities can authorize concessions for fish farming at sea.

12.1.1 Aquaculture in Sicily

Over the past few years there has been a continuous and intense development of aquaculture for fish like seabream, seabass, dentex, greater amberjack and recently, for red tuna.

At present, Sicily holds an important position on an international and Mediterranean level as far as plants and productive capacity are concerned. Innovations are active and numerous, stimulated by an increased popularity of Sicilian produce on the market and a greater integration with the fishing industry which is penalized by restrictive measures imposed by the need to protect fishing resources.

Moreover, cheaper competition from countries along the Mediterranean shores and from other parts of the world makes a policy of diversification necessary, concerning specific production and size and the choice of certification of product quality, as a competitive instrument, and as a guarantee of safety for consumers.

Favourable environmental conditions aid the development process which has solid historical roots connected with the farming of euryhaline species existing in the salt lakes near Trapani, reaching very advanced technological levels. At the beginning, the plants were along the Western coasts of the island, with structures mainly on the shores, but later they increased and were to be found along all the coasts. First they were on the shores, but then in open sea: floating cages are positioned in protected areas and off- shore, and have the advantage of reduced costs of investment and management because of improved quality of the product and a reduced conflict with other coastal activities.

In Sicily, 24 plants are active in sea and salty water and 18 have been requested. They are to be found along the shores of the whole island, in areas defined as urban, where the population increases significantly during the summer season and where many other coastal activities take place.

12.2 Sea-Farming Activity Planning

Intensive fish farming at sea in cages or enclosures or in shore plants creates an impact on the environment in the surrounding areas, damaging them according to the intensity of the activity and depending on the ecological and physical characteristics of the environment (Findlay and Watling, 1994; GESAMP, 1996; Gowen and Bradbury, 1987; Gowen et al., 1988; Hargrave, 1994). Polluting substances derive from unused fodder, catabolytes and stools of the organisms, eventual substances used for therapies, products used to avoid fouling (Gowen and Bradbury, 1987; Ackefors and Enell, 1994; Wu, 1995; Axler et al., 1996; Kelly et al., 1996) (Fig. 12.1).

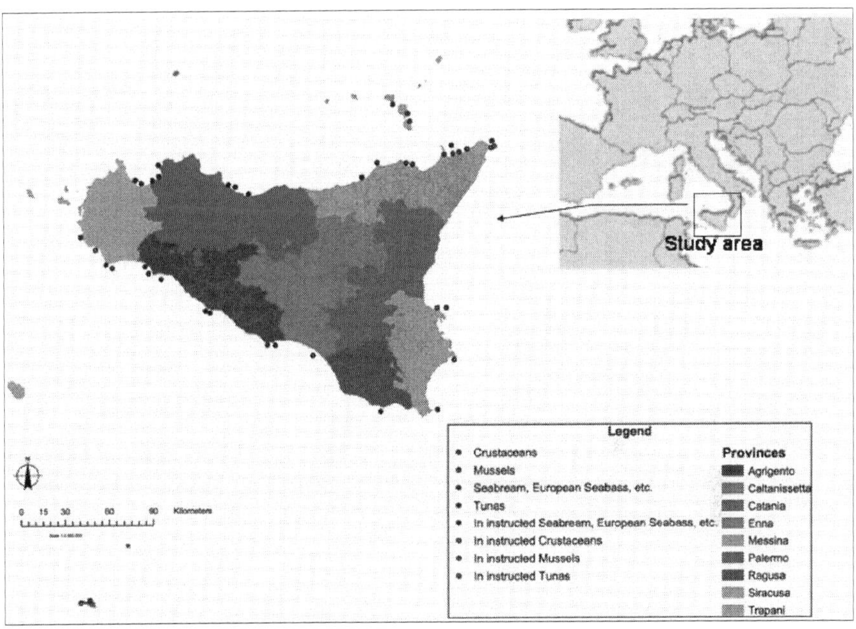

Fig. 12.1 Position of the existing and planned plants along the Sicilian coasts and the distribution of species and provinces (See also Plate 19 on Page 401 in Color Plate Section)

The impact of aquaculture activity is highly specific to the site and standard solutions cannot be predicted in each case – however the following suggestions can be made:

A number of variables to be considered, the method of data acquisition for each type of variable, and the limits these values have to be restricted to.

Sea fish- farming in densely populated areas like the Sicilian coast, which is already used for many other activities, can create conflicts and competition. Before starting up, it is vital to verify any possible interactions with other uses of the sea or the shore as a resource (Ackefors and Enell 1990; Ackefors and Grip 1995).

These uses can be: coastal local fishing (Wildsmith, 1992), industrial and local industries, maritime transport etc (Katranidis et al., 2003). Therefore sea fish- farming should be integrated with other activities in order to allow a balanced socio-economic development of the whole local population. Synergy can be found in the induced, either in the construction or management phases. For example, sea fish-farming can find valid support in local fishing and sometimes it can be a stimulus to its own balanced development. It can represent one of the few solutions to a total or partial conversion imposed for some kinds of fishing (swordfish or drag fishing).

Since sea fish-farming can have possible effects on the environment, even at a certain distance, the Regione Siciliana has published "Rules to mitigate the impact of aquaculture" for sea fish-farming industries in order to encourage the development of sustainable fish-farming at sea to protect the environment and local economy. These rules contain information needed to evaluate potential impact on environmental resources, ecological sustainability and social acceptability before giving permission for starting up a plant. The proponent must fill in forms giving information including a description of the socioeconomic context in which the activity will be.

Table 12.1 shows the activities which could interfere with aquaculture and the minimum distances to be observed, proposed by the guidelines prepared by the Regione Siciliana, and the potential environmental risks.

The model proposed by the Regione Siciliana is an instrument of preventive measures against impact, to ensure that official bodies can make decisions based on better information than is available now, and bearing in mind economic downfalls, the possibility of already existing activities or the possibility of increasing their potential, and the ways in which the negative effects on the environment may be avoided or reduced to a certain extent (Rosenthal, 1994, 1997).

The aim of the guidelines is to suggest the choice of site to whoever is planning aquaculture activities at sea or on the shore, to give Management instruments to constantly control that the limit of environmental acceptability is observed. The use of the area which is to be used is also described and all this should protect the health

Table 12.1 Minimum distances to be observed by off-shore plants (MDA = Minimum Distance Allowed)

Type of activity in neighbouring areas	MDA (Km)	Connected risks
Protected Areas (in general)	5	Alloctone introduction
Protected areas (Naturalistic or archaeological)	3	Alloctone introduction and /or Visual impact
Urban areas	3	Contamination and increased Impact
Industrial agricultural activities	5	Contamination and increased Impact
Recruiting areas of fish species	1	Alloctone introduction
Aree dedicate a peculiari attività di pesca	1	Over-fishing
Fish organism breeding areas	8	Contamination and increased Impact
Mollusc breeding areas	3	Increased impact
Sport and/or recreational areas	1.5	Visual impact

Source: Territorial and Environmental Ministry.

of the population with the aim of evaluating environmental compatibility of the area and safeguard the interests of other simultaneous activities (O'Sullivan, 1992; Garrett et al., 1997; Midlen and Redding,1998).

For a correct management of the coast, the development of sea fish- farming is ruled by Law N°.146/94 (EEC Directive 85/337 and 97/11) and Law 152/99 (EEC Directive 91/271 and 91/276). Before any use of environmental resource, the planner must give the competent Authorities information concerning the state of the area to be used and the potential effects on the environment. After the Directive 97/11/UE, Aquaculture and Sea-farming also have to undergo an Environmental Impact Assessment (EIA), and this Directive applies if the plant is new or if it is an extension of existing activities.

The EIA is an efficient instrument to ensure that decisions to develop activities are made on the basis of available information concerning the environmental reality, subjected to exploitation.

The approach suggested by the Strategic Environmental Evaluation, adopted in 2001 (SEA) 2001/42/EC ensures that the consequences of projects are identified during planning and after they start functioning, thereby allowing a greater integration of planned activities in the environmental context in which they take place. The guidelines proposed by the Regione Siciliana also derive from rules and regulations observed in many countries in the world, where aquaculture represents an important quota of the Gross Domestic Product (GDP) .The regulations of the Scottish Environment Protection Agency – SEPA), the Canada Environment and USA Agencies belonging to the National Oceanic and Atmospheric Administration have also been taken into consideration.

12.3 Geographical Information Systems (GIS)

There are many environmental and territorial variables that need to be evaluated in order to define the suitability of the sea for off-shore breeding activities. This is why geographic information systems are important to conduct an analysis between environmental components and breeding activity. They have proved to be useful in the development and increase in planning programmes in aquaculture activity (Meaden and Kapetsky, 1991; Kapetsky and Travaglia, 1995; Aguilar-Manjarrez, 1996; Ross, 1998), thanks to the possibility of visualizing, cataloguing and comparing physical, biological and economic variables of the analyzed ecosystems.

Research concerning the use of this technology in aquaculture has studied aspects of the scale (continental, national, regional and local) and of the various species bred (molluscs, shellfish, fish), contributing greatly to the diffusion of possible applications of geographic information systems in aquaculture (Aguilar-Manjarrez and Ross, 1995; Nath et al., 2000)

It is possible to use geographic information systems near coastal areas, and this emphasizes the characteristics of space-type analysis instruments and planning (Populus et al., 1997), thanks to the possibility of elaborating information using various data (oceanography, ecology, economy, etc.) The GIS is one of the few instruments able to clearly express functions, uses and interactions of the ecosystem (Kapetsky e Aguilar-Manjarrez, 2002). The results, due to the fact that they concern various disciplines, make it possible to initiate revision processes of legislative aspects and/ or rules in force in the areas studied.

A geographic information system has been created in order to verify the corresponding distances between sea- breeding plants and neighbouring areas suggested in the document prepared by the Regione Siciliana. By using data relative to the characteristics of the continental and marine environment, it has been possible to describe the environmental context of sea-breeding activities present in Sicily.

Data relative to the description of the marine environment have been furnished by the Sea Defence System (Si.Di.Mar)[1] – Ministry of the Environment and Defence of the Territory- whereas geographical information concerning the coastal environment belong to the I&CLC 2000 CORINE Land Cover[2] project, produced on a national level by APAT (Agency for the Protection of the Environment and for Technical Services). Geographical data relative to the distribution of fish-farming activities along the coast were furnished by the Regione Siciliana – Territorial and Environmental Ministry.[3] Thanks to this information it has been possible to localize off-shore fish farms already functioning and planned ones which could be functioning over the next few years.

Since the studied area is vast, and elements present on the territory are not homogenous, it is easy to realize that a map of the use of the territory required a definition of the homogenous group categories. This is why we adopted the classification system proposed by the Corine Land Cover Project. The CORINE cartography classification is based on 44 classes distributed on three hierarchical levels- the details and thematic contents make it especially suitable for environmental analysis on a regional and national scale. In this case we used the III level which contains 44 classes.

Preliminary operations of collection and reorganization of map information have allowed us to realize a system capable of measuring the resources Sicilian aquaculture is in conflict with, and they have been compared with those in the regional document reported in Table 12.1.

12.4 Results

The check concerned the relationship between aquaculture activity and marine biotopes in the census carried out by the SiDiMar data bank and the Corine Land Cover map . In particular, the minimum distances allowed (MDA) were verified relative to the types of activity present in areas next to the plant. The results of the measurements are reported in Table 12.2, where the effective distance from the biotope taken into account can be seen, as can the distance suggested by the guidelines of the Regione Sicilia.

Table 12.2 MDA measurements from marine (Benthos distance) and from land ecosystems (CORINE distance)

				Sea Environment			Terrestrial Environment			
Column I	Column II	Column III	Column IV	Column V	Column VI	Column VII	Column VIII	Column IX	Column X	Column XI
Fish Farming Species	CODE LEVEL III	Label Level3	Marine biotopes	Distance Benthos meters	Distance provide meters	Difference metres	Distance Corine meters	Distance provide meters	Difference meters	Total Results
Been requested crustaceans	123	Port areas	Mosaic of Posidonia live and dead	0	1000	-1000	2100	5000	-2900	
Been requested seabream, seabass etc	111	Continuous urban fabric	Cimodocea nodosa	0	1000	-1000	800	3000	-2200	
Been requested seabream, seabass etc	112	Discontinuous urban fabric	Cimodocea nodosa	1800	1000	800	2400	3000	-600	
Been requested seabream, seabass etc	112	Discontinuous urban fabric	Cimodocea nodosa	600	1000	-400	1900	3000	-1100	
Been requested seabream, seabass etc	112	Discontinuous urban fabric	Cimodocea nodosa	1400	1000	400	1100	3000	-1900	
Been requested seabream, seabass etc	123	Port areas	Cimodocea nodosa	300	1000	-700	1600	5000	-3400	
Been requested seabream, seabass etc	123	Port areas	Mosaic of Posidonia live and dead	1800	1000	800	4000	5000	-1000	
Been requested seabream, seabass etc	211	Non-irrigated arable land	Cimodocea nodosa	2600	1000	1600	3200	5000	-1800	
Been requested seabream, seabass etc	211	Non-irrigated arable land	Posidonia on dead matte	0	1000	-1000	1900	5000	-3100	
Been requested seabream, seabass etc	221	Continuous urban fabric	Posidonia on sand	200	1000	-800	1400	5000	-3600	
Been requested seabream, seabass etc	242	Complex cultivation patterns	Posidonia on rock	1700	1000	700	2300			
Been requested seabream, seabass etc	242	Complex cultivation patterns	Mosaic of Posidonia alive and dead	0	1000	-1000	2100			
Been requested seabream, seabass etc	323	Sclerophyllous vegetation	Posidonia on sand	1100	1000	100	1300			
Been requested seabream, seabass etc	332	Bare rocks	Posidonia on rock	1600	1000	600	1700			
Been requested istruttoria tuna	112	Discontinuous urban fabric	Posodonia on biologic substrate	900	1000	-100	1400	3000	-1600	
Been requested istruttoria tuna	211	Non-irrigated arable land	Posidonia on rock	22300	1000	21300	3700	5000	-1300	
Been requested istruttoria tuna	321	Natural grasslands	Posidonia on matte	500	1000	-500	1200			
Been requested istruttoria tuna	323	Sclerophyllous vegetation	Posidonia on matte	200	1000	-800	400			
Mussel	512	Water bodies	Posidonia on sand	2200	1000	1200	0			
Mussel	521	Coastal lagoons	Cimodocea Nodosa	1000	1000	0	0			
Seabream, Seabass etc	111	Continuous urban fabric	Posidonia on matte	900	1000	-100	1500	3000	-1500	
Seabream, Seabass etc	111	Continuous urban fabric	Posidonia on sand	2200	1000	1200	700	3000	-2300	
Seabream, Seabass etc	112	Discontinuous urban fabric	Posidonia on sand	4000	1000	3000	4600	3000	1600	
Seabream, Seabass etc	112	Discontinuous urban fabric	Posidonia on matte	600	1000	-400	1100	3000	-1900	
Seabream, Seabass etc	112	Discontinuous urban fabric	Cimodocea nodosa	500	1000	-500	1200	3000	-1800	
Seabream, Seabass etc	123	Port areas	Cimodocea nodosa	1300	1000	300	2600	5000	-2400	
Seabream, Seabass etc	222	Fruit trees and berry plantations	Posidonia on sand	21600	1000	20600	1700	5000	-3300	
Seabream, Seabass etc	222	Fruit trees and berry plantations	Posidonia on sand	2200	1000	1200	1200	5000	-3800	
Seabream, Seabass etc	223	Olive groves	Cimodocea nodosa	1100	1000	100	1600	5000	-3400	
Seabream, Seabass etc	223	Olive groves	Posidonia on sand	26500	1000	25500	900	5000	-4100	
Seabream, Seabass etc	242	Complex cultivation patterns	Cimodocea nodosa	900	1000	-100	2700			
Seabream, Seabass etc	243	Land principally occupied by agriculture	Cimodocea nodosa	1100	1000	100	3900			
Seabream, Seabass etc	323	Sclerophyllous vegetation	Cimodocea nodosa	0	1000	-1000	800			
Seabream, Seabass etc	323	Sclerophyllous vegetation	Posidonia on matte	2200	1000	1200	3300			
Seabream, Seabass etc	323	Sclerophyllous vegetation	Posidonia on rock	2700	1000	1700	1300			
Seabream, Seabass etc	323	Sclerophyllous vegetation	Posidonia on rock	400	1000	-600	1100			
Seabream, Seabass etc	331	Beaches, dunes, sands	Cimodocea nodosa	500	1000	-500	1700			
Seabream, Seabass etc	511	Water courses	Mosaic of Posidonia live and dead	800	1000	-200	1400			
Tuna	112	Discontinuous urban fabric	Cimodocea nodosa	1600	1000	600	3000	3000	0	
Tuna	112	Discontinuous urban fabric	Posidonia on rock	1000	1000	0	2200	3000	-800	
Tuna	223	Olive groves	Posidonia on matte	5500	1000	4500	5300	5000	300	
Tuna	323	Sclerophyllous vegetation	Posidonia on sand	600	1000	-400	800			

Legend	
	Plant observes the suggested distances
	Limit condition
	Discordance between measurements

In Table 12.2, from left to right, one finds: species bred; the Corine code, the type of land closest to the classified fish-farm according to the III detail level; the type and distance of the marine biotope closest to the plant; the minimum allowed distance suggested by the regional guidelines for the studied marine biotopes, the difference between the values in columns V and VI; the distances of the plants from the Corine Land Cover in column VIII. Column IX reports the minimum allowed distances between the various types of activity suggested by the regional Table and those found in the Corine Land Cover (empty spaces, culumn IX, Table 12.2).

The correspondence with the specific requirements of the Regione Siciliana are expressed by a numeric value and also by the colour of the squares: green represents agreement between measurements, i.e: where the plant observes the suggested distances: red shows discordance; yellow shows measurements on the borderline of suggested distances (Table 12.2) (Fig. 12.2)

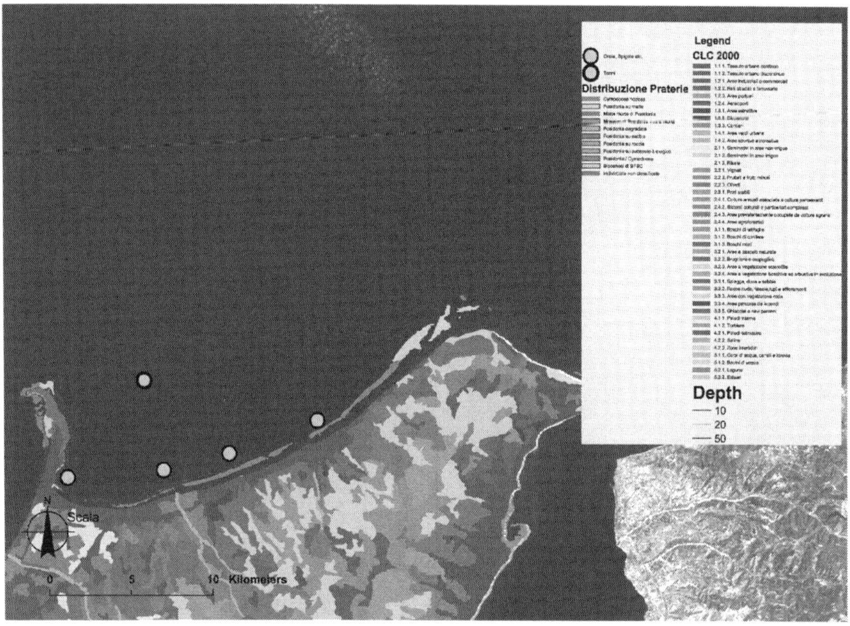

Fig. 12.2 Detail of plants and biotopes (circled in *yellow-green*) distributed along the Tyrrhenian coast in the province of Messina (brown in Fig. 12.1.) (See also Plate 20 on Page 402 in Color Plate Section)

12.5 Conclusions

Thanks to the recent institution of the Regional Agency for the Protection of the Environment (Italian ARPA), it is now possible to predict periodical check-ups on the activity of fish farming at sea. The documents needed to obtain permission from the State property office are deposited with the competent Authorities, and allow us to have information concerning initial environmental conditions of each site used for fish-farming with a certain reliability. Information concerning changes undergone by the neighbouring areas over periods of time, however, is scarce. This makes it necessary to define monitoring plans and to have better and more available information. An evaluation of the variables which are present on a vast scale is an extremely important aim of the specific programme to lessen impact. This approach means a substantial change in environmental monitoring programmes which are at present the duty of the fish farms.

Adoption of referral protocols allows a definition of an environmental standard, consequent performance measures and indications on any eventual unacceptability of impact.

One possible development could be a "sea regulation plan" where areas suitable for off-shore breeding are shown, as well as types of organisms which can be bred and surfaces available on request.

12.6 Appendix

1. The CORINE Land Cover is recognized on a European level as a basic instrument for the definition of territorial policies of various European Commission services, such as DG- Regional Policy, DG- Environment and DG- Agriculture besides the AEA and network belonging to the European Topic Centres (ETCs). The applications of the Corine Land Cover (CLC) in planning are many, including the elaboration of plans for an integrated management of the coastal areas ( ).
2. The SiDiMar data Bank collects data from regional observation points for the marine environment. Besides the oceanographic data, the SiDiMar collects data concerning geographic distribution of benthonic environments, such as Cymodocea nodosa, Posidonia oceanica, at SFBC depths; characterized by sands of a certain diameter, indicating a biocenosis, living at mobile depths , characterized by a specific list of species repeated in time and space. The SiDiMar is still the only data bank which collects data relative to the marine environment on a national level, in GIS form.Verified and valid data are transferred into the SiDiMar data bank, connected on-line to the ANPA and Regions, and are available on internet, so that they can be of benefit.
3. The data Bank belonging to the Territorial and Environmental Dept. of the Regione Siciliana- Maritime Property Office, on the basis of the granted permissions, contains the GIS information concerning the following: Size of the areas occupied by the structures, their precise geographical location, breeding plants and types of organisms bred, besides information concerning the firms. This data base is accompanied by a data bank relative to requests for new authorizations for off-shore fish farming which are preceded by the words "being processed" in the Geographic information system, followed by the type of fish farming.

Bibliography

Ackefors, H., Enell, M., (1994). The release of nutrients and organic matter from aquaculture systems in Nordic countries. *J. Appl. Ichthyol.* 10 (4), 225–241.

Ackefors, H., Enell, M., (1990). Discharge of nutrients from Swedish fish farming to adjacent sea areas. *Ambio* 19, 28–35.

Ackefors, H., and Grip, 1995. The Swedish model for Coastal zone management, Report 4455. Swedish Environmental Protection Agency, Stockholm, 83p.

Aguilar-Manjarrez, J., (1996). Development and evaluation of GIS-based models for planning and management of coastal aquaculture: a case study in Sinaloa, Mexico. *PhD Thesis*. Institute of Aquaculture, University of Stirling, Scotland, UK.

Aguilar-Manjarrez, J., and Ross, L.G. 1955. Managing Aquaculture Development: The role of GIS in environmental studies for aquaculture. GIS World. 8: (3), 52–66.

Axler, R., Larsen, C., Tikkanen, C., McDonald, M., Yokom, S., Aas, P., (1996). Water quality issues associated with aquaculture: a case study in mine pit lakes. *Water Environ. Res.* 68 (6), 995–1011. European Environmental Agency. Environmental assessment report No 10. Copenhagen. Europe′s environment: the third assessment.

Cecchi, G., Munafò, M., Sambucini, V., (2004). La cartografia Corine Land Cover a supporto della pianificazione ambientale a scala regionale e nazionale. GEOMATICA Standardizzazione, interoperabilità e nuove tecnologie. *Atti 8ª Conferenza Nazionale ASITA*. Roma, 14–17 December 2004. http://www.asita.it/8/Pdf/1125.pdf

Fernandes, T.F., Eleftheriou, A., Ackefors, H., Eleftheriou, M., Ervik, A., Sanchez-Mata, A., Scanlon, T., White, P., Cochrane, S., Pearson, T.H., Read, P.A., (2001). The scientific principles underlying the monitoring of the environmental impacts of aquaculture. *J. Appl. Ichthyol.* 17(4), 181–193.

Findlay, R.H., Watling, L., (1994). Toward a process level model to predict the effect of salmon net-pen aquaculture on benthos. In *Modelling Benthic Impacts of Organic Enrichment from Marine Aquaculture* (Hargrave, B. T., ed.). Canadian Technical Report of Fisheries and Aquatic Sciences, Canada, pp. 47–48.

Folke, C., Kautsky, N., (1992). Aquaculture with its environment: Prospects for sustainability. *Ocean Coastal Manag.* 17(1):5–24.

Garrett, E., Spencer dos Santos, C.L., Jahncke, M.L., (1997). Public, animal, and environmental health implications of aquaculture. *Emerg. Infect. Dis.* 3, 453–457.

GESAMP (IMO/FAO/UNESCO/WMO/WHO/IAEA/UN/UNEP Joint Group of Expert on the Scientific Aspects of Marine Pollution). (1996) Monitoring the Ecological effects of Coastal aquaculture Wastes. *GESAMP Reports Studies* 57, 38 pp.

Gowen, R.J., Bradbury, N.B., (1987) The Ecological Impacts of Salmonid Farming in Coastal Waters: A Review. *Oceanogr. Mar. Biol. Ann. Rev.* 24, 563–575.

Gowen, R.J., Brown, J.R., Bradbury, N.B., McLusky, D.S., (1988) Investigations into benthic enrichment, hypernutrification and eutrophication associated with mariculture in Scottish coastal waters (1984–1988). *Report* to the Highlands and Islands Development Board, Crown Estate Commissioners, Nature Conservancy Council, Countryside Commission for Scotland and Scottish Salmon Growers Association. 289 pp.

Gowen, R., Bradbury, N.B., (1987). The ecological impact of salmonid farming in coastal waters: a review. *Oceanogr. Mar. Biol. Ann. Rev.* 25, 563– 575.

Hargrave, B.T., (1994). A benthic enrichment index. In *Modelling Benthic Impacts of Organic Enrichment from Marine Aquaculture* (Hargrave, B. T., ed.). Canadian Technical Report of Fisheries and Aquatic Sciences, Canada, pp. 79–91.

Kapetsky, J.M., Travaglia, C., (1995). Geographical information systems and remote sensing: an overview of their present and potential applications in aquaculture. In: Nambiar, K.P.P., Singh, T. (Eds.), *AquaTech '94: Aquaculture Towards the 21st Century*. INFOFISH, Kuala Lumpur, pp. 187–208.

Kapetsky, J.M.C.D., Aguilar-Manjarrez, J., (2002). GIS for the development and management of coastal aquaculture: present applications and new opportunities. Aquaculture Europe 2002. Seafarming today and tomorrow. *EAS Special Publication* 32, 55–60.

Katranidis, S., Nitsi, E., Vakrou, A. (2003). *Coastal Management*, 1 January, 31(1), 37–53(17).

Kelly, L.A., Stellwagen, J., Bergheim, A., (1996). Waste loadings from a fresh-water Atlantic Salmon farm in Scotland. *Water Res. Bull.* 32(N5), 1017–1025.

Meaden, G.J., Kapetsky, J.M., (1991). Geographical Information Systems and Remote Sensing in Inland Fisheries and Aquaculture. *FAO Fisheries Technical Paper*, No. 318.

Midlen, A., Redding, T., (1998). *Environmental Management for Aquaculture*. Chapman & Hall, London. 223 pp.

Munday, B., Eleftheriou, A., Kentouri, M., Divanach, P., (1992). *The interactions of aquaculture and the environment*. A bibliographical review. Commission of the European Communities. Brussels: Directorate General for Fisheries.

Nath, S.S., Bolte, J.P., Ross, L.G. and Aguilar-Manjarrez, J. 2000. Applications of Geographical Information Systems. (GIS) For Spatial. Decision Support in Aquaculture, Aquacultural Engineering. 23: 233–278.

O'Sullivan, A.J., (1992). Aquaculture and user conflicts. Aquaculture and the environment. *Spec. Publ. Eur. Aquacult. Soc.* 16, 405–412.

Pearce, D.W., Markandya, A., Barbier, E.B., (1989). *Blueprint for green economy*. London: Earthscan Publications.

Populus, J., Loubersac, L., Prou, J., Kerdreux, M., Le Moine, O., (1997). Geomatics for the management of oyster culture leases and production. In: D. Green and G. Massie (ed), *Symposium on GIS and Computer Mapping for Coastal Zone Management*. University of Aberdeen, 29–31 August.

Rampacci, M., (2000). Censimento Nazionale Impianti Piscicoltura Specie Eurialine del Consorzio Unimar. http://www.unimar.it/Sites/Site0004/Documents/00041766000039.pdf

Rosenthal, H., (1994). Aquaculture and the environment. *World Aquaculture* 25(2):4–18.

Rosenthal, H., (1997). Environmental issues and the interaction of aquaculture with other competing resource.

Ross, L.G., (1998). The use of Geographical Information Systems in Aquaculture: A Review. Paper presented at I Congreso Nacional de Limnologia, Michoacan, Mexico. November 1998.

Wildsmith, B., (1992). Aquaculture conflicts in Atlantic Canada. *World Aquaculture* 23(2), 151–162.

Wu, R.S.S., (1995). The environmental impact of marine fish culture: towards a sustainable future. *Mar. Pollut. Bull.* 31 (4–12), 159–166.

Chapter 13
Soft Copy Photogrammetry to Measure Shore Platform Erosion on Decadal Time Scales

Uwe Dornbusch, Cherith Moses, David A. Robinson, and Rendel Williams

Abstract Shore platform erosion is considered to be a driving mechanism for cliff recession on many rocky coasts and a threat to the stability of cliff defence structures, yet the spatial pattern of platform erosion, as well as the rate of lowering, is poorly documented. Measurements based on techniques such as the Micro-Erosion Meter or portable laser scanner, though highly accurate for short time scales and specific locations, are difficult to extrapolate in space and time. They also fail to measure meso-scale changes such as block removal. This chapter describes and illustrates a photogrammetric method of quantifying spatial and temporal changes on shore platforms.

13.1 Introduction

Shore platforms are usually backed by cliffs, which in urbanised areas are often protected by engineering structures such as seawalls and groynes. Where the cliffs are not protected, shore platform erosion combines with wave attack at the foot of the cliff to control cliff-line retreat (Trenhaile, 1997, 2005). Even when protective structures are in place, shore platforms continue to erode, as is evidenced by scour marks around the base of the structures. Quantification of platform erosion is important on natural cliffed coastlines, in order to help predict cliff-line retreat, and on cliffed coastlines that are protected, in order to help estimate the life expectancy of the engineering structures.

This chapter describes a new method for measuring erosion across the whole intertidal width of shore platforms on engineering time scales. The accuracy of the method is demonstrated for a shore platform cut in chalk in Southeast England.

U. Dornbusch (✉)
Department of Geography, University of Sussex, Chichester Building BN1 9QJ, UK
e-mail: uwe_dornbusch@ymail.com

This chapter was published earlier in the Journal of Coastal Conservation 11:193–200 (2007).

D.R. Green (ed.), *Coastal and Marine Geospatial Technologies*, Coastal Systems and Continental Margins 13, DOI 10.1007/978-1-4020-9720-1_13,
© Springer Science+Business Media B.V. 2010

13.2 Existing Methods for Measuring Shore Platform Erosion

To date, efforts to measure rates of platform erosion have concentrated mostly on small-scale spatial and temporal changes. Instruments, such as the Micro-Erosion Meter (Robinson, 1976) and the portable Laser Scanner (Williams et al., 2000), provide highly accurate micro-scale measures of surface downwearing (in millimetres and usually for a maximum of only a few years, with the exception of Stephenson and Kirk (1996)). Unfortunately, the small number and selective location of measurement sites in relation to the frequent large size and varying topography of the shore platforms makes spatial and temporal generalisation of the results difficult. In addition, the instrument legs need to be in contact with fixed reference points, usually in close (<0.1 m) proximity to the measurement surface. This makes it impossible to measure erosive processes like block removal as the reference points are liable to be eroded at the same time as the measurement surfaces. Other shortcomings include the difficulty of measuring surface lowering in confined spaces, such as narrow runnels (gullies), partly because of the size of the instruments, especially the laser scanner, and partly because of the narrow range of heights over which they will operate.

Conventional surveying techniques such as total station surveys can cover larger areas with a high positional and vertical accuracy. However, they are very time consuming when the platforms are wide and the complexity of the micro-relief requires that a high density of points be measured. A dense network of runnels crosses some shore platforms at right angles to the shoreline, and in extreme cases the platform surface rises or falls by over a metre in distances of less than a metre. Also, the repeated submersion of the platforms by the tide hinders the surveying, especially close to low water mark where the platform is exposed only for a short time. In addition, historic data based on total station surveys of sufficient spatial density are generally lacking.

Developments in softcopy photogrammetry and automatic Digital Elevation Model (DEM) extraction (e.g. Ackermann, 1996) allow a much more detailed representation of the topography of shore platforms. Unlike other photogrammetry software where elevation information can be obtained only for points that are several image pixels apart, the software package used in this study (PCI Geomatics Orthoengine 9.1) can perform automatic DEM extraction for each image pixel. In addition, ground control points for the aerotriangulation can be collected with a GPS in sufficient number over large areas and with a high accuracy that allows for very stable bundle adjustments. Using air photographs also has the advantage that historic air photographs are often available.

The authors are aware of only one previous attempt to use air photographs to assess changes in shore platform topography (Stephenson, 2001). This compared photographs of the platforms at Kaikoura in New Zealand taken 52 years apart, but the erosional losses could not be measured due to the small scale of the air photographs and lack of suitable ground control points.

13.3 Site Selection, Data and Method

The chalk coast of Southeast England provides an ideal area for the investigation of platform erosion. Observations and previous studies (e.g. Williams et al., 2000) indicate that the platforms are eroding and the high pixel contrast between the white chalk, the dark flint that it contains, and the seaweed cover is ideal for the image correlation and feature-matching algorithms employed by the software to extract elevation information from stereo pairs of photographs. The coast also includes stretches of platform where sea defences of different ages are present and others where they are absent. In addition, the cliffs are located where dense urban development on the clifftop provides numerous, high-quality ground control points (Fig. 13.1). The test site showing erosion rates and patterns presented in this study is located on the Peacehaven frontage (at "Bastion Steps") where defences were built between 1975 and 1977. A second test site nearby is described in Dornbusch et al. (2008). At both sites, the cliffs and platforms are composed of Newhaven Chalk of Santonian and Campanian age. The mean tidal range is 4.5 m and the coast is very exposed to southwesterly waves travelling up the English Channel.

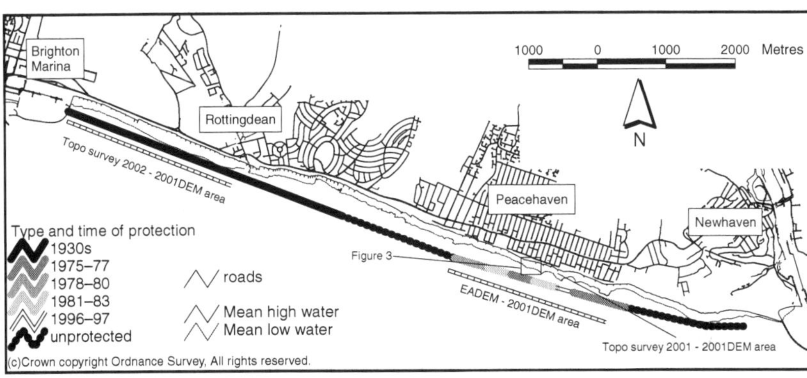

Fig. 13.1 Location of the test area between Brighton and Newhaven, Southeast England, showing schematically the location of seawalls and groynes, and their dates of construction. The location and extent of the DEM test sites are also shown

13.3.1 Data

An Annual Beach Monitoring Survey (ABMS) of the coast of Southeast England is currently carried out by the Environment Agency. Aerial photographs at a scale of 1:5,000 are taken each spring around a spring low tide.

The earliest air photographs selected for examination were flown at the start of the ABMS in spring 1973. Because the flight path lay up to 300 m landward of the cliff line, the cliff toe is sometimes obscured in the photographs. The original black and white negatives have been scanned at 14 μm especially for this study, providing a ground resolution of ~7 cm. Calibration information for the camera, which had a focal length of 152.05 mm, is no longer available. To provide the minimum internal orientation parameters it has been assumed that there was no offset in the principal point and no distortion. The fiducial mark coordinates were calculated by measuring the relative distances between the corner fiducial marks on 3 randomly selected photographs. The resulting DEM is referred to as the 1973DEM.

The latest photographs studied were flown on May 5, 2001. These have been scanned for the Environment Agency at 21 μm, providing a ground resolution of ~11 cm. A full set of camera calibration parameters is available. The resulting DEM is referred to as the 2001DEM.

Ground control points were collected landward of the shore platform using static differential GPS referenced to the Environment Agency GPS network that is tied into the Ordnance Survey National Grid. The GPS point accuracy thus achieved is about ±2 cm in the horizontal and ±2 cm in the vertical. The ground control points were chosen on the basis that they were clearly visible as far as possible on both the 1973 and 2001 photographs. Most were corners of roadside drains and rectangular manhole covers aligned parallel to the flight path.

13.3.2 Method

The photogrammetric analysis was carried out for a 10.5 km long coastal strip between Brighton and Newhaven using PCI Geomatics Orthoengine 9.1. For both sets of photographs, approximately 140 ground control points were used and approximately 100 automatic tie points generated for the shore platform. The root mean square (rms) error for the ground control points and tie points in the bundle adjustment was <0.05 m Automatic DEM extraction was carried out without interpolating between failed cells. This approach has the advantage that only measured cells form part of the DEM and therefore surface elevation comparison is performed solely on measured values, but has the disadvantage that the DEM has holes in it. The resulting geocoded DEMs were exported in grid format to ArcView. All subsequent analyses (elevation comparison between different data sets) were carried out using ArcView 3.2a.

13.3.3 Data Control

To determine whether the photogrammetric software used produces elevation data for each image pixel at sufficient vertical resolution to measure changes over the 28 year period, the 2001DEM was compared to three different data sets: one photogrammetric survey carried out for the Environment Agency and two ground surveys. Due to a lack of alternative elevation data, the quality of the 1973DEM can be assessed only in relation to the 2001DEM, as described in the results section.

The photogrammetric control survey consists of a set of elevation points provided by the Environment Agency at a stated vertical accuracy of ±0.2 m that form part of a coarse DEM (EADEM) of the entire coast of Southeast England based on the same air photographs of 2001. The advantage of this data set is that it uses the same photographs, yet the number, accuracy and distribution of ground control points are different, as are the software and personnel carrying out the photogrammetry, thus providing an independent control. Points on the shore platform were collected only close to low water at the time of the survey and at a mid-platform position.

Figure 13.2a compares the 2001DEM and EADEM along a 2.5 km long stretch of the Peacehaven frontage. More than half (57%) of the 2700 points common to both DEMs show elevation differences of the order of ±0.2 m The difference between the mean of the points in the two DEMs is 0.02 m, indicating a low systematic error (vertical offset between the two surfaces). Points with significantly higher elevation differences often occur close to the cliff edge or on the cliff face (above 25 m OD in Fig. 13.2a), where small changes in horizontal position can have a significant effect on the elevation. In open areas and across the shore platform the differences are generally less than ±0.2 m. Other points with large elevation differences are associated with the undercliff promenade (at ~5.5 m OD in Fig. 13.2a). Here the error introduced by small positional differences either at the edge of the seawall

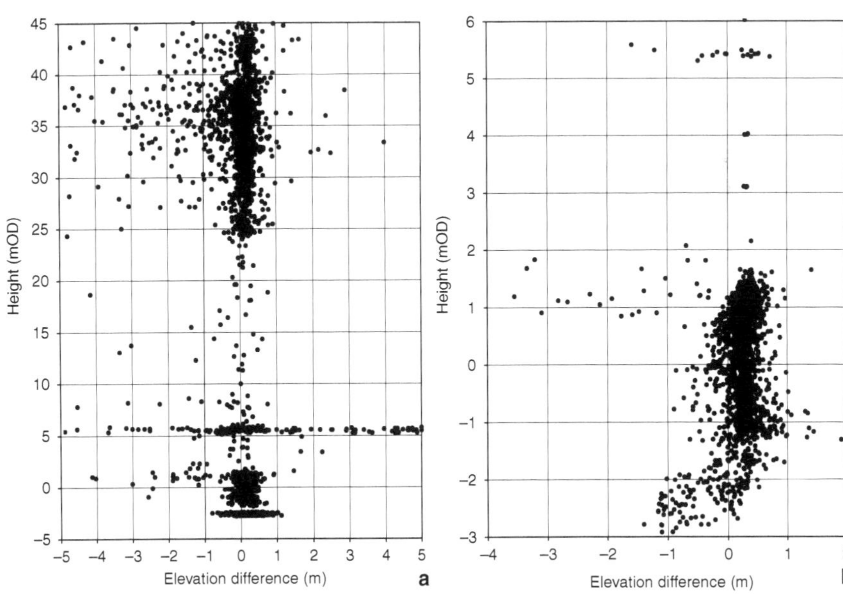

Fig. 13.2 Scatter plot showing (**a**) the elevation difference between the EADEM and the 2001DEM in relation to the EADEM elevation, (**b**) the elevation difference between the total station survey and the 2001DEM in relation to the total station elevation

or at the cliff/promenade junction is increased by the fact that the automatic DEM extraction process performs poorly on the featureless concrete surface.

Two more sets of data control points were obtained from independent ground surveys. In March 2001, Alain Hénaff carried out a conventional total station survey on the shore platform at Peacehaven (Fig. 13.1) over a period of 3 low tides, capturing 2262 points within an area of ~13,000 m^2 (i.e. 1 point every 5 m^2). His main aim was to delimit topographic features on the platform, such as steps and runnels. This was achieved by recording point pairs in close proximity at the top and bottom of the features. The survey was carried out in local coordinates, but in 2005 these were transformed into Ordnance Survey National Grid coordinates using a DGPS survey of the only 5 points in the original survey that could still be identified in the field. The rms-error of the transformed data is assumed to be ± 0.02 m in both the horizontal and vertical.

Comparison between the survey data and the 2001DEM for 2026 points not located on the seawall or groynes gives a standard deviation of 0.17 m, indicating that 60% of all points have an elevation difference smaller than ± 0.2 m (Fig. 13.2b). Variations at 5.5 m OD and between 1 and 2 m OD relate to problems close to the seawall and the groynes. Here, the DEM values record the upper surface of the structures whilst the total station points record the elevation of the toe of the structures. Similarly, the increasing negative differences below -2 m OD are associated with deep runnels at the seaward end of the platform. These are captured by the total station as closely spaced point pairs with large elevation differences, but not by the 2001DEM with its poorer horizontal resolution. If, for example, the 2001 DEM has one pixel of 0.2×0.2 m in extent with a height of -2 m and the total station survey records 2 points of elevation -2.1 m and -3 m within that area, Fig. 13.2 will show two points, one with an elevation of -2.1 m and a difference of -0.1 m and one with an elevation of -3 m and a difference of -0.9 m.

Brighton and Hove City Council carried out another total station survey in May 2002 prior to repair work on the sea defences between Brighton and Rottingdean (Fig. 13.1). Profiles seaward of the seawall were surveyed every ~70 m with points recorded at regular intervals of ~5 m The comparison of 591 points common to this survey and the 2001DEM on the chalk platform in the same area showed that 90% of the points have an elevation difference smaller than ± 0.2 m.

The comparison of elevations in the 2001DEM with those derived by other photogrammetric methods and conventional ground surveys shows that a DEM of the chalk shore platform can be produced from the aerial photographs with a horizontal ground resolution of ~0.1 m and an elevation error of $<\pm 0.2$ m.

Due to a lack of independent control data on the platform for the 1973DEM the only control has to be based on areas of unchanged topography on the cliff top and the assumption that differences in elevation between the 2001DEM and 1973DEM on the shore platform are either negligible ($<\pm 0.2$ m) or negative because the platform can only have remained unchanged in elevation or been lowered.

13.4 Comparison Between 1973 and 2001

Figure 13.3 shows the changes in platform elevation at Peacehaven over the 28 years of survey. Most of the middle and lower part of the platform, together with unchanged areas on the cliff, show elevation differences of the order of ±0.2 m, which would indicate that the 1973DEM has a very similar error in the elevation values compared to the 2001DEM. On the middle and lower part of the platform, erosion is concentrated in lines perpendicular to the shore, due to the development or deepening of runnels. The concentration of these features in a band on the lower part of the platform is because backward erosion of a chalk layer has created a ~1 m high step, which is incised by numerous runnels. However, the most significant feature is the band of erosion that reaches a maximum at distances between ~5 and 15 m seaward of the seawall, where elevation losses well in excess of 1 m and with a maximum of 2.2 m are measured. Elevations close to the foot of the seawall cannot be compared due to the landward position of the flight path in 1973. In addition to the erosion parallel to the seawall, erosion has also been concentrated around the groynes.

Construction drawings (Stammers, 1982) show that the foundations for the seawall and the groynes were dug more than a metre below the surface of the platform. The excavations were filled with concrete up to the level of the platform surface, but because of subsequent platform erosion now form a plinth. The erosion immediately next to the groynes can therefore be measured in the field by measuring downward from the top of the plinth. Field measurements along the eastern side of the eastern groyne shown in Fig. 13.3 range from 0.6 m near the tip to 1.5 m near the middle of the groyne. These measurements match closely those found using the DEM comparison (0.55 and 1.6 m respectively) shown in Fig. 13.3.

At the test site there is very little shingle that could have contributed to the platform erosion so that the measured erosion seems to have been caused by the heavy machinery used in the construction of the concrete groynes and seawall. Stammers (1982) shows two photographs (Figs. 10 and 11) of cranes and bulldozers in front of the seawall and around the groynes in the locations that record the highest rates of erosion. The good spatial correlation between the areas of highest erosion and that of highest construction traffic suggests that the latter either eroded part of the chalk during the construction or fractured and weakened the chalk, making it susceptible to later erosion.

For the period 1973–2001, the maximum erosion of 2.2 m is equivalent to an annual average rate of erosion of 7.8 cm. The average for the whole platform extending seawards between the two groynes shown in Fig. 13.3 (24,228 m^2) is 0.36 m, which is equivalent to a rate of only 0.013 m per year. Using only the area enclosed by the two groynes (4612 m^2) the average lowering is 0.87 m over the 28 years or 0.03 m per year. These averages may conceal important year to year variations that the long interval between the surveys used does not identify. In particular, erosion in the vicinity of the groynes and seawall is likely to have increased significantly immediately following their construction.

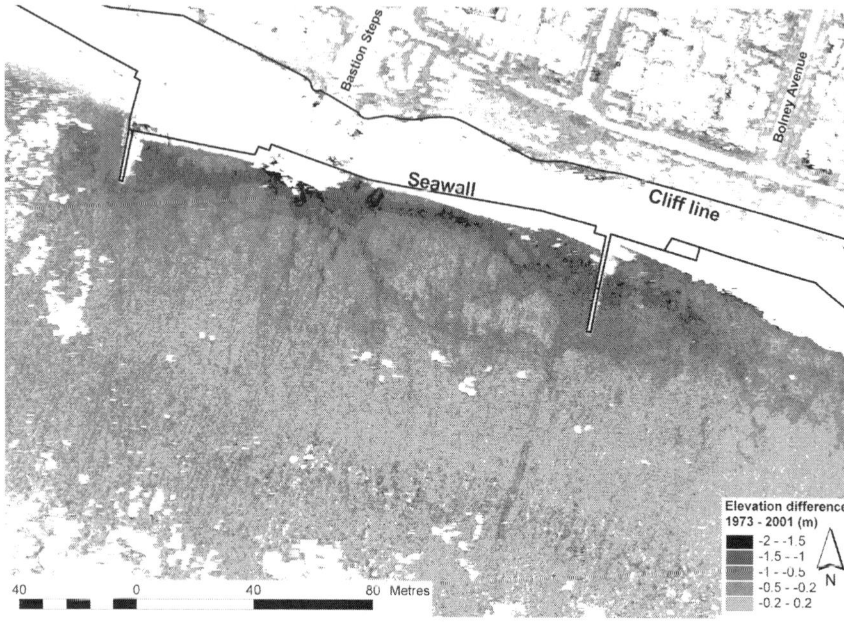

Fig. 13.3 Elevation difference between the 1973DEM and 2001DEM. The *white* area has no data because no elevation values were obtained from either or both DEMs. The cliff top shows the outline of streets with the street verges and the grass area immediately behind the cliff line representing areas with no elevation change between 1973 and 2001

13.5 Discussion

This chapter is concerned to evaluate the use of photogrammetry in platform erosion studies. The comparison of digital elevation models created from high resolution air photographs with independent data shows that the method can produce almost continuous DEMs with grid sizes of < 0.2 m and vertical elevation errors with standard deviations < 0.2 m. Apart from the quantification of erosion, independent of whether this occurs through gradual downwearing or block removal, this method is the first to create a true spatial representation of changing erosion that can then be linked to possible causes. While the middle and lower platform shown in Fig. 13.3 is likely to show natural changes associated with the deepening of runnels and the backward erosion of steps created by individual chalk beds, the magnitude and pattern of erosion on the upper platform would indicate enhanced erosion influenced at least in part by the construction of coastal defences. Further studies are planned of unprotected cliff sections and a recent study on erosion patterns and causes around engineering structures (Dornbusch et al., 2007) has demonstrated the applicability of the method described in this chapter.

13.6 Conclusion

On shore platforms similar to those discussed and using air photographs with a scale of at least 1:5,000, modern softcopy photogrammetry software combined with GPS ground control point collection allows the creation of almost continuous digital elevation models with ground resolutions of <0.2 m and vertical accuracies of a similar magnitude. Surface comparisons can be carried out reliably to detect elevation changes in excess of 0.2 m even for isolated objects only a few pixels in size. The use of air photographs allows access to valuable data sets that go back several decades so that changes in the topography of platform surfaces can be measured over engineering time scales.

Acknowledgements The authors would like to thank the Environment Agency for access to the air photographs and their DEM data, Brighton and Hove City Council for access to their survey data between Brighton and Rottingdean, and the MAST III funded ESPED project for access to Alain Hénaff's survey data at Peacehaven. This study was funded by the Beaches At Risk INTERREG III project.

References

Ackermann F (1996) Techniques and strategies for DEM generation in: Greve, C. (ed), Digital Photogrammetry: an addendum to the manual of Photogrammetry. *American Society of Photogrammetry and Remote Sensing*, Washington DC, pp 135–141.

Dornbusch U, Moses C, Robinson D and Williams R (2008) Soft copy photogrammetry to measure shore platform erosion on decadal time scales. *Journal of Coastal Conservation*, 11:193–200.

Dornbusch U, Robinson D, Williams R and Moses C (2007) Chalk shore platform erosion in the vicinity of sea defence structures and the impact of construction methods. *Coastal Engineering*, 54(11):801–810.

Robinson LA (1976) The micro-erosion meter technique in a littoral environment. *Marine Geology*, 22:M51–M58.

Stammers RL (1982) Coastal defence engineering in East Sussex – Part II. *Municipal Engineer*, 109:353–359.

Stephenson WJ (2001) Shore platform width – a fundamental problem. *Zeitschrift für Geomorphologie*, 45(4):511–527.

Stephenson WJ, Kirk RM (1996) Measuring erosion rates using the micro-erosion meter: 20 years of data from shore platforms, Kaikoura Peninsula, South Island, New Zealand. *Marine Geology*, 131:209–218.

Trenhaile AS (1997) *Coastal Dynamics and Landforms*. Clarendon Press, Oxford.

Trenhaile AS (2005) Modelling the effect of waves, weathering and beach development on shore platform development. *Earth Surface Processes and Landforms*, 30(5):613–634.

Williams RBG, Swantesson JOH, Robinson DA (2000) Measuring rates of surface downwearing and mapping microtopography: The use of micro-erosion meters and laser scanners in rock weathering studies. *Zeitschrift für Geomorphologie*, Supplementband 120, pp 51–66.

Chapter 14
Regionalizing Coastal Zones with Geospatial Tools for Integrated Coastal Zone Management

Jorge I. Euán Avila and Alfonso Cuevas Jiménez

Abstract Integrated coastal zone management (ICZM) requires clearly defined geographical regions. Geospatial tools in geographic information software provide systematic approaches for defining coastal zones and the limits of a nested hierarchy of management zones or units. In this chapter, a comprehensive set of approaches and their implementation are presented using several tools and data such as proximity, topography/bathymetry, watershed analysis, ecotones, hidrogeological data, cultural and administrative features as well as climate, sea surface temperature, ocean color and biodiversity distributions and condition. A simple, illustrative example is provided for the coastal zone of the Yucatan Peninsula in Mexico.

Keywords Regionalization · Management · Coastal boundary · Spatial planning · Remote sensing and GIS

14.1 Introduction

Coastal zones are experiencing increasing strain from urbanization, overfishing, mineral extraction, changing sediment load, coastal deforestation, pollution, tourism, global warming and sea level rise. This strain threatens the future of myriad ecosystems and presents a challenge to coastal zone management (Kay and Alder, 2005; Lindeboom, 2002). Population growth, property values, poverty, ineffective control of development and a lack of integration between management and planning can lead to resource overuse, encourage environmental degradation, and heighten coastal conflicts (Kay and Alder, 2005; Cicin-Sain and Knetch, 1998; Ongkosongo, 1992).

J.I. Euán Avila (✉)
Departmento de Recursos del Mar, Centro de Investigacion y de Estudios Avanzados del IPN, A.P. 73, CORDEMEX, 97310, Merida, Yucatan, Mexico
e-mail: euan@mda.cinvestav.mx

D.R. Green (ed.), *Coastal and Marine Geospatial Technologies*, Coastal Systems and Continental Margins 13, DOI 10.1007/978-1-4020-9720-1_14,
© Springer Science+Business Media B.V. 2010

Fig. 14.1 Nine examples of coastal geomorphology in Mexico (*left to right* and *top to bottom*): Cancún, Quintana Roo; Progreso, Yucatán; Huatulco, Oaxaca; Acapulco, Guerrero; Puerto Vallarta, Jalisco; Manzanillo, Colima; San José de los Cabos, Baja California Sur; Mazatlán, Sinaloa; and Hecelchakán, Campeche (See also Plate 21 on Page 402 in the Color Plate Section)

Mexico's coastal zone is extremely diverse (Fig. 14.1), which allows it to support a heavy human activity load. Most human activities in this zone are extractive, such as fishing and oil drilling, or service-related, such as tourism and transportation (Hernandez and Kempton, 2003; Capurro et al., 2002; Rivera-Arriaga and Villalobos, 2001). The intensity and extent of these activities have begun to overtax the zone's ecosystems leading to impacts such as smaller catches, high levels of water pollution, habitat degradation and loss, reduced visual quality, urban sprawl and loss of public recreational areas. If the pace of current use continues, many human activities may become unsustainable in the long-term.

This same trend is occurring in coastal zones worldwide. Reversing it will require an integrated management approach based on scientific data, the participation of users and administrators and application of sustainability principles (Holland and Bernal, 2002; Chua, 1997; GESAMP, 1996; UNCED, 1992). Developing an integrated coastal zone management (ICZM) program for the large areas covered by coastal zones involves selection of the general boundaries to be used in developing an initial planning framework. A systemic or ecosystemic approach is best adopted in this situation in that it allows landscape characteristics to be properly evaluated and effectively aggregates the effects of human activities, thereby facilitating sustainable resource management (Yáñez-Arancibia and Day, 2004; Franklin, 1997). This regional planning approach to ICZM aids in identifying and verifying compatibility between current activities and zone resources, potential land uses and conservation measures (IMCRA, 1998). Both academics and practitioners in many countries have acknowledged its usefulness in regionalizing and managing these zones. However, establishing the geographic boundaries of regions and dividing them into smaller segments based on common biotic and abiotic attributes and processes poses serious challenges for gathering, integrating, updating, visualizing and accessing data, especially in countries with a limited capacity for researching large areas. Current technologies can bypass this limitation and provide an initial step toward regionalizing coastal zones in tropical areas: satellite remote sensing provides large amounts of detailed data; geographic information systems are effective for integrating and analyzing this data; and the Internet is extremely useful for accessing further data and publishing results.

The regionalization process creates opportunities for land-use planning, including the possibility of assessing the protection status of coastal environments. For this reason, it can be useful in (1) determining each region's level of protection; (2) calculating ecosystem integrity (i.e. ecosystem health); (3) establishing risks to and limitations on long-term viability; (4) identifying management alternatives; (5) developing a framework for management of fisheries and other coastal resources (e.g. beaches and wetlands) that integrates potential and/or current activities (e.g. tourism and aquaculture); (6) facilitating selection of sites and structural and non-structural controls; and (7) recognizing organizational/institutional needs.

14.2 Coastal Zone Boundaries

Regionalizing any area requires clearly-defined boundaries. In the case of coastlines, the most basic approach is to define boundaries based on distance from one axis along the coastline, called a longshore axis, and another (normal) axis, called an "on-off axis" (OECD, 1993). A more effective and useful approach, however, is to use one or more polygons to define the area to be regionalized. These can be established in a number of ways: (a) proximity or buffering based on a reference line; (b) topography/bathymetry; (c) transition lines between ecosystems (ecotones); (d) estimation of watershed boundaries; (e) identification of aquifer seepage and discharge influence; and (f) government administrative boundaries (Table 14.1).

Proximity. This method involves establishing a boundary line at an equal distance from the object in question, in this case, a shoreline (SL). Boundaries established in this way are somewhat arbitrary in that they cut through ecosystems, cross administrative boundaries and disregard topographic and bathymetric data. However, this approach is quite useful in scenarios where data is lacking and economic activities and/or human populations are clearly concentrated within the boundaries. This constitutes a simple use of geospatial analysis systems because it allows testing with different boundaries and monitoring of enclosed surfaces and other attributes. The attribute that best complies with additional selection criteria can then be adjusted or selected. Coastal contours can be obtained from vector-based topographic maps or

Table 14.1 Themes and criteria that can be used in establishing coastal zone boundaries

Themes	Variables/criteria	Findings
Proximity	Buffer zone on both sides of coastline	Shoreline (SL) used as a benchmark and buffer zone calculated at a given distance (d). SL can be obtained through vectors (.shp, .dxf or .e00) or converted to vectors using raster maps. http://www.ngdc.NOAA.gov/mgg/shorelines/gshhs.html
Topography/ Bathymetry	Elevation and depth	DEMs such as ETOPO data from the Shuttle Radar Topography Mission (STRM). In Mexico, DEMs are available from the INEGI (1:250000). Navigation charts. http://glcf.umiacs.umd.edu/index.shtml
Watersheds	Boundaries of watersheds/microwatersheds linked to coastal zone	DEM data and a watershed-analysis program to generate watershed and microwatershed boundaries, along with runoff pattern, slope and aspect data. Watersheds must be selected based on criteria such as size and distance to SL.
Ecotone	Land cover type and condition	Global land use maps from Global Land Cover Facility (GLCF). Multispectral image classification. http://zulu.ssc.nasa.gov/mrsid/mrsid.pl
Human activities	Type and extent of land uses, population distribution, and gross domestic product	Global land use maps from Global Land Cover Facility (GLCF). http://glcf.umiacs.umd.edu/services/ colwidth="3pc" landcoverchange/ Land use maps, population and housing statistics, and national economic data from INEGI. http://www.inegi.gob.mx/geo/default.asp
Administrative boundaries	Municipal, state or national boundaries, EEZ	Maps with municipal, state, and international divisions
Combination	Combining different methods	Management of geospatial tools in a co-registered database

nautical charts in .SHP, .DXF, or .E00 formats while global data can be obtained at http://www.ngdc.NOAA.gov/mgg/shorelines/gshhs.html.

Topography/bathymetry. Topography often serves as a natural boundary for ecosystem development and establishment of productive or recreational activities, which is why several authors have suggested using elevation as a criterion to define the terrestrial limits of coastal zones. Bathymetry is used to determine marine area limits, usually using a given depth associated with light penetration and vegetation growth, although other criteria can also be used. Digital elevation model (DEM) data can be obtained from the Global Land Cover Facility's (GLCF) Internet page (http://glcf.umiacs.umd.edu/index.shtml) in resolutions of 1, 3 and 30 seconds of arc. The Shuttle Radar Topography Mission (SRTM) gathered elevation data to generate the most complete digital topographic database of the planet currently available. In Mexico, the National Institute of Statistics, Geography and Informatics (Instituto Nacional de Estadística Geografía e Informática - INEGI) provides DEMs at a scale of 1:50,000.

Watersheds. The watershed concept is among those that best captures the idea of an ecosystemic approach to ICZM because water flowing off land masses predominantly runs into the oceans, transporting natural and anthropogenic materials (e.g. metals, nutrients, PCBs, viruses and petroleum products, among others). Encompassing this inland influence on coastal ecosystems is vital due to the influence of runoff in these ecosystems and any consequent impacts on the potential for long-term, sustainable development. The area covered by management actions must therefore incorporate activities far removed from coastlines. Defining watersheds and microwatersheds is relatively simple with a DEM and a boundary identification program. Watershed analysis modules are often included in geospatial analysis software, and are effective for identifying watersheds as well as runoff networks, accumulated flows and slopes. DEMs are the primary data source for watershed analysis.

Ecotones. An ecotone is the transition area between ecosystems. Ecosystems such as mangrove swamps, rocky coasts, salt marshes, coastal lagoons, estuaries, delta systems, salt flats, beaches, etc. are associated with the coastal zone, and the presence and condition of these ecosystems are determined by the extent to which the influence of the marine environment extends inland; when an ecosystem boundary is far inland, marine influence is considered to be minimal. Management actions in a coastal zone will influence these associated ecosystems, making ecotones a very practical way of delimiting coastal zone management boundaries. Ecotones are shown on thematic maps, or multispectral images can be used in classification processes to identify coastal ecosystems and their boundaries. The Global Land Cover Facility (GLCF) provides data derived from remote sensing systems that aid in better understanding environmental systems at local and global scales. CIRCA project images are also good sources for useful coastal landscape classifications (Tucker et al., 2004). Using data from these sources, coastal zones can be delimited and marine-influenced infrastructure can be identified, including wetlands, urban areas, agricultural fields, inland water bodies, transport networks, etc.

Hydrogeological basins (aquifers). Subterranean hydrological, or hydrogeological basins are spaces under the earth's surface in which water accumulates. Surface water, with its nutrient and anthropogenic material load, filters into this groundwater, or aquifer, a portion of which discharges into the ocean. These discharges can pose a threat to coastal waters and affect human activities in the coastal zone, meaning groundwater movements and flow directions can be important when establishing coastal management zone boundaries. However, hydrogeological flow networks can be difficult to document, and available data in many countries is often insufficient for use as an alternative source.

Infrastructure and socioeconomic activities. Actions and policies implemented in coastal zones are guided by the need to regulate activities in these areas in an effort to minimize impacts, reduce conflicts and optimize natural resources use. Understanding the spatial distribution of socioeconomic activities allows planners to suggest the most suitable boundaries for management, while defining coastal zone boundaries based on infrastructure and socioeconomic activities is an effective way of ensuring that these are covered by management policy. Analysis of aerial images is the most effective way of documenting land use and changes in it. Boundary definition benchmarks in coastal zones can include maritime traffic, recreation and tourism, subsistence fishing and population, among other factors.

Administrative boundaries. Administration of coastal areas, wetlands and marine territory is usually carried out at the federal level, whereas most coastal zone socioeconomic activities are commonly governed at the municipal level. Using the boundaries of municipalities, states and/or countries on a coast or those containing areas which influence a coast is a way of including governance in management planning. One consideration is that political boundaries vary widely and frequently include areas with no relation to a coastal zone, meaning very restrictive geographic definitions of inland coastal zones need to be applied to exclude areas irrelevant to coastal zone management. A country's Exclusive Economic Zone (EEZ) can be used to define an administrative boundary in marine territory.

Combinations. Combining a number of these approaches in defining coastal zone boundaries is a way of creating an area that best fits management needs. This can be done using geospatial tools in a co-registered database.

14.3 Coastal Regionalization

Forman (1990) defines a region as a mosaic of ecosystems, the stability of which is the key element of regional sustainability. Using this definition, it is important to identify the boundaries of areas that apparently operate as an entire region, and can tolerate changes while still functioning. In coastal zones, these regions are defined by factors such as climate, coastal oceanographic processes, geomorphology/geology and biota, among others (Table 14.2).

Climate. Climate is one of the most important factors defining a region's natural development. It incorporates elements such as ambient temperature, precipitation, and extreme events such as cyclones. Monthly average temperatures are available from weather stations (http://www.ncdc.noaa.gov/oa/ncdc.html), the advanced very high resolution radiometer (AVHRR) or moderate resolution imaging spectroradiometer (MODIS) satellite images. In Mexico, the best source for average monthly precipitation for long periods is the National Water Commission (Comisión Nacional del Agua – CNA). A national geographic data base for Mexico is available through the National Commission on Biodiversity Knowledge (Comisión Nacional para el Conocimiento de la Biodiversidad – CONABIO) (http://www.conabio.gob.mx/informacion/geo_espanol/doctos/cart_; linea.html). Hurricane route maps, including the geographic coordinates of each hurricane's route, are available at http://hurricane.csc.noaa.gov/hurricanes/.

Table 14.2 Themes and criteria that can be used in establishing coastal regions

Themes	Variables/criteria	Findings
Climate	Ambient temperature, precipitation, and extreme events. Climate classification maps (e.g. Kôppen)	Monthly/annual averages based on AVHRR or MODIS satellite images for temperature and average precipitation are available from NASA. Hurricane route maps with geographic coordinates for each route path are available from NOAA. In Mexico, the CONABIO climate map is useful: http://www.conabio.gob.mx/informacion/geo_ espanol/doctos/cart_linea.html
Oceanography	Temperature, primary productivity and overall color of ocean and coastal water masses. Other factors such as wave, current and tide energy	MODIS and SEAWIFS data are useful for analyzing variations in coastal water temperatures and primary productivity. Numerically-based circulation patterns can add to this analysis. Tide gauge data for Mexico are available from the UNAM and the Centro de Investigacíón Científica y de Educacíón Superior (CICESE), Ensenada. Wind data can be obtained from the CNA
Geomorphology	Juts, crags, barrier islands, reefs, canyons, lagoons, shelves, slopes, ocean plains, continental elevation	Global DEMs provide data on all these variables. In Mexico, the CONABIO and National Atlas of Mexico, published by the UNAM's Institute of Geography, have produced a map called *Geomorfología Marina* which defines some aspects of the shelf, slope, plains, etc. Topographic maps commonly produced in many countries can be used. These sources can be complemented with data from topographic (average elevation, range and rugosity) and bathymetric maps
Coastal geology	Soils, rocks and sediments.	INEGI geological map (1:1000000). http://www.inegi.gob.mx/geo/default.asp
Biota	Typical flora and fauna of main mangrove, reef and submerged aquatic vegetation (SAV) areas	AVHRR data processed with the vegetation index indicates the best-conserved vegetal masses and possible annual variations. SPOT or ETM Images can help to improve coastal vegetation classifications. In Mexico, land use and vegetation maps are available from the INEGI, and species distribution data from the CONABIO

Coastal Oceanography. Data for the surface temperature of water masses have been produced by the AVHRR and can also be taken from current MODIS images. Primary productivity based on ocean color can be calculated using data from the Coastal Zone Color Scanner (CZCS), SeaWifs and MODIS, while ocean surface wind circulation patterns can be obtained from QuickScat. Ocean floor topography, such as depth, slopes and shelf dimensions, can be obtained from digital elevation models. Valavanis (2002) provided an excellent list of online sources for this type of data. Marine hydrodynamics have been studied and monitored using remote sensors which produce spatial data, as well as other instruments (Hellweger et al., 2004).

Geomorphology/geology. Some physical characteristics of coastal areas can be obtained from DEMs and hydrographic studies, while global data can be taken from ETOPO and STRM maps. Useful geomorphological characteristics for coastal zone analysis include beaches, barrier islands, reefs, canyons, lagoons, shelves, slopes, ocean plains and continental elevation, among others. These can be complemented with topographic (average elevation, range and rugosity) and bathymetric maps (see above). Coastal geology data (e.g. soils, rocks and sediments) can be taken from national geological maps. In Mexico, the INEGI, the CONABIO and the National Atlas of Mexico, published by the UNAM's Institute of Geography, provide a marine geomorphology map that defines some aspects of the marine shelf, slope and plains.

Biota. Data on the typical flora and fauna of principal coastal ecosystems, such as mangroves and other non-submerged vegetation, can be estimated using AVHRR, MODIS, ETM, SPOT and global or national landcover vegetation maps. The normalized vegetation index (NDVI) can be used to define the best-conserved vegetal masses and their possible seasonal variations using MODIS data that can be obtained from the Global Land Cover Facility (http://glcf.umiacs.umd.edu/data/modis/ndvi/index.shtml). For submerged aquatic vegetation (SAV) and reefs, multispectral scanner (MSS), Landsat TM and ETM, SPOT, QuickBird and compact airborne spectrographic imager (CASI) data have proven useful in highly transparent, shallow areas. In Mexico, the INEGI's forest inventory and vegetation and land use maps provide the best description of coastal cover at this level.

The geospatial tools included in GIS software can facilitate integration of multiple data types and implement the procedures for delimiting regions. The present study involved DEM analysis and processes to select, merge, and combine vector layers, and group pixels in raster images using classification methods. Classification procedures are available in image analysis programs and are divided into supervised and unsupervised processes. Popular procedures are k-means and Isodata, as well as methods employing fuzzy logic (Tso and Mather, 2001; Lillesand and Kiefer, 1987). Different land uses are currently being explored using multicriteria analysis based on weights and standardized factors in an effort to determine the suitability of sites or zones with different risk levels due, for example, to flooding or use conflicts.

14.4 Results

Initially, the proximity concept was applied to the DEM map (Fig. 14.2A) to produce a shoreline buffer zone 20 km offshore and inland (Fig. 14.2B). A 50 m elevation mask was then applied to exclude areas above this level (Fig. 14.2C) and all subwatersheds totally or partially within the 20-km buffer were highlighted (Fig. 14.2D). After this preliminary processing, the regionalization process was begun.

Combination of the watershed and climate maps produced a map with 12 climatic regions which also included municipal, state and international boundaries to identify regions shared by different administrative entities (Fig. 14.3A). Land uses (Fig. 14.3B) and vegetation density (Fig. 14.3C) within the regions were then overlaid to complete the regional data. From this peninsula-wide view, the Carmen climatic region (Aw1) and its subwatersheds (Fig. 14.3D) was chosen as a sample region. In a process similar to that applied to the coastline as a whole, this region map was then fine-tuned by selecting the watersheds totally or partially within the 20 km shoreline buffer zone (Fig. 14.3E) and overlaying the NDVI values grouped into five ranges as well as municipal and state boundaries (Fig. 14.3F).

The final map (Fig. 14.3F) clearly shows that five municipalities share the Carmen region and indicates what ecosystems each municipality contains and their

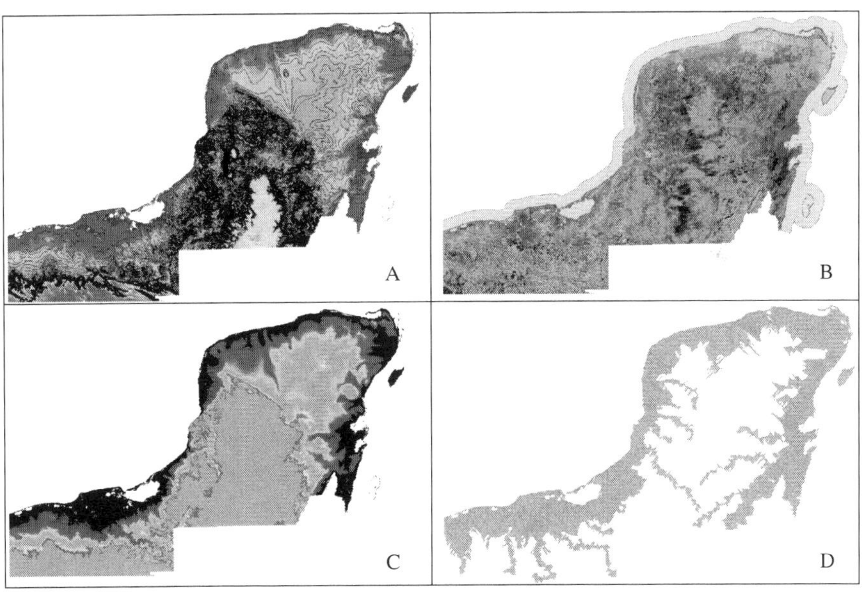

Fig. 14.2 (**A**) Digital elevation map of the Yucatan Peninsula within Mexico; (**B**) Proximity, shoreline buffer 20 km offshore and inland; (**C**) Area below 50 m elevation; (**D**) Hydrological subwatersheds totally or partially within the 20 km buffer (See also Plate 22 on Page 403 in the Color Plate Section)

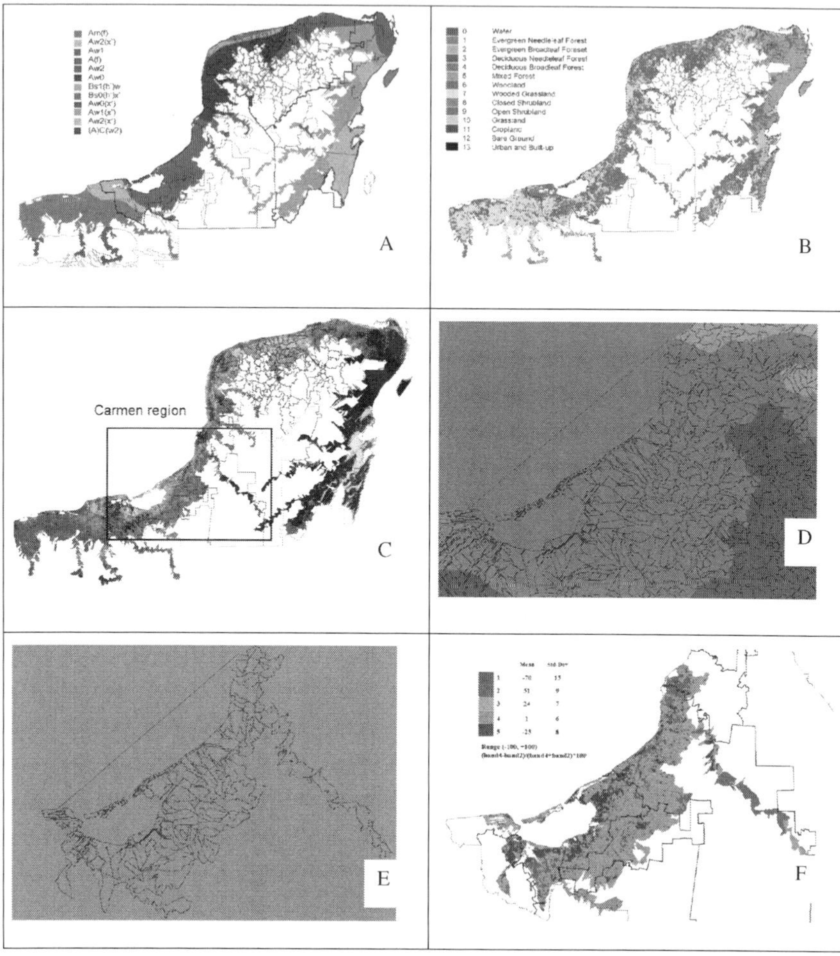

Fig. 14.3 (**A**) Shoreline and government administrative boundaries overlaid with climatic regions contained within watershed boundaries; (**B**) Shoreline and government administrative boundaries overlaid with land uses contained within climatic regions and watershed boundaries; (**C**) Shoreline and government administrative boundaries overlaid with NDVI values contained within climatic regions and watershed boundaries; (**D**) Close-up of Carmen climatic region (Aw1), including sub-watersheds; (**E**) Carmen climatic region showing selected watersheds totally or partially within 20 km SL buffer; (**F**) Carmen climatic region overlaid with five NDVI classes within watershed boundaries, as well as municipal and state boundaries (See also Plate 23 on Page 404 in the Color Plate Section)

condition. Using this map, Campeche municipality was separated out as an example to show how vegetation density and land use in the portion of a region within a particular municipality can be isolated for a more detailed analysis (Table 14.3) and at the same time management considerations should take into account that this portion of land belongs to a large environmental area.

Table 14.3 Campeche municipality showing vegetation density and land use in portions of Carmen region within municipality boundaries. Categories: VL-very low, VH-very high, H-high, M-medium, and L-low (Color key is same as Fig. 14.2B and 14.2F)

	Vegetation density				Land cover and use		
Municipality	Type	Cells	km^2	Municipality	Type	Cells	km^2
Campeche	1-VL	18	1	Campeche	2	1743	109
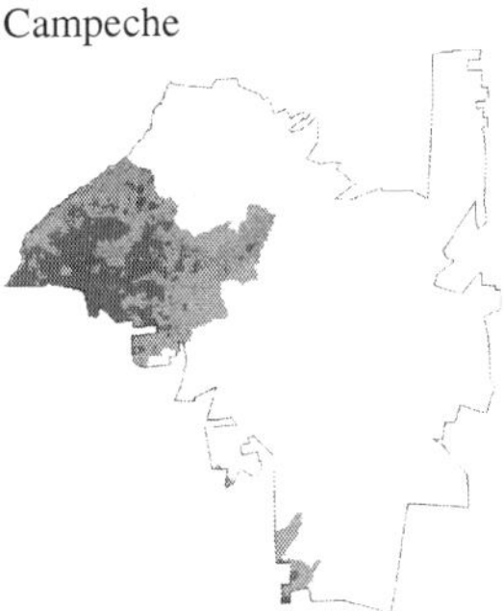	2-VH	3285	205	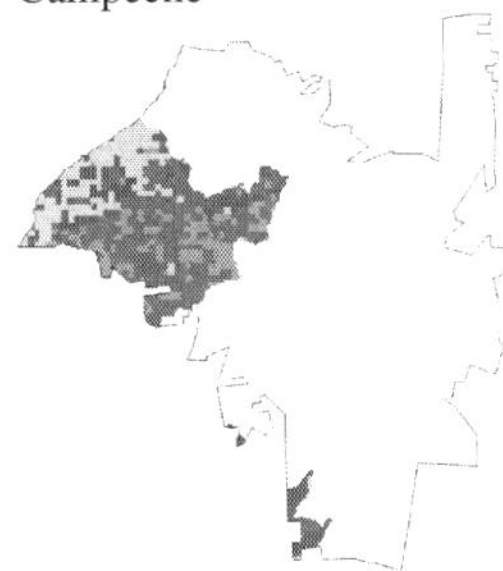	4	4974	311
	3-H	4884	305		6	1685	105
	4-M	1907	119		7	2029	127
	5-L	335	21		11	3	0

14.5 Conclusions

Coastal management regions for the Yucatan Peninsula in Mexico were established using environmental data and GIS programs incorporating geospatial tools. Using a digital elevation map, a shoreline buffer was calculated using a 20 km proximity criterion and then a coastal zone defined based on topography, watersheds and subwatersheds. The coastal zone was regionalized by overlaying climatic data on watersheds. Vegetation data allowed the map to be segmented by even small areas within which the NDVI index was applied as a measure of regional ecosystem integrity and land use as a measure of economic activity. The use of government administrative divisions makes it possible to identify the environmental conditions of particular municipalities within the regional context, and assess needs for collaboration between other administrative entities to implement environmental protection or restoration actions based on an ecosystem approach. Geospatial tools are effective for identifying the risks and limitations implied by present and potential uses of the coastal zone. Multicriteria approaches are currently in use for evaluation of specific activities. By assisting in activity site and management tool selection, the regional vision applied here may also have management applications for resources and activities beyond the municipal level, such as fisheries, beaches, wetlands, tourism and aquaculture. Other biotic, abiotic, socioeconomic and political data can also be incorporated into this regional approach. Making the results of this type of analysis available to economic sectors, administrators and other interested parties will provide them with a more thorough knowledge of the coast and its resources. If used effectively, this data can encourage development of human activities which are more compatible with the environment and with each other. Regionalization is also a potentially vital tool for integrated coastal zone management because it can contribute to establishing criteria for future development, as well as modifying regulations, identifying data shortfalls, determining zoning needs and furthering the extremely complex decision making processes involved in managing coastal resources.

References

Capurro, L., Euán, J., and Herrera, J. (2002). Manejo sustentable del ecosistema costero de Yucatán, *Avance y Perspectiva* vol. 21, julio-agosto 2002, CINVESTAV.

Chua, T.E. (1997). The essential elements of science and management in coastal environmental management. *Hidrobiología*, 352(1–3), 159–166.

Cicin-Sain, B., and Knetch, R.W. (1998). *Integrated Coastal and Ocean Management: Concepts and Practices*. Island Press, Washington, DC; Covelo, California, 517p.

Forman, R.T.T. (1990). Ecologically sustainable landscapes: the role of spatial configuration, in: Zonneveld IS and Forman RTT (eds) *Changing Landscapes: An Ecological Perspective*. Springer-Verlag, New York, pp. 261–278.

Franklin, J.F. (1997). Ecosystem management: an overview in Ecosystem Management, in: Óbice MS and Haney Λ (cds) *Applications for Sustainable Forest and Wildlife Resources*. Yale University Press, New Haven, CT, 361p.

GESAMP (1996). The contributions of science to coastal zone management. IMO/FAO/UNESCO-IOC/WMO/WHO/IAEA/UN/UNEP, Joint Group of Experts on the Scientific Aspects of Marine Environmental Protection, *Rep. Stud., GESAMP* (61): 66p.

Hellweger, F.L., Schlosser, P., Lall, U., and Weissel, J.K. (2004). Use of satellite imagery for water quality studies in New York Harbor. *Estuarine, Coastal and Shelf Science*, 61(3): 437–448.

Hernandez, A., and Kempton, W. (2003). Changes in fisheries management in Mexico: Effects of increasing scientific input and public participation, *Ocean & Coastal Management*, 46: 507–526p.

Holland, G., and Bernal, P. (2002). Framework of cooperation, chapter 10, in: Field JG, Hempel G and Summerhayes CP (eds) *Oceans 2020 Science Trends, and the Challenge of Sustainability*, Island Press, Washington, DC, 257–281p.

IMCRA. (1998). Interim Marine and Coastal Regionalisation for Australia: An ecosystem-based classification for marine and coastal environments, Version 3.3, Ed. R Thackway and ID Cresswell, Reserve Systems Section, Biodiversity Group, Environment Australia, Commonwealth Department of the Environment.

Kay, R., and Alder, J. (2005). *Coastal Planning and Management*. Second Edition, Taylor & Francis, London and New York, 380p.

Lillesand, Thomas M. and Kiefer, Ralph W. (1987). Remote Sensing and Image Interpretation, Second Edition, Wiley; 721p.

Lindeboom, H. (2002). The coastal zone: an ecosystem under pressure, chapter 3, in: Field JG, Hempel G and Summerhayes CP (eds) *Oceans 2020: Science trends, and the Challenge of sustainability*, Island Press, Washington, DC, 49–84 pp.

OECD (1993). *Coastal Zone Management: Integrated Policies*. Head of Publication Services, Organization for Economics Co-operation and Development (OECD). Paris, France.

Ongkosongo, O.S.R. (1992). Human activities, environmental problems and management of the North Coast of West Java, Indonesia, with emphasis on the Jakarta Bay. In: Coastal System Studies and Sustainable Development. Proceedings of the COMAR Interregional Scientific Conference, *UNESCO Technical Papers in Marine Science* No. 64.

Rivera-Arriaga, E., and Villalobos, G. (2001). The coast of Mexico: approaches for its management, *Ocean & Coastal Management*, 44: 729–756.

Tso, Brandt and Mather, Paul M. (2001) Classification Methods for Remotely Sensed Data. Taylor & Francis, 332p.

Tucker, C.J., Grant, D.M., and Dykstra, J.D. (2004). NASA's global orthorectified landsat data set, *Photogrammetric Engineering & Remote Sensing*, 70(3): 313–322.

UNCED (1992). Protection of the Oceans, all kinds of Seas, including enclosed and semi-enclosed seas and coastal areas and the protection, rational use and development of their living resources. In *Report of the United Nations of the Environment and Development*. Chapter 17, Rio de Janeiro, 3–14 June.

Valavanis, Vasilis D. (2002). Geographic Information Systems in Oceangraphy and Fisheries Sciences. Taylor & Francis Books, Ltd; 240p.

Yáñez-Arancibia, A., and Day, J.W. (2004). The Gulf of Mexico: towards an integration of coastal management with large marine ecosystem management, *Ocean & Coastal Management*, 47: 537–563.

Chapter 15
Applying Geospatial Technologies to Weedmat Monitoring and Mapping: The Ythan Estuary, NE Scotland

David R. Green

Abstract This chapter examines the potential for using archival colour vertical aerial photography to map the spatial location, distribution, extent and changes over time in weedmat coverage for the Ythan estuary, North East Scotland, UK. Archival colour aerial photography scanned into a computer, and processed using image processing techniques resulted in a series of mosaiced images for visual interpretation of the location and extent of weedmats for each year of photography selected. Input of the imagery into a Geographical Information System (GIS) enabled the derivation of maps showing change and quantitative estimates of the areal coverage of macro-algal weedmats for each year. It is concluded that macro-algal weedmats can easily be differentiated on colour aerial photography from background mud and sand deposits, but it is not possible to reliably separate individual macro-algal species from each other, or to distinguish weedmat densities without the availability of additional co-incident field data.

Keywords GIS · Remote sensing · Weedmat · Archival · Aerial photography

15.1 Introduction

This chapter outlines a practical working methodology for the use of multi-temporal archival panchromatic and colour aerial photography and GIS for the extraction of multi-temporal spatial information (location and extent) of macro-algal weedmats in the Ythan Estuary, North East Scotland. The reasons for undertaking this work are threefold. Firstly, casual observations made over time at the estuary suggest that the extent of weedmats in the estuary has changed and generally increased over time. To establish whether this is the case or not requires multi-temporal observations of the

D.R. Green (✉)
Department of Geography and Environment College of Physical Sciences, Centre for Marine and Coastal Zone Management (CMCZM), University of Aberdeen, AB24 3UF Scotland, UK
e-mail: d.r.green@abdn.ac.uk

D.R. Green (ed.), *Coastal and Marine Geospatial Technologies*, Coastal Systems and Continental Margins 13, DOI 10.1007/978-1-4020-9720-1_15,
© Springer Science+Business Media B.V. 2010

location and extent of the weedmats in the estuary. Secondly, whilst ground survey techniques are obviously crucial in confirming observations, time available and the costs of undertaking the work are usually prohibitive, especially on a repeat basis. A cost-effective alternative is therefore required. Finally, establishing a standardised approach to weedmat monitoring and mapping is required in order to compare coverage for each year as well as to establish a baseline for future studies that may involve modelling.

15.2 Background

Around the world, macro-algal weedmats[1] have increasingly become a topic of interest over time. Despite this, according to Edwards (1999) relatively little has been written about the remote sensing of macro-algal weedmats and/or the assessment of macro-algal biomass. However, studies by Budd et al. (1980–1983, and 1985), Coulson and Budd (1979), Collier et al. (1996), and Baily et al. (2002) have used aerial photography to monitor and map algal weedmats. The latter utilized a combination of specially flown colour infrared (CIR) aerial photography, photogrammetric techniques and GIS software to map intertidal vegetation, specifically *Spartina*, *Ulva*, and *Enteromorpha* in Langstone, Portsmouth, and Chichester Harbours on the south coast of England. Aerial photo-interpretation of both single and stereo CIR photography at a scale of 1:10,000 was used to map the areas of vegetation from 1998 to 2000. Young et al. (2000) also made use of specially flown CIR stereo aerial photography at a scale of 1:7200 combined with Differential Global Positioning System (DGPS), ground-truth and ancillary data to map distributions of eelgrass and green macro-algae (*Ulva* and *Enteromorpha*) in the USA. They also used a 35 mm hand-held camera with CIR film as a demonstration of an inexpensive way to map spatio-temporal distributions of green macro-algae in a coastal estuary. Green et al. (1998) demonstrated the practicality of using a 35 mm camera, film, and filter combination on a large-scale model aircraft to monitor and map macro-algal weedmats using a digital image processing system. Natural Environment Research Council[2] used archival aerial photography and other airborne imagery (1991–1994), including Airborne Thematic Mapper (ATM[3]) imagery for the monitoring of algal weedmats for UK estuaries.

A significant feature of the Ythan Estuary in North East Scotland, and well-documented in the past by Raffaelli and Nicol (1988), Raffaelli and Plomer (1989), and Raffaelli and Way (1996), has been the presence of benthic macro-algae

[1] Macro-algae or macroscopic algae are more commonly referred to as seaweed and are large plant-like structures found in coastal and estuarine waters around the World

[2] NERC, 1997. Webpage (no longer available)

[3] In the UK, the Airborne Thematic Mapper (ATM) is a scanning sensor for collecting multispectral data in the visible, near/short/ thermal infrared (http://www.nerc.ac.uk/arsf/Pages/Instruments/ATM.htm)

(*Enteromorpha, Ulva, and Chaetomorpha*) resulting in the formation of extensive "green" mats over the intertidal flats of the estuary. Studies by Raffaelli et al. (1988, 1989, 1996), used a combination of aerial photographs dating from the 1940s to the late 1990s and field work, to (a) identify the location and distribution of the weedmats, (b) to establish the possible cause(s) of the weedmats (e.g. increasing amounts of nitrogenous fertiliser being applied to the Ythan catchment over time), and (c) to determine the impact upon invertebrates in the underlying estuarine sediments, shorebird populations and their behaviour (see for example: Lewis and Kelly, 2001; Dunne, 2003).

The methods for repeat environmental data acquisition, monitoring and mapping are well documented in Raffaelli and Plomer (1989) and Raffaelli and Way (1996) and include: field surveys; photo-interpretation of specially flown vertical aerial photographs acquired from light aircraft flying at 1000–1500 ft; and projection of colour photographic slide transparencies for manual tracing and area assessment with the aid of mm graph chapter and a planimeter.

Since the work by Raffaelli et al. (1988, 1989, 1996), geospatial technologies such as remote sensing, Geographical Information Systems (GIS), Global Positioning Systems (GPS), the Internet and mobile GIS have all developed very quickly for use on desktop and, more recently, portable computers. There is now considerable potential using this technology to repeat and check the results from previous studies; capacity to build up digital environmental databases comprising geo-rectified and mosaiced photography and vector map layers that can be used as a baseline for future monitoring and mapping exercises; and to derive quantitative information e.g. distance and area measurements more quickly. Some justification for an approach that uses geospatial technology is provided by Young et al. (2000: II-286) who state that "the rapid development of benthic macroalgae on estuarine mudflats during the summer growing season makes difficult the accurate documentation of spatial and temporal distributions of such algae from ground surveys alone".

15.3 Study Area

The Ythan Estuary (57° N, 2° W) lies approximately 20 km to the North of Aberdeen on the east coast of Scotland. It is a tidal estuary that extends about 8 km inland to Ellon, and it drains a predominantly agricultural catchment. The estuary averages 300 m wide, with an average depth at high water of 2.5 m with a tidal range of less than a meter to 3 m or more (Fig. 15.1).

15.4 Objectives

The overall objective of this work was to devise a practical remote sensing- and GIS-based methodology to extract information (location and extent) about macro-algal

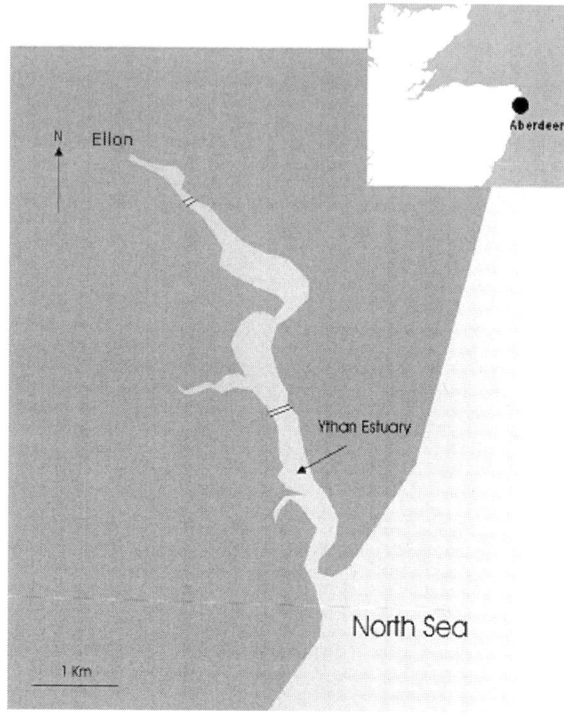

Fig. 15.1 Map of the Ythan Estuary study area

weedmat coverage in the Ythan Estuary derived from multi-temporal archival aerial photography of varying scales acquired from different sources. More specifically to:

- Use archival aerial photography;
- See if it was possible to detect and map macro-algal weedmats from the aerial photography without the opportunity to undertake ground-truthing;
- Establish a practical procedure by which it was possible to derive recommendations to improve future opportunities for monitoring and mapping the weedmat coverage in future years;
- Identify any temporal trends in the location, distribution and areal extent of the macro-algal weedmats;

  and

- Create an environmental database as the basis for both the development of a desktop GIS and an online GIS-based Internet public information system for the Ythan Estuary.

The practical work involved the following tasks:

- Acquisition of the aerial photographic data sets from different sources;
- Inspection and selection of the aerial photographs deemed to be most suitable for the work; scanning of the aerial photographic datasets;

- Geo-correction of the scanned aerial photography using a desktop scanner; Mosaicing of the geo-corrected aerial photography in a digital image processing system;
- Manual and on-screen photo-interpretation, classification and mapping of the multi-temporal aerial photography within GIS software.

All of the aerial photographs available were scanned using a desktop scanner at 150 dpi in order to minimise the filesizes of the raster images. The images were geo-corrected using between 4 and 16 GCPs per image, a polynomial transformation (orders 1 and 2), and resampling using the nearest neighbour algorithm. Where possible, the overall RMSE (Root Mean Square Error) was reduced to <1. Geo-referencing utilized digital Ordnance Survey (OS) Land-Line Plus 1:2,500–1:10,000 scale map tiles for the study area downloaded from the Edina Digimap service (http://edina.ac.uk/digimap). Each image was geo-referenced to the Transverse Mercator, Airy Spheroid, OSGB 1936 Projection/Spheroid/Datum.

In order to conserve computer disk storage space and to aid in the display monitor and memory requirements each mosaic was "cropped" using a digitised boundary of the high water mark (HWM) of the Ythan Estuary catchment area derived from Ordnance Survey (OS) Landline data. The HWM boundary was digitised and used as a "cookie cutter" to reduce the extent of the mosaics to the boundary of the estuary.

Using a combination of the pan and zoom tools, units of macro-algal weedmats, identified with the aid of a photo-interpretation key, were delineated by tracing the boundaries with the aid of the default on-screen interactive digitising tools. Minor edits to the digitised boundaries were then completed once all the units had been mapped. This was done so as to minimise the human errors associated with on-screen digitising. In order to maintain consistency in the interpretation and digitising of the weedmat units the work was carried out by a single individual.

15.5 Interpretation

Frequently a minimum mappable unit (mmu) is defined as the basis for determining a "cut-off" point or "threshold" to help decide which units to digitise and which to ignore. This is defined in terms of spatial dimensions of the smallest possible unit e.g. weedmat unit that can be visually identified and its boundaries mapped. In this study it depended upon the scale of the photography used and ranged from approximately 3–5 m. Careful consideration also has to be given to exactly where a boundary line is drawn around the feature of interest. The use of the zoom tool with onscreen digitising helps the interpreter to map small units, the decision to include being based upon whether it looks like weedmat and its size in relation to ease of digitising and knowledge of the appearance of weedmat on the ground.

To some extent the choice of which unit to map also depends upon:

- Original photographic scale;
- Quality of the aerial photography/mosaic;

- Degradation of the original photographic image e.g. steps undertaken in the creation of the digital image used for onscreen digitising and interpretation i.e. scanning, mosaicing and colour balancing;
- Quality of display monitor and screen display resolution used during onscreen digitising and interpretation process;
- Image sharpness and the contrast between the features of interest and the background or surrounding area or surface; sometimes difficult at the edges of photographs;
- Image magnification and zoom; some features are more easily identified when zoomed out (homogeneity and context) whilst others benefit from zooming in (detail);
- Eyesight of the interpreter;
- Knowledge and experience of the interpreter.

Additional help in the interpretation can be provided through experience, references to contextual information, and the examination of hardcopy aerial photographs that generally have a sharper image than their scanned digital equivalent.

Background information about the features and environment of interest can provide important assistance in the interpretation of aerial photography. In the case of weedmats, typically *Enteromorpha* forms vast mats that lie on the surface of estuarine mud- and sand-flats. These mats can either be very fine coverings (thin) or several cms deep (thick). Free-floating mats of *Enteromorpha* may also be deposited on other vegetation types. *Ulva* is frequently associated with *Enteromorpha, and* inhabits the middle to low intertidal zone (or eulittoral to sublittoral zone) and grows in sheltered environments but may also be found in brackish water, particularly estuary environments. *Ulva* may often be found attached to rocks in the middle to low intertidal zone but can also survive in water as deep as 10 m in calm and protected harbours. They are also often found in abundance where there are enhanced nutrient supplies (sewage outfalls), metal contamination, where wave shearing forces are low and herbivory is reduced. Young et al. (1998: II-39) note that "in late summer, substantial quantities of green macroalgae often break loose from the substrate and are deposited in other habitats of the intertidal area". Additional information is also provided from knowing about the type and distribution of sediment. It is known that some weedmat species are more closely associated with certain sediment types (Orr, 2002). Where the density of the weedmats is such as to allow underlying substrate to show through this may alter the visual appearance of the weedmats. For example, where there are mussel beds or seaweed then this may make the weedmat may appear visually darker. In this study help in this respect is provided by the acquisition of several sediment maps for the estuary that were digitised as separate GIS layers (Orr, 2002; Chan, 2003) (Fig. 15.2). Knowledge of the hydrography of the estuary is also useful, and some information on this was forthcoming from Stove (1978). Bearing this information in mind, the general location of weedmats in each mosaic also seemed to correspond to the known "habitat conditions" i.e. sheltered environments such as are found in harbours, the middle to low intertidal zone, and calm shallow water.

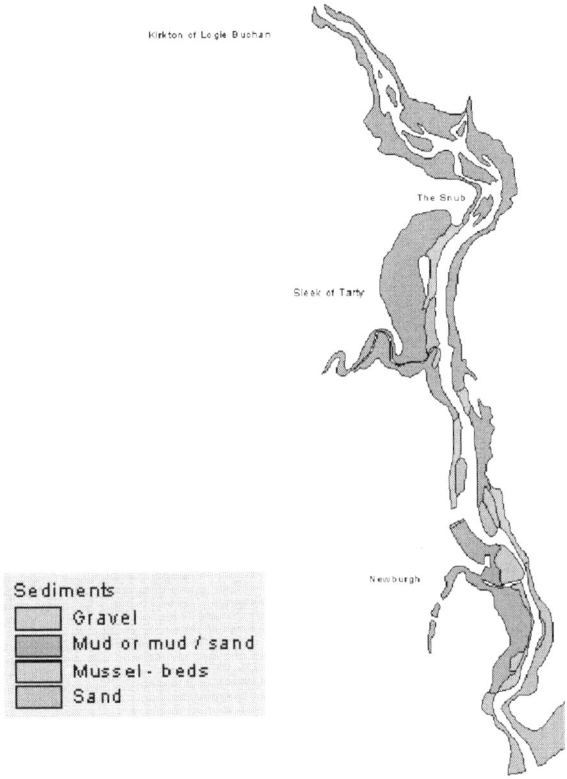

Fig. 15.2 A sediment map of the Ythan Estuary

15.5.1 Methods

One major constraint of this work was that operationally it had to be completed in approximately three weeks. This included selecting, scanning, geo-correcting, and mosaicing the aerial photography for four dates (1989, 1992, 1994, and 2000), photo-interpretation of each aerial photographic set, import into a GIS, identification, delineation and digitising of the weedmat unit boundaries, and completion of the GIS database including documentation (metadata) (Fig. 15.3).

15.5.2 Interpretation Key

An approach that is frequently used to aid in the photo-interpretation process is a photo-interpretation key providing a visual guide for the interpreter. An Internet search for examples of keys relating to macro-algal weedmats did not, however, yield any useful examples. A practical alternative, therefore, was to devise a photo-interpretation key for the work being undertaken. This approach assumes that

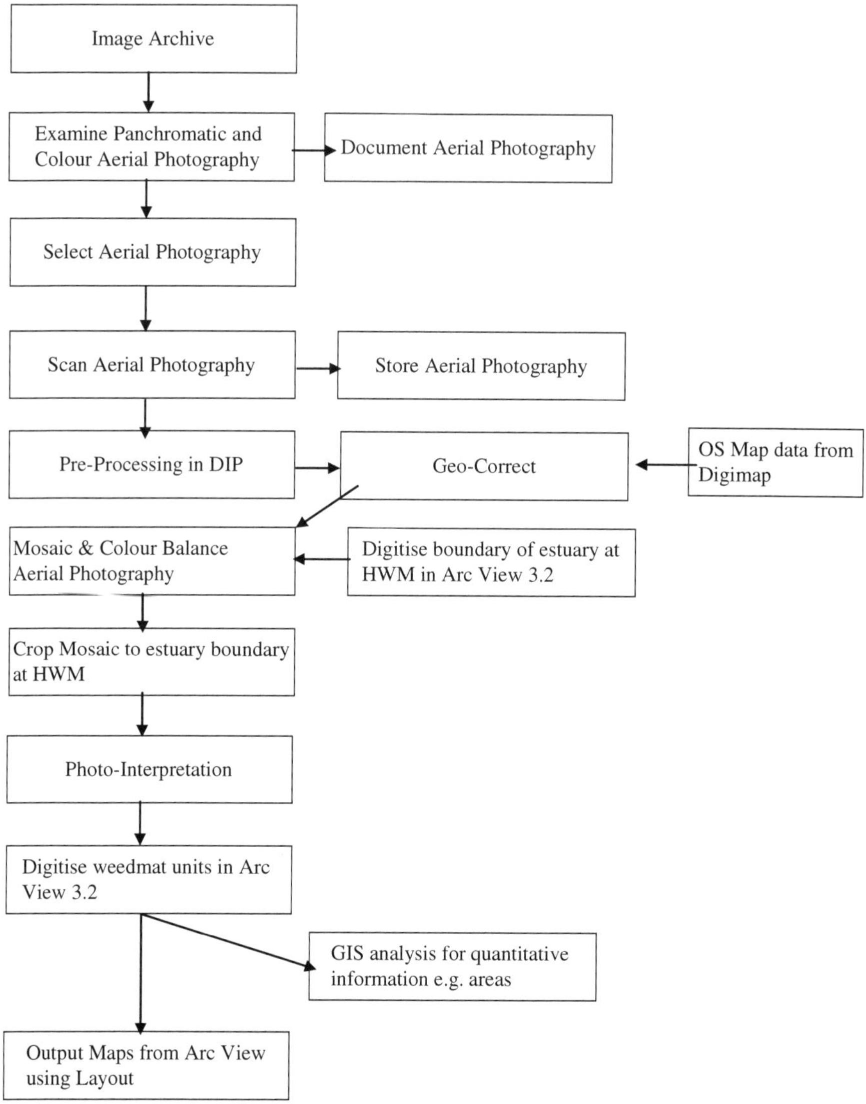

Fig. 15.3 The methodology for weedmat mapping

macro-algal weedmats are likely to be generally very similar in their appearance on all of the aerial mosaics and through time. For example, weedmats generally appear a distinct "emerald green" on colour photographs in contrast to their bottom sediment surroundings (sand and mud: pale brown to grey brown). The interpretation key was constructed by locating examples of the different classes (dense and sparse weedmat) to be interpreted, where additional information provides some certainty as to the correct identity and representativeness of the class.

15.5.3 Spectral Profiles

The use of spectral profiles for each different class to be mapped can also be very helpful when interpreting an image. In effect this provides a simple way by which it is possible to establish the visual colours of the different classes. Unfortunately, once again, an Internet search yielded no good examples of spectral data for macro-algal weedmats. As a practical alternative, therefore, the spectral profile tool in the digital image processing software was used to generate some basic profiles of the examples of weedmat units as an aid to classifying the images.

15.6 Comparisons and Problems

One of the main objectives of this work was to identify any temporal trends in the location, distribution and areal extent of the macro-algal weedmats. This was based upon comparisons between measurements derived in this study and those of Raffaelli et al. (1988, 1989, 1996). Using the display and overlay functionality of a GIS it was possible to undertake the following:

- Check the quantitative area measurements of Raffaelli et al. (1988, 1989, 1996);
- Overlay maps from both sources for the same year to compare both the interpretations and quantitative measurements.

Weedmat maps by Raffaelli and Way (1996) for the South Quay area at Newburgh, whilst very generalised, could be compared by digitising the original maps and inputting them into a GIS. It was possible to locate four ground control points in these maps allowing them to be geo-corrected, sufficient to facilitate overlay with our geo-corrected aerial photographs and mosaics as well as with maps for each date. This allows some basic comparisons to be made between different years and potentially forms a basis within a GIS to produce a "change map" using different colours and shadings that can be viewed separately or as part of a simple animation.

15.7 Practical Constraints

Some practical constraints included:

- The type, quality, scale and extent of coverage of the archival aerial photography: not all of the photography had been flown for the current purpose, and much of it was of relatively poor quality, variable scale, condition and there was incomplete coverage by the flight lines. Other photography was, by comparison, high quality, large scale colour aerial photography specially flown for the task;

- Although some of the photography had been flown around approximately the same time each year, some had not, and furthermore there were no overflights throughout the "growing season";
- Scanning was undertaken using a low-cost desktop scanner: although physically possible to scan aerial photography using this low-cost hardware, it would have been more appropriate to source and use a photogrammetric scanner. However, such devices are often not available and are expensive to purchase. Therefore, there may be loss of information and distortions present in the scanned photography that will affect the interpretation;
- Geometric correction of some of the photographic frames was found to be difficult because of the lack of easily identifiable ground control points (GCPs) on both the aerial photographs and the reference maps. This is a common problem in estuarine (and other) areas. One solution is to undertake fieldwork using a portable GIS, GPS, and digital map base. If it is possible to commission the aerial photography, then it may also be possible to place ground control points on the ground visible in the imagery. Alternatively, aerial photography now often automatically includes GPS co-ordinates;
- Colour balancing results from Erdas Imagine were not uniform and were often relatively poor compared to those achieved using another software product, PanaVue (http://www.panavue.com). Differences between photographs within a single flight can occur due to sun-sensor-ground angles, sensor view angle, and small atmospheric changes. Between flight differences can also result from the use of different film batches, cameras and photographic processing. Although some results were not as good as had been hoped for in providing a "colour balanced" mosaic, on the whole it did not create a serious problem when interpreting the mosaic on-screen;
- Although cutting the filesize of each mosaic aided in the digitising procedure speeding up the display of the image backdrop for digitising, in practice the full image mosaic provided more context for the weedmat interpretation and information extraction.

15.8 Discussion

Macro-algal weedmats are relatively easily identified on good quality colour aerial photography acquired at scales of 1:3000–1:5000, providing that the weedmat density is quite high (Fig. 15.4). Identification is significantly aided by the distinct contrast with the background or substrate. However, identification is less straightforward on smaller resolution, archival and panchromatic photography. Comparison with larger scale photographs and ground-based 35 mm photography confirms this. These findings concur with those of Raffaelli et al. (1988, 1989, 1996) regarding

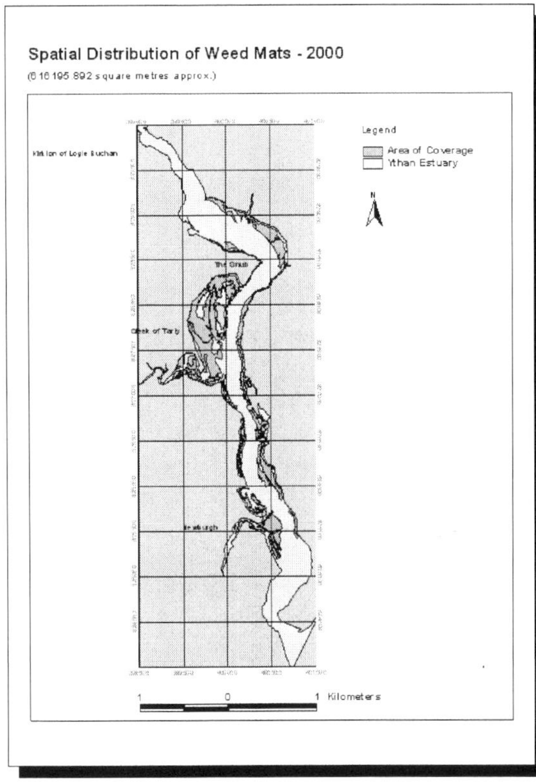

Fig. 15.4 An example GIS map showing the spatial distribution of weedmats

the visual recording of weedmats with a density greater than 1 kg/m^2, although it was not possible to confirm this observation through fieldwork in this study. Where weedmats are not so dense and the units are sparse or small, or the weedmats are submerged at the time of the overflight it is far less easy to identify the weedmats and particularly to ascertain their unit boundaries, extent and exact ground coverage.

Despite the potential problems outlined above it is possible to conclude the following: the spatial distribution and location of the weedmats in the Ythan Estuary is very similar for each year. This is to be expected given the likely causes of weedmats e.g. pollution, shelter, sediment substrate etc.; as might be expected there also appear to be differences in the areal weedmat coverage from year to year.

As found by Raffaelli et al. (1988, 1989, 1996) there is a trend over time. However, although such similarities and trends appear to exist it is difficult to quantitatively establish the degree to which the estimates are similar or accurate. Possible sources of differences and error are:

- The different interpretations (choice of the minimum mappable unit (mmu), identification and mapping of the weedmat units (boundary detail));
- The comparability of the method used to derive weedmats by each interpretation;

- The accuracy and origin of the image rectification used as the basis for mapping;
- The quality of the aerial photography used.

Thus, although quantitative estimates of weedmat area can be practically and easily obtained, it is relatively difficult to establish whether they are accurate estimates and how much error is associated with each of the different components of the methodology used in both the current and previous studies. In the current study, for example, there was no opportunity to gather ground-truth information for the dates of aerial photography used. In future studies there is a need to ensure that coincident ground-truthing can be undertaken to aid in the detailed interpretation of the aerial photography.

Aerial photo-interpretation of either prints or digital scanned images may not be easy. Young et al. (1998), for example, noted a problem in being able to distinguish macro-algal weedmats when they co-exist with other vegetation. Baily et al. (2002) noted problems due to variability in the CIR photography from year to year, atmospheric/weather conditions, processing, time, and film deterioration, all of which inhibit interpretation. The appearance of vegetation on a photograph may also vary as a function of the angle of light and the position of the photograph when it was taken. They also point to a quotation from Daels and Antrop (1978) who observe that interpretation should only be made from within a single stereopair and not across different years. Furthermore, an important point is the subjective nature of the interpretation due to differences in image quality and the skill and experience of the observer. Boundary definitions vary as change in density of cover can be gradual. Additionally, Daels and Antrop (1978) note that vegetation mapping requires photographs taken at the peak of the growing season, in good weather, at low spring tide. In reality, however, a perfect combination of these factors is not always possible.

The application of digital image processing techniques to the analysis of colour aerial photography is not particularly straightforward and the classification of the weedmats is not always successful or repeatable. Young et al. (1999: 1176) observe that "the use of digital image processing techniques alone to classify. ... was problematic ... because macroalgae, for example, share the spectral characteristics of other classes in the visual portion of the electromagnetic spectrum". A solution proposed by Young et al. is "to use these techniques as a first step in the classification process, followed by a visual examination of the imagery, and to edit the results in a traditional approach to photo-interpretation". Alternative approaches might be to use contextual information in the classification process.

15.9 Recommendations

To extend the work undertaken here, it is suggested that further acquisition of aerial photography should take the following considerations into account:

Table 15.1 Seasonal succession of bloom macro-algae

Macro-Algae	From	To	Peak
Enteromorpha spA	May	November	May–June
Enteromorpha spB	May	November	July–September
Chaetomorpha linum	Mid-May	November	July/August
Ulva lactuca	Late May	Mid-October	September

Source: Adapted from Raffaelli et al. (1998, p. 108)

- For comparisons with previous data and information derived from aerial photographs overflights should be acquired at approximately the same time and at the same scale as previously using the same camera setup, film type and processing;
- Some aerial overflights are also made for other times during the season to try to ascertain whether or not there is an optimum period for aerial photographic acquisition to assist in the identification and mapping process. The optimum periods for such overflights could be established by examining previous aerial photography and a range of other information sources such as graphs. A summary of these graphs suggests the following timetable for aerial overflights (Table 15.1);
- Colour Infrared (CIR), colour negative (Young et al., 1999) and/or multispectral aerial photography is considered for future studies, the former offering the opportunity to maximise the identification and separability of the macro-algal communities from the substrate. Other sources of remotely sensed imagery are considered as alternatives to traditional aerial photography offering better spatial and spectral resolution helping to assist in weedmat interpretation. These include Compact Airborne Spectrographic Imager (CASI), Airborne Thematic Mapper (ATM), Light Detection and Ranging (LIDAR), high-resolution satellite imagery and airborne digital photography; Most recently, high resolution satellite imagery has been considered as a way forward.
- Collection of additional field or ground-based data and information coincident with the aerial photography (or other imagery) to assist in the photo-interpretation process.

15.10 Future Studies

A number of additional studies are now underway as part of a long-term objective to develop an environmental database for the Ythan Estuary and to continue to examine the problem of weedmat monitoring and mapping in the Ythan Estuary. These studies are designed to overcome some of the limitations identified in the current study, and an established methodology for using geospatial technologies to derive such information. These studies include mapping weedmat coverage from more recent aerial photography (Chan, 2003; Dunne, 2003), ground-truthing of the weedmats (Orr, 2002), and the use of mobile mapping technology to map additional geographical information (Chan, 2003); for example, sediment distribution. Future studies of the weedmats will also attempt to utilise a wider range of remotely sensed imagery including imagery from airborne platforms such as CASI and LIDAR, and satellites for which it will be possible to organise coincident ground-truthing. There are already plans to fly multispectral digital camera and video sensors mounted on a microlight over the Ythan.[4] In part this will help to provide a regular and flexible source of remotely sensed imagery that will replace the aerial photography flown to date using a traditional aerial camera mounted on a light aircraft.

In the longer term, the digital datasets generated in this study will provide a useful resource and form a baseline database for the Ythan Estuary. Already the weedmat datasets have been used as the basis for several studies of bird distribution[5] (Dunne, 2003). The map datasets will also be utilized within an online mapserver to provide public access to environmental information about the Ythan Estuary and its surrounds.

15.11 Summary and Conclusions

The current study was primarily initiated to develop a practical methodology for the derivation of weedmat maps for the Ythan Estuary using existing archival remotely sensed imagery. This work also provided a basis upon which to compare and contrast the extensive work already undertaken by Raffaelli and Nicol (1988), Raffaelli and Plomer (1989), Raffaelli and Passoni (1992), and Raffaelli and Way (1996), by creating comparable maps based upon independent interpretations, and also enabling the digital capture of the datasets first created by these studies for integrated analysis and further comparison. Input of the datasets to a GIS also provided an opportunity to integrate the weedmat maps with other datasets to create a series of environmental datasets for future analyses and to assist in the interpretation and mapping exercise. The practical methodology developed will also be useful as a baseline for subsequent work in the Ythan that will utilise new sources of airborne and spaceborne remote sensing, the results of which can easily be visualised in ways that might be useful in the wider context of a public information system.

Whilst general comparisons between the different studies of the Ythan weedmat problem can be made, the results of any detailed comparisons must be interpreted very carefully because assessments of the coverage have been derived from photographs of varying scales, quality and date, and by different methods and interpreters. In order to monitor and map the macro-algal weedmat coverage accurately in the future, and to determine if there is a genuine change over time, will require very careful planning and execution of any aerial overflights. Ideally this will

---

[4]personal communication with G. Wright – Macaulay Institute, Aberdeen, Scotland, UK (2003)

[5]personal communication with I. Patterson – University of Aberdeen, Scotland, UK (2003)

necessitate special and frequent overflights. This is a matter for serious consideration in the development of future monitoring and mapping studies whereby it is possible to ensure appropriate aerial photography, scale and resolution for the task; and appropriate timing of the overflights at one or more points during the season. For example, aerial photographs need to be acquired throughout the season, as well as late on in the season when the macro-algal coverage is known to peak.

To this end, however, the current and past studies do provide a very useful basis upon which to develop a practical methodology for the future that can be used as a baseline for monitoring and mapping strategy for the Ythan Estuary.

In conclusion, the combination of digital data capture, conversion and input capabilities, digital image processing and GIS technology offers considerable potential to create a baseline GIS-based system as the basis for future multi-temporal monitoring and mapping of environmental change of macro-algal weedmats in the Ythan Estuary. The proposed methodology for mapping weedmats is greatly simplified, both timewise and practically, once the initial investment in the technology has been made, and in the long-term, is far more efficient in terms of the number of steps required to undertake the mapping. It also helps to make the method of working for further studies far more consistent, objective and standardised.

It is clear, however, that whilst it is possible to utilise archival imagery with the aid of computer technology for studies for which it was not intended, any future work must ensure that the type of remotely sensed imagery, its timing and acquisition are more closely matched to the requirements of the task. Based upon the current experience it is possible to develop an appropriate strategy that will overcome the limitations of using archival imagery thereby providing a solid foundation for more accurately estimating the spatial distribution, amount and trends in macro-algal weedmats in the future using remote sensing.

Although this study initially only set out to develop a practical methodology to extract information about macro-algal weedmats from archival aerial photography, further consideration suggests that any future studies should also take into account other sources of information to aid in the extraction of information. For example, another way to aid in the location and mapping of the weedmats in the context of their geographical location and setting would be to utilise a Digital Elevation Model (DEM). Draping aerial photography and map data over the DEM would help to provide another way of visualising the data. Although the topography of the Ythan Estuary and its surrounds is not dramatic, the use of vertical exaggeration, zoom and pan tools in a GIS provides an opportunity to examine the environment from a different perspective and to help in the interpretation. The availability of both LIDAR data and CASI imagery would also aid in the monitoring and mapping of weedmats in the future. Another more analytical type of approach would be to examine the impact of changing agricultural regimes on sediment load and deposition, and both water quality and temperature and its impact on weedmat growth, as a basis for developing a modelling approach to weedmat location and distribution. At the time of this study such information was not available and this would therefore be a topic for future investigation.

References

Baily, B., Collier, P., Fontana, D., and Pearson, A. (2002): Mapping the inter-tidal vegetation of southern English coastal harbours for pollution monitoring and habitat conservation using false colour infrared photography. Journal of Coastal Conservation. Volume 8(1):77–86.

Chan, K. (2003): Sediment and Macro-Algae Relationship of the Ythan Estuary: A Remote Sensing and GIS Approach. Unpublished M.Sc. Dissertation, Department of Geography, University of Aberdeen, Scotland, UK, 60p.

Collier, P., Fontana, D., and Pye, K. (1996): Aerial FCIR Image Interpretation in Chichester and Langstone Harbours 1994. Report for the Environment Agency. University of Portsmouth.

Coulson, M.G., and Budd, J.T.C. (1979): Survey of the Inter-tidal Vegetation of Chichester and Langstone Harbours. Report to Southern Water Authority. 15p.

Daels, L., and Antrop, M. (1978): The use of a temporal analysis of FCIR images for the evaluation of the vegetation damage in industrial areas. Proceedings of the International Symposium on Remote Sensing for Observation and Inventory of Earth Resources and the Endangered Environment, Freiburg, Germany, pp. 2083–2093.

Dunne, B. (2003): The Relationship Between Algal Mats and Wading Bird Populations in the Ythan Estuary, North East Scotland. Unpublished M.Sc. Dissertation in Ecology, University of Aberdeen.

Edwards, A.J. (1999): Applications of Satellite and Airborne Image Data to Coastal Management: Seventh Computer-Based Learning Module (BILKO for Windows), UNESCO, Paris, France, 185p.

Green, D.R., King, S.D., and Morton, D.C. (1998): Small-scale airborne data acquisition systems to monitor and map the coastal environment. Proceedings of the 5th International Conference: Remote Sensing for Marine and Coastal Environments, San Diego, California, 5–7 October 1998, 1479–1486.

Lewis, L.J., and Kelly, T.C. (2001): A short-term study of the effects of algal mats on the distribution and behavioural ecology of estuarine birds. Bird Study, Vol. 48:354–360.

Orr, G. (2002): The Distribution of Enteromorpha Alga Mats in Relation to the Sediment Facies of the Intertidal Zone of the Lower Reaches of the Ythan Estuary. Unpublished Undergraduate Dissertation. Department of Geography. University of Dundee, Scotland, UK, 75p.

Raffaelli, D., and Nicol, A. (1988): The Distribution of Enteromorpha Weed in the Ythan Estuary. Unpublished Report. 4p.

Raffaelli, D., and Plomer, F. (1989): The Distribution of Weed Mats in the Ythan Estuary in 1989. Unpublished Report. NCC 3p.

Raffaelli, D., and Way, S. (1996): Long-Term Changes in the Ecology and Usage of the Ythan Estuary. Scottish Office Report. 19p.

Raffaelli, D.G., Raven, J.A., and Poole, L.J. (1998): Ecological impact of green macro-algal blooms. Oceanography and Marine Biology: An Annual Review. Vol. 36: 97–125.

Stove, G.C. (1978): The Hydrography, Circulation and Sediment Movements of the Ythan Estuary. Unpublished PhD Thesis. University of Aberdeen.

Young, D.R., Cline, S.P., Specht, D.T., Clinton, P.J., Robbins, B.D., and Lamberson, J.O. (2000): Mapping spatial/temporal distributions of green macroalgae in a Pacific Northwest Coastal Estuary via small format color infrared aerial photography. In Proceedings of the Sixth International Conference: Remote Sensing for Marine and Coastal Environments, Charleston, South Carolina, May 1–3, 2000. Veridian ERIM International, Ann Arbor, Michigan, ISSN 1066-3711, Vol. 2, pp. II-285–II-291.

Young, D.R., Specht, D.T., Clinton, P., and II Lee, H. (1998): Use of color infrared aerial photography to map distributions of eelgrass and green macroalgae in a non-urbanized estuary of the Pacific Northwest, USA. In Proceedings of the Fifth International Conference on Remote Sensing for Marine and Coastal Environments, Vol. II:37–45, ERIM International Inc., Ann Arbor, pp. II-37–II-45.

Young, D.R., Specht, D.T., Robbins, B.D., and Clinton, P.J. (1999): Delineation of Pacific Northwest SAVs from aerial photography: natural color or color infrared film. In Proceedings: From Image to Information. 1999 ASPRS Annual Conference, Oregon Convention Center, Portland, Oregon, May 17–21, 1999. pp. 1173–1178.

Chapter 16
"Ythanview" – Visualizing an Estuary and Virtual Fieldwork at the Ythan Estuary, Scotland, UK

David R. Green and Katarzyna Bojar

Abstract This chapter describes an ongoing research project, "YthanView", to provide online access and visualization tools for geospatial data and information characterizing the Ythan Estuary in northeast Scotland. The project addresses requirements for new and innovative ways to store, catalogue, access, and visualize a wide range of terrestrial and coastal data and information for a small estuarine environment, and to make it more widely available for teaching, research and consultancy activities. Building upon work currently being undertaken at the University of Aberdeen, to facilitate wider access to estuarine data and information resources, the project also considers ways to introduce undergraduate and postgraduate students to coastal and estuarine fieldwork, to reduce staff/student contact hours, and to establish closer links with academic research and consultancy as part of its aims. Several examples illustrate the different ways in which visualization can be used to display, explore, and communicate geospatial information for an estuarine environment.

Keywords GIS · Visualisation · Coastal management · Ythan estuary · Ythanview

16.1 Introduction

The main aim of this chapter is to present the role of coastal landscape visualization using different spatial data applied in multimedia cartography and GIS in the context of Integrated Coastal Zone Management (ICZM). The objectives are to demonstrate the creation of a multimedia environment with three-dimensional (3D) visualization of the landscape in the form of VRML environments, fly-throughs and cartographic animations with dynamic elements. Furthermore, to

D.R. Green (✉)
Aberdeen Institute for Coastal Science and Management (AICSM) and the Centre for Marine and Coastal Zone Management (CMCZM), University of Aberdeen, Aberdeen, AB24 3UF, Scotland, UK

D.R. Green (ed.), *Coastal and Marine Geospatial Technologies*, Coastal Systems and Continental Margins 13, DOI 10.1007/978-1-4020-9720-1_16,
© Springer Science+Business Media B.V. 2010

present the use of the Internet as a distributed computing environment for spatial data visualization in cartography and GIS. Finally, to show different methods of creating cartographic animations using different software. A small project was created in ESRI's ArcView 3.3. GIS software to present maps with hyperlinked ground photography. This approach provides a more realistic and up-to-date view of a study site e.g. an estuarine environment, and also a useful aid in visualization studies and for modelling the effects of environmental change.

The visualization work described here is part of an ongoing research project, YthanView, designed to provide online access and visualization tools for geospatial data and information characterizing the Ythan Estuary and surrounds in north-east Scotland, UK. The YthanView project was initially developed to address the need to examine new and innovative ways to store, catalogue, access, and visualize a wide range of terrestrial and coastal data and information for a small estuarine environment, and to make such data and information more widely available for teaching, training, research and consultancy activities. The YthanView project evolved from other work currently being undertaken in the Department of Geography and Environment in the Centre for Marine and Coastal Management (CMCZM), OceanLab and the Aberdeen Institute for Coastal Sciences and Management (AICSM), to facilitate wider access to estuarine data and information resources. YthanView also provided the opportunity to address a growing need to find new and innovative ways to introduce undergraduate and postgraduate students to a new subject area (i.e. coastal and estuarine studies) and fieldwork, whilst seeking to reduce staff/student contact hours, as well as establishing closer links with research and consultancy work being undertaken by staff.

The development of YthanView has taken advantage of the wide range of currently available geospatial and online Internet technologies to provide a framework to collect, store, analyze, visualize and communicate geospatial data and information. Using environmental databases established from a variety of existing sources, a virtual fieldwork experience for a coastal and marine study area, the Ythan Estuary, is also being developed. Provision of an online Internet resource, including virtual fieldwork exercises, laboratory practicals, assignments, both extends and broadens current course delivery opportunities. Innovative ideas such as the use of guest speaker videos, MS Powerpoint presentations, and greater use of remote sensing, GIS and mobile technologies for fieldwork are also explored as ways of enhancing the student experience.

Practically YthanView has provided the means to develop an online resource and project base to assist in the reduction of teaching contact hours for the Marine Resource Management (MRM) degree programme and project work; to enhance project work resources currently available to MRM students for undertaking project work; to develop a virtual fieldwork experience for MRM students with the aim of replacing existing fieldwork as well as complementing half-day and full-day field excursions; and to provide online virtual fieldwork modules for the Ythan Estuary. A number of examples are provided to illustrate the different ways in which visualization tools can be used to display, explore, and communicate geospatial data and information for an estuarine environment.

16.2 The Role of Geospatial Technologies in Marine and Coastal Management

In the last 10 years the geospatial technologies (e.g. GIS, remote sensing, cartography, digital mapping, GPS and the Internet) have found an increasing role in all aspects of monitoring, mapping, and modelling of the coastal environment: the coastal zone, terrestrial, and marine environments, and in the wider context of integrated coastal zone management. Many different applications of the technologies used on their own or integrated can be cited (e.g. Green and King, 2003b, 2003d, c, d, 2002a, b; Green et al., 1998; Green, 1995).

16.3 Data Collection

Different geospatial technologies have been used to gather data at a wide variety of different spatial and temporal scales. For example, GPS and mobile technologies have been used to gather data at the local scale (Green, 2005; Green and King, 2003a). Airborne and spaceborne remote sensing have been used to gather a wide range of environmental data to study coastal environments ranging from beaches to estuaries, pollution, marina and port management, as well as ship monitoring. Whilst satellite sensors have, until relatively recently, been largely used to gather data providing information about patterns and distributions over large areas (refs) aerial photography and more recently airborne sensors such as the Airborne Thematic Mapper (ATM), CASI (Compact Airborne Spectrographic Imager), and LIDAR (Laser Induced Detection and Ranging) offer much higher spatial resolutions. In combination, remote sensing provides a wide range of sources of environmental data for studying coastal and marine areas at many different spatial and temporal scales (Green and King, 2003b).

16.4 Storage, Access and Processing

The advent of the digital era in information technology (IT) means that most environmental data now in use are in a digital format convenient for storage, processing and analysis within a computer system. GIS software provides a toolbox in which much geospatial data is now handled and processed. Whilst remotely sensed data is still typically processed within a Digital Image Processing (DIP) system, increasingly there are very few differences between digital image processing and GIS software as more geographic data are being handled in integrated software capable of handling both vector and raster data formats. Software labelled as GIS, however, is more commonly used for handling a wide range of geographical data.

16.5 Display

An important starting point, as well as an endpoint in the processing of data into information is to display or visualise the data, usually in the form of a graphic e.g. a graph, picture or map. Visualisation can be either 2D or 3D. Both GIS and DIP software provide a suite of visualisation tools to allow for the display of data in a variety of different ways e.g. as a graph, a photograph, a remotely sensed image, or a digital terrain model. This enables exploratory data analysis as well as communication of information in both a static and dynamic form, and the creation of virtual field environments that can be useful resources for students and stakeholders involved in coastal management Commercial software is also enhanced with a wide variety of freeware and shareware in the form of utility software and plugins.

16.6 Integration

Access to spatial information using GIS and/or Internet-based information systems facilitates the integration of many different layers of spatial data and information. Providing access to a seamless source of data and information for the coastal environment, for example, provides the basis to bring together coastal, terrestrial, and marine datasets into a single portal or work environment. Already the advantages of bringing together the land and marine environment for coastal areas has been demonstrated by the ICZMAP project (http://www.iczmap.com/) co-ordinated by the Ordnance Survey (OS), the UK Hydrographic Office (UKHO), and the British Geological Survey (BGS). Another related project, Integrated Coastal Hydrography (ICM) (http://www.coastalhydrography.com/), aims to provide metadata for the hydrographic data of UK coastal waters.

16.7 Spatial Data Infrastructures (Data Quality, Standards, Metadata), and Data Models

To facilitate the efficient exchange and access to geospatial datasets, consideration must necessarily be given to the use of data and metadata standards, data models and spatial data infrastructures (SDI). Much work has been carried out in recent years and the justification for these considerations is clearly documented in the work of the ISO, the OGC, and the FGDC amongst others. A number of data models have been developed for coastal and marine data. The ESRI Marine Data Model is one such example (http://dusk.geo.orst.edu/djl/arcgis/). Another data model developed for Protected Areas Management, including marine protected areas, is the NATURE-GIS project (http://www.gisig.it/nature-gis). NATURE-GIS is a demonstrator project for INSPIRE (http://inspire.jrc.it/home.html) (**IN**frastructure for **SP**atial **I**nfo**R**mation in **E**urope initiative). An integral component of this project is the development of a set of guidelines for the use of geographic data and information in protected areas management, including consideration of the different potential uses of geographic information, the need for metadata, and the role of data models. The importance of marine and coastal data standards, metadata, and spatial data infrastructures is also being studied by the MOTIIVE project (http://www.motiive.net) which is directly targeted at the coastal and marine stakeholder.

16.8 The Role of Visualisation

Scientists visualize data for a wide range of different purposes, from exploring unfamiliar datasets to communicating insights revealed by visual analyses. Visualization is a process for representing large amounts of data as abstract images to better understand the meaning of data or information. Visual images help users to develop and share insights and can help them to be more creative. Today, visualization of spatial data is no longer restricted by technical constraints. Modern computer and multimedia techniques extend cartographic maps using pictures, video clips, animation and sound.

Multimedia is the combination of basic types of media such as text, graphics, pictures, sound and video. It is intended to expand the channels of information available to the end-user, as well as providing an accessible tool with considerable potential for developing display interfaces to geographical data and information. It offers different ways to view data that has been generated and stored by spatial resources packages. Interactive maps, using hotspots and buttons provide access to underlying data and information, as well as metadata, and allow for the map display to link to other information offering an enhanced spatial information resource. Multimedia technology changes the visualization of spatial data. The map, the traditional presentational form of spatial data, is complemented by other media such as pictures, animation, sound and video. Each of these additional media has particular advantages for communicating information.

Within the context of Coastal Zone Management (CZM), one of the emerging data management tools is GIS. Together with cartography, GIS can easily be used to visualize spatial data used in CZM. A GIS can display the Earth in the form of realistic, three-dimensional perspective views through then use of animations that convey information more effectively, and to wider audiences than traditional, two-dimensional static maps. Obtaining, handling, and processing data is often difficult in a 3D environment. The primary data and information required for building and maintaining a 3D GIS includes orthorectified imagery, and digital terrain models (DTMs), 3D features (vectors) and non-spatial attribute information associated with a 3D feature. Digital terrain modelling is a powerful tool in GIS analysis and visualization, and the storage, display and analysis of data about the terrain surface is one of the most widely used areas of GIS functionality (Wise, 2004). Grid-based DTMs are directly compatible with satellite data and can provide a useful basis for terrain analysis (Longley et al., 2001).

Visualization techniques can also be employed to give better information and add interpretive value. Colours and textures can be used to show changing contour heights. Using sun angles and lighting on one side of the image, hill-shading and other illumination models can also be used to highlight topographic relief. Other visualisation techniques can be used to enhance a DTM using composites of imagery and data drapes. Moreover, perspective, integrated oblique views and fly-throughs can also enhance visual interpretation. Viewshed map displays can also be used to determine the visibility of objects from different locations.

The tools to create cartographic animations are now becoming more widely available. Animation of images can be used to create the illusion of movement depicting a trend or change. Animation may also include sound (narration, music, or sound effects) resulting in a multimedia animation, which can be further incorporated into a hypermedia presentation. In the context of the spatial data handling process, animation is a prominent form of dynamic presentation. It is an effective method of visual communication that allows for the representation of very complex processes. The importance of animation is that it creates a link between reality and the abstraction process in the map. It means that animation can easily deal with real data as well as abstract and conceptual data. Animations can be used not only to tell a story, or explain a process, but also have the capability to reveal patterns, relationships, or show trends (Cartwright, 1999). Frame-based animations include special effects between frames such a fades, wipes or swipes and dissolves (e.g. tools within the Erdas Imagine suite of software). Cast-based animation, also called tweening or key-frame animation, is based upon the concept of the cell, an individual layer or frame of animation. Using a computer automatically allows the creation of a specified number of frames between two key frames with objects displayed proportionally in each frame. Tweening also enables a smooth transition in some characteristics of an object or its shape.

The "look" of an animated map is accounted for not only by the visual variables and typography, but also by the dynamic variables. The dynamic variables can be seen as additional tools to design animation because the pace (fast/slow) and character (smooth/abrupt) of the illusory motion created in an animation depends on how a set of dynamic variables are used. A single static graphic, whose visual characteristics noticeably and regularly change, such as flashing point symbols can be used to depict the phenomenon's distribution. In this instance, a dynamic variable (duration) merely reinforces the visual variables (the value etc.) that comprise the point symbol. Another application of the dynamic variables is to emphasize attributes or relationships among attributes of symbolized cartographic features.

16.9 Internet

The World Wide Web (WWW) or the Internet provides a powerful basis for an information system by merging the techniques of networked information and hypertext. Moreover it has considerable potential to increase the applications of mapping and GIS. Distributed Geographic Information (DGI) systems based on a combination of network communications and GIS technologies provides access to geospatial data and information in a variety of forms, including maps, images, datasets, analysis operations and reports. More advanced services such as spatial modelling and spatial data analysis are being used by only a few online systems (Bossomaier and Green, 2001).

16.10 Ythan Estuary Context and Study Area

At the University of Aberdeen, the Ythan Estuary (57°N, 2°W; 14 km north of Aberdeen) and the adjoining sand dune system, the Sands of Forvie National Nature Reserve (NNR), has long been an area of academic research and fieldwork. General studies about the geography of the area have been complemented by those on hydrography, sediment, bird habitat, ecology and macro-algal weedmats (e.g. Stove, 1978; Raffaelli et al., 1989, 1998; Raffaelli, 2000; Chan, 2003; Dunne, 2003; and Orr, 2003). Datasets for the area include topography of the sand dunes as well as vegetation maps and other information such as footpaths, and archaeological features.

The Ythan Estuary is one of Britain's smallest estuaries, tidal and extends for approximately 8 km in the general direction of Ellon. It drains a catchment of about 650 km^2 of intense arable agricultural land and has a maximum width of 620 m at the Sleek of Tarty (Fig. 16.1) and at its narrowest point is approximately 300 m wide. The narrow tidal estuary of the River Ythan is part of the Forvie National Nature Reserve (NNR), which also includes the Forvie Peninsula and Forvie Links. The Sands of Forvie National Nature Reserve (NNR) has also been designated as a RAMSAR site, a Site of Special Scientific Interest (SSSI), an EU Special Protection Area (SPA), and an EU Special Area of Conservation (SAC), where designations were assigned to the estuary because of the birds species it supports.

The average depth at high water is 2.5 m with a tidal range of between less than a meter to 3 m or more. Elevations within the catchment range from sea level to 200 m and the average annual rainfall varies from 700 mm on the east coast to 900 mm on the highest parts of the catchment The low water channel is about 71 ha and in the intertidal area there are 115 ha of mudflat and 70 ha of mussel beds and sand. The salinity of the estuary varies seasonally with water in the lower reaches ranging from 30–34%, the middle reaches from 5 to 35% and the upper reaches 0–25%.. The Forvie Peninsula and Forvie Links are predominantly made up of sand, largely consisting of deglaciated deposits from the last Ice Age (approximately 10,000 years ago) washed down from the glaciers of the present day Cairngorms to the shelf

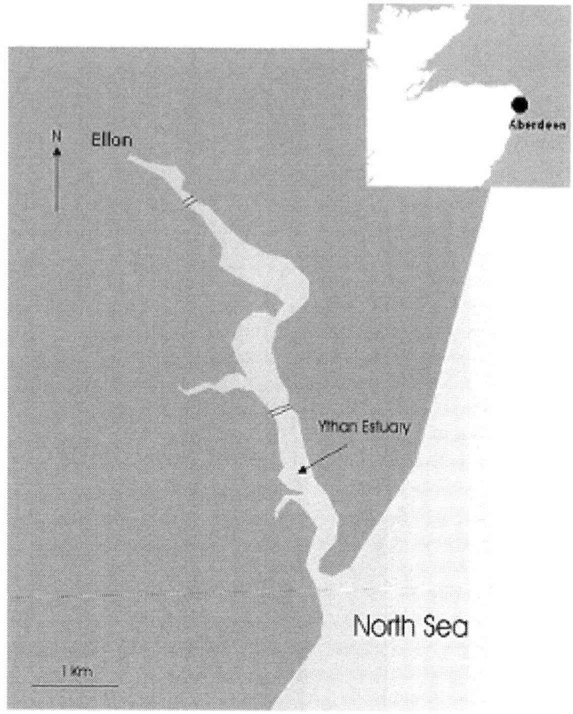

Fig. 16.1 Location of the Ythan Estuary in North East Scotland, UK

of the North Sea. The southern part of the estuary is underlain by glacial fluvio-glacial, estuarine and beach terraces, and the northern part of the Forvie peninsula is underlain by a till covered plateau due to the glacial past. The nature of the dune system means that vegetation within the Sands of Forvie exists in a highly dynamic state and is dependent on active physical processes. These processes are wind, wave and tidal. The Sands of Forvie has active mobile boundaries and older more stable dune systems. This has provided a range of habitats identified from coastal sands, bogs to heath-grasslands. The distribution of vegetation throughout the Sands of Forvie is controlled by the nature of the dunes and their associated processes.

The intertidal area supports a wide variety of wading birds which use the estuary for roosting, feeding and overwintering such as the pink-footed geese, *Anser brachyrhynchus,* which feed on surrounding farmland in winter, and also for the sandwich Tern, *Sterna sandvicensis*, which breed within the dune system. Moreover, the Ythan's mudflats support several types of macro-algae namely, *Enteromorpha spp. (E. prolifera, E. intestinalis), Chaetomorpha linum, Ulva lactuca* and *Cladophora spp.* (Gorman and Raffaelli, 1993; Raffaelli et al., 1989, 1998; Raffaelli, 2000; Green and King, 2002a, b)

16.11 Background

The background to the study described in this chapter began with a number of studies about the monitoring and mapping of macro-algal weedmats. All of these studies sought to make use of a combination of remote sensing and Geographical Information Systems (GIS) to monitor, map and analyse geographical data for an environmental application. These studies are described in Green (1995), Green et al. (1998), Green and King (2002b, 2005a). Given the significant interest expressed over time by a wide range of stakeholders in studying the Ythan Estuary and the surrounding Sands of Forvie sand dune system, it was considered an important objective to try to bring together the outcome of the past, present and future research studies, including the datasets into a digital archive so as to provide greater educational, training, research and consultancy access to potentially useful datasets. As a result, the YthanView project was developed.

16.12 Ythanview

YthanView is an ongoing research project designed to provide improved access for educators, researchers and students at the University of Aberdeen to geospatial datasets for the Ythan Estuary and surrounding area (in particular the Sands of Forvie National Nature Reserve (NNR)) in Northeast Scotland, UK. This project originated, in part, with the educational and research work currently undertaken by the Centre for Marine and Coastal Zone Management (CMCZM – http://www.abdn.ac..uk/cmczm) including an internal University of Aberdeen Learning and Technology Unit (LTU) proposal approved in 2004, and the research work programme currently being undertaken within the Aberdeen Institute for Coastal Science and Management (AICSM – http://www.abdn.ac.uk/aicsm). YthanView was also designed to complement other work currently being undertaken on the Ythan Estuary and surrounds by other researchers within the University e.g. OceanLab (http://ww.abdn.ac.uk/oceanlab), and some of the research interests of the Macaulay Institute (http://ww.mluri.ac.uk), as well as the recently established East Grampian Coastal Forum. The project is co-ordinated by David R. Green (d.r.green@abdn.ac.uk) under the auspices of the Centre for Marine and Coastal Zone Management (CMCZM), the Aberdeen Institute for Coastal Science and Management (AICSM), and OceanLab (http://www.abdn.ac.uk/oceanlab) at the University of Aberdeen. To date the project has gathered together a number of geospatial datasets, information and references for work on the Ythan Estuary that involve the application of the geospatial technologies including: remote sensing (aerial, airborne and satellite platforms), Geographical Information Systems (GIS), Global Positioning Systems (GPS), cartography and digital mapping, as well as databases, decision support systems, and the Internet. These will form the basis of the resources to be made available to staff and students. Current categories of interest are:

- *Map/GIS* datasets (ESRI shapefile format (.shp))
- *Imagery* (usually in the form of scanned colour and panchromatic aerial photographs, airborne imagery, and satellite imagery (SPOT and Landsat) – .jpg and ERDAS Imagine .IMG format)
- *DEMS* (ERDAS Imagine.IMG format)
- *Photographs* (single and panoramic photographs – TIFF, GeoTIFF, and ERDAS Imagine.IMG format)
- *Dissertations* (B.Sc. and M.Sc. dissertations – .DOC (MS-word) or .PDF (Acrobat Reader))

Initially stored on an internally accessible network drive, these will subsequently be provided as a searchable online catalogue and database with online map access using map and image server technology, as well as password protected access to the datasets.

16.13 Visualisation

As part of the YthanView project, some work was carried out to investigate how a variety of geospatial (image and map) datasets could be brought together using "off-the-shelf" information technology tools to facilitate visualisation of the Ythan Estuary and surrounding environment.

Many sources of data were available for use in this project. The data sets that were used came in the form of map data, DTMs, remotely sensed imagery, photographs and digitized shape files of algal and sediment cover. A wide range of commercially available off-the shelf software products were used to visualise the datasets available for the Ythan Estuary and Sands of Forvie. These included Erdas Imagine 8.5, ArcView 3.3, Bryce 5, and Microsoft Internet Explorer using a number of different plugins for 3D, video and animation. Only a small selection of examples output is shown here in Fig. 16.2a, b and c. Figure 16.2a shows an interactive hypermedia animation with several forms of interactivity generated with Erdas Imagine 8.5 using the ParallelGraphics Cortona® VRML Client used to view VRML model. In this example, Ordnance Survey (OS) GRID Digital Elevation Model (DEM) at a scale of 1:50000 (exaggeration 2; level of details 4%) is overlaid with raster Ordnance Survey map data at a scale of 1:50000; an aerial colour photographic mosaic, and a polygon cover (.shp) for 1989 macroalgal weedmat cover. It is saved in VRML in Erdas Imagine 8.5. This application has potential for the analysis of weedmat distribution and land-use/ land-cover analysis dependent upon surface relief. The second example, Fig. 16.2b, is an ArcView 3.3. project using raster Ordnance Survey (OS) map data at a scale of 1:50,000 with hotlinks to 21 ground truth colour photographs at 1280 × 960 pixels resolution saved in TIFF format and hot linked into a series of photographic arrays showing the direction in which the photograph was taken. Potential applications include a visual presentation of tourist attractions

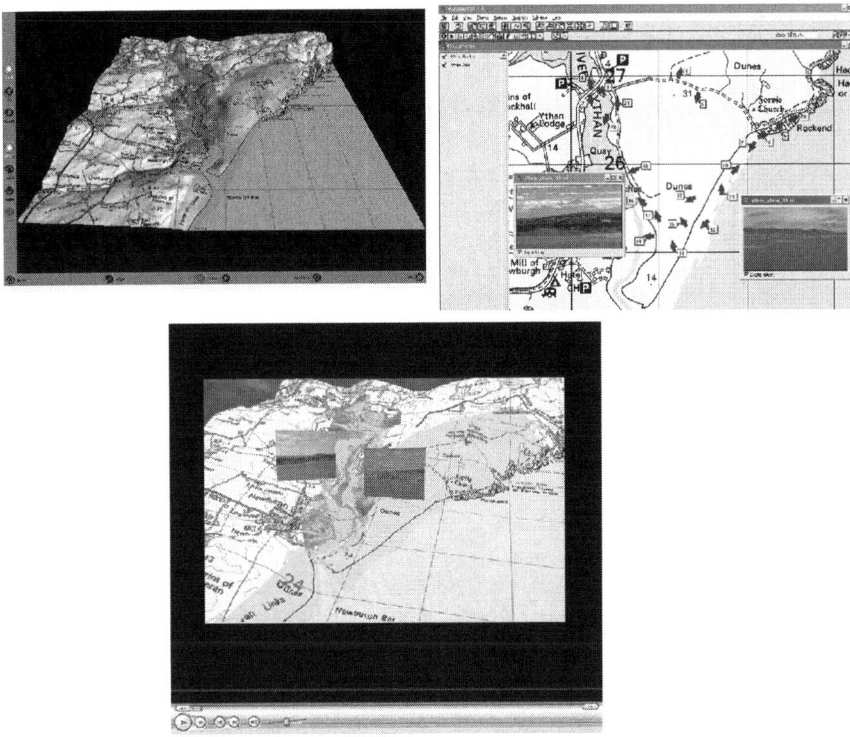

Fig. 16.2 (**a**) An interactive hypermedia animation with several forms of interactivity generated with Erdas Imagine 8.5 using the ParallelGraphics CortonaR VRML Client used to view VRML model. (**b**)An ArcView 3.3. project using raster Ordnance Survey (OS) map data at a scale of 1:50,000 with hotlinks to 21 ground truth colour photographs at 1280 × 960 pixels resolution. (**c**) The third example is a video clip (11.4 MB file size, with a duration of 25 s, comprising 375 frames, at 15 frames per second, using the DivX High Definition Profile, at a resolution of 768 × 512) generated by the Bryce 5 software (See also Plate 24 on Page 405 in the Color Plate Section)

in the context of sustainable coastal tourism. Figure 16.2c, the third example, is a video clip (11.4 Mb filesize, with a duration of 25 s, comprising 375 frames, at 15 frames per second, using the DivX High Definition Profile, at a resolution of 768 × 512) generated by the Bryce 5 software. The imagery contained in the video was created using a combination of Erdas Imagine 8.5 and Virtual GIS, VRML 1.0, and Corel Photo Paint 9. This is an example of cast-based, process animation, using motion and trajectory with some elements of thematic animations (comparing distribution) and aerial animations (fly-through) (Lobben, 2000), and a non-temporal animation with elements from animation with successive build-up (displaying themes in sequences) and the animation with changing representation (a simulated fly-through).

16.14 Discussion

Multimedia cartography and GIS can play a very important role in the process of visualisation of spatial data. Recent technical advances in digital cartography and its convergence with the Internet have fostered the development of interactive visualization of geospatial information on-line. Multimedia as an interaction with multiple forms of media is supported by the computer that is both the tool of multimedia and its medium. Without means of creation or distribution, the current interactive form of multimedia would not be able to exist.

In this study, the use of multimedia cartography and GIS is presented in the context of coastal landscape visualization. Image draping, photorealistic rendering, virtual worlds and static images viewed in the Internet browser were the main methods used in the process of visualization of geospatial data in monitoring and mapping coastal areas. Multimedia presentations can play an important role in ICZM only if they provide appropriate information about the issues of concern with a better understanding of geospatial matters; and also allow one to perform analyses and to forecast future demand on coastal resources.

Although data and information are essential components of ICZM, they may be limited by availability. However, good multimedia presentations should allow one to identify some of main resources/components such as: land area, agricultural land, build-up area, forests, water resources, marine resources or wild life resources and also infrastructure represented by roads, railways, water supply amongst others. In this case, 3D visualization is a useful tool for annotating 2D images by providing a simulation of, and the potential to, explore the landscape. On this basis the identification of key indicators of existing conditions to the past and present state of the coastal environment is possible.

3D modelling where data is draped over a DEM together with the creation of different types of animation for data sourced from different years not only provides better understanding of spatial phenomenon but also enables comparisons, analyses and the potential to discover and isolate reasons for landscape change. Multimedia presentations, especially those with navigational tools, allow one to identify the impacts of the main land and marine uses, after analyses of appropriate components. Coastal and marine ecological or landscape value can be influenced by urbanization and settlement, tourism and recreation, industry, fisheries and aquaculture, transportation, agriculture and forestry.

Virtual environments created with databases should allow identification of coastal resources under stress or at risk, and their level of vulnerability or risk of degradation. Although natural risks such as seismic activity, flooding, tsunami, landslides and volcanic eruptions are not necessarily easy to predict, areas most exposed to risk can be identified. Forecasting future demands on coastal resources can also be generated with animation showing changes of the phenomenon over the time. GIS can also be used to show, for example, where marine features are in relation to terrestrial features and the geographical position of a coastline can be mapped over a period of time. Using animation techniques it is possible to model sea-level rise scenarios and potential impacts on the coast. However, forecasts should be based both on projections of existing trends and on forecasts of the activities expected to affect the natural systems in the future.

Virtual worlds generated in this study allow one to perform land-use analysis depending on the surface relief. Analysis of the research area using sunlight illumination tools can provide additional information about erosion and the distribution of vegetation. Moreover, the relationship of estuarine channels and sandbanks can also be more readily visualized and analysed.

Using animations created in the Bryce software, the distribution of weedmats and their changes can be better understood and additional analyses using sediment coverage can be made. All these changes can also be easily compared. Furthermore, analysis of the coastal line can give the view of erosion impacts and changes in vegetation can be analysed.

Multimedia cartography and GIS can be very effective visual information tools for costal monitoring, including detailed monitoring of coastal landscapes, habitats, sediment processes, such as erosion, deposition with the consequence effects on the position of the shoreline and mapping vegetation changes, especially dune and marsh systems.

Due to the main ideas of this research, multimedia information system was created corresponding to the needs of expert and non-expert audience. These multimedia presentations could be presented on online and offline systems in an attractive and user friendly way using utilized an online mapserver. It also means that in the future the access to this kind of geospatial information about the Ythan Estuary derived from multimedia virtual database can be used by the public users as a regional tourist information system and also be used in scientific research and analyses.

Future multi-temporal monitoring and mapping of environmental changes in the Ythan could be also integrated with archival and new datasets to generate GIS based Internet environment. Implementation of other available data from different years into the project such as: satellite images, aerial photography covering other parts of the study area, panoramic photographs and possibly videography could result in creating new virtual worlds with VRML models, fly-throughs and other animations. Moreover, datasets concerning geology and bathymetry together with DEM for bathymetry of the sea and the estuary could enhance multimedia visualization and enable more complex analyses. Besides, viewsheds layers overlaid over DTMs could allow additional GIS analyses.

New capabilities and visualization methods could also be explored within visualization software packages used in this study. Also due to both software and processing limitations other professional landscape programs, depending on their availability, could be applied to produce different multimedia presentations and enhance a higher degree of realism in the final output.

However, the whole process of generation virtual environments requires appropriate computer requirements such as: hardware, high speed processor, very good graphic card, and huge amount of storage space. The final result will also depend on available datasets, their resolutions, compatible file formats and visualization software packages.

This multimedia presentation of Ythan Estuary enhanced with other new 3D visualizations placed over the Internet or stored in the form of multimedia CD could be used to introduce new users to the characteristics of the region, particularly to its nature, environmental changes and risks, tourist attractions and even the local economic infrastructure including ecotourism aspects.

16.15 Summary and Conclusions

This chapter has briefly outlined an innovative and ongoing project, YthanView, at the University of Aberdeen. Initially developed to examine new ways to store, catalogue, access, and visualize a wide range of terrestrial and coastal data and information for a small estuarine environment, and to make such data and information more widely available for teaching, research and consultancy activities, YthanView has also provided a number of other educational opportunities. One of these is new ways to introduce undergraduate and postgraduate students to a new subject area (i.e. coastal and estuarine studies) and fieldwork, and to establish closer links between education and research and consultancy work being undertaken whilst at the same time seeking to reduce student contact hours.

The development of YthanView has sought to take advantage of the wide range of current geospatial and online Internet technologies available to provide a framework to collect, store, analyze, visualize and communicate geospatial data and information. Using the environmental databases established via existing sources, a virtual fieldwork experience for a coastal and marine study area, the Ythan Estuary, is now being developed. Provision of an online Internet resource, including virtual fieldwork exercises, laboratory practicals, and assignments, will both extend and broaden current course delivery offerings. The use of guest speaker videos and greater use of remote sensing, GIS and mobile technologies for estuarine fieldwork are also being explored as ways to enhance the student experience. Practically, YthanView has provided the means to develop an online resource and project base to assist in reduction of teaching contact hours for the University of Aberdeen's Marine Resource Management (MRM) degree programme and project work; to enhance project work resources currently available to MRM students for undertaking project work; to develop a virtual fieldwork experience for MRM students with the aim of replacing existing fieldwork, as well as complementing half-day and full-day field excursions; and to provide online virtual fieldwork modules for the Ythan Estuary, NE Scotland.

This is an ongoing project, and is currently being expanded and enhanced to develop an online map and image-based information system and resource for the Ythan Estuary, one that can provide access to data and information for a wide range of coastal and marine stakeholders.

References

Bossomaier, T.R.J., and Green, D.G., 2001. *Online GIS and Spatial Metadata*. Taylor & Francis, London.

Cartwright, W., 1999. Extending the map metaphor using web delivered multimedia. *International Journal of Geographical Information Science*, Vol. 13, No. 4, pp. 335–353.

Chan, K., 2003. Sediment and Macro-Algae Relationship of the Ythan Estuary: A Remote Sensing and GIS Approach. *Unpublished M.Sc. Dissertation*, Department of Geography, University of Aberdeen, Scotland, UK, 60p.

Dunne, B., 2003. The Relationship Between Algal Mats and Wading Bird Populations in the Ythan Estuary, North East Scotland. *Unpublished M.Sc. Dissertation* in Ecology, University of Aberdeen.

Gorman, M., and Raffaelli, D., 1993. The Ythan Estuary. *Biologist*. Vol. 40, No. 1, pp. 10–13.

Green, D.R. 1995. Preserving a fragile environment: integrating technology to study the Ythan Estuary. *Mapping Awareness* 9(3, April 1995):28–30.

Green, D.R., 2005. *Going Mobile: Mobile Technologies and GIS*. URISA Quick Study Guide. URISA, 60p.

Green, D.R, and King, S.D., 2002a. *Coastal Remote Sensing Guidelines*. URISA Quick Study Guide. URISA. 2002. 36p.

Green, D.R., and King, S.D., 2002b. Ythan Estuary – The Distribution of Weedmats in the Intertidal Zone. *Report to Formartine Partnership*, University of Aberdeen, Aberdeen.

Green, D.R., and King, S.D., 2003a. Pencil out, stylus in: geospatial technologies give fieldwork a new dimension. *Geography*. Vol. 89(1):58–70.

Green, D.R., and King, S.D., 2003b. (Eds.) *Coastal and Marine Geo-Information Systems: Applying the Technology to the Environment*. Coastal Systems and Continental Margins. Vol. 4. Kluwer, The Netherlands. 580p.

Green, D.R., and King, S.D., 2003c. Application of new technologies to coastal management. In: Taveiro-Pinto, F. (Ed.) *Good Practices in Coastal Zone Management and Coastal Defence*. Chapter 6. pp. 65–83. ISBN: 972-752-057-X. Hydraulics and Water Resources Institute, Porto, Portugal.

Green, D.R., and King, S.D., 2003d. Advances in coastal GIS modeling. In Lakhan, V.C. (Ed.) *Advances in Coastal Modeling*. Chapter 21, pp. 553–580, Elsevier, Amsterdam.

Green, D.R., 2005a. Applying Geospatial Technologies to Weed Mat Monitoring and Mapping: The Ythan Estuary, NE Scotland. *Zeitschrift fur Geomorphologie* (Special Issue) Volume 141:197-212.

Green, D.R., and King, S.D., 2005b. GIS-based estuarine management systems. In: Bartlett, D. and Smith, J. (Eds.) *GIS for Coastal Zone Management*, Chapter 18. CRC Press, Boca Raton, FL, pp. 239–255.

Green, D.R., King, S.D., and Morton, D.C. 1998. Small-scale airborne data acquisition systems to monitor and map the coastal environment. *Proceedings of the 5th International Conference: Remote Sensing for Marine and Coastal Environments*, San Diego, California, 5–7 October 1998, 1479–1486.

Lobben, A., 2000. Classification and application of cartographic animation. *The Professional Geographer*. Vol. 55(3): 318–328.

Longley, P., Goodchild, M.F., Maguire, D.J., and Rhind, D.W., 2001. *Geographic Information Systems and Science*. John Wiley and Sons. Chichester. 454p.

MOTIIVE. Marine Overlays on Topography for Annex II Valuation and Exploitation. Data Harmonisation and Integration for Coastal and Marine Stakeholders. http://www.motiive.net.

Orr, G., 2003. The Distribution of Enteromorpha Alga Mats in Relation to the Sediment Facies of the Intertidal Zone of the Lower Reaches of the Ythan Estuary. *Unpublished Undergraduate Dissertation*. Department of Geography. University of Dundee, Scotland, UK. 75p.

Raffaelli, D., 2000. Interactions between macro-algal weedmats and invertebrates in the Ythan estuary, Aberdeenshire Scotland. *Marine Research.* Vol. 54: 71–79.
Raffaelli, D., Hull, S., and Milne, H., 1989. Long-term changes in nutrients, weedmats and shore-birds in an estuarine system. *Chroniques: Advances in Marine Biology.* Vol. 30:259–270.
Raffaelli, D.G., Raven, J.A., and Poole, L.J., 1998. Ecological impact of green macro-algal blooms. *Oceanography and Marine Biology: an Annual Review.* Vol. 36: 97–125.
Stove, G.C., 1978. The Hydrography, Circulation and Sediment Movements of the Ythan Estuary. *Unpublished PhD Thesis.* University of Aberdeen.
Wise, S.M., 1998. Digital Terrain Models – Traps for the Unwary. http://www.ccg.leeds.ac.uk/geocomp/1998/80/gc_80.htm. Accessed 2004.

Chapter 17
Analysis of Long-Term Changes of a Sandy Shoreline Utilising High-Resolution Aerial Photography

Tobias Dolch

Abstract The Wadden Sea is the shallow sedimentary coastal zone of the south-eastern North Sea. It is a highly dynamic system where shorelines are continuously changed by wind, waves, currents and tides. These morphodynamics were surveyed at the island of Sylt which is located in the northern Wadden Sea. The base of this long-term study is a series of high-resolution aerial photographs from 1928 to 2003. The analyses of the georeferenced aerial photographs and the quantification of shoreline changes were conducted with a Geographic Information System (GIS).

The northern part of the island of Sylt turned out to be the most interesting area where different coastal processes can be observed. The west shore of northern Sylt is subject to severe erosion as it is exposed to strong westerly winds, a consistent westerly wave wash and strong currents. Therefore, a considerable coastal retreat was observed here. In contrast to this, the north shore turned out to be an accumulation area. However, the east shore of northern Sylt is rather inactive due to its sheltered position as well as solid coastal protection constructions, such as dykes and petrified coastal sections.

In order to compensate the coastal retreat at the west shore, sand replenishment is carried out almost annually since 1972. By this, the loss of sand gets compensated and the shoreline could be fixed. At the same time, the annual accumulation rates at the north shore increased considerably. Due to the sand replenishment at the west shore, a lot of unconsolidated material is artificially supplied to the depositional system. These sediments get transported to some extent to the north shore of Sylt where they are deposited.

High-resolution aerial photographs and their analysis with GIS proofed to be very suitable to monitor coastal processes and the efficiency of coastal protection measures, like sand replenishment. This technique is a powerful method to detect erosion "hotspots". It can be especially recommended for long-term and local surveys at a smaller spatial scale.

T. Dolch (✉)
Alfred Wegener Institute for Polar and Marine Research, Wadden Sea Station Sylt, Hafenstr. 43, 25992 List, Germany
e-mail: Tobias.Dolch@awi.de

D.R. Green (ed.), *Coastal and Marine Geospatial Technologies*, Coastal Systems and Continental Margins 13, DOI 10.1007/978-1-4020-9720-1_17,
© Springer Science+Business Media B.V. 2010

Keywords Shoreline changes · GIS · Aerial photographs · Wadden Sea · Sand replenishment

17.1 Introduction

Aerial photographs are an important data source for monitoring and a Geographic Information System (GIS) represents an appropriate tool to analyse these data. Coastal monitoring with aerial photographs and GIS is a common technique carried out in different environments (Li et al. 1998, Moore and Griggs 2002, Rankey and Morgan 2002, Hapke 2005). This method is particularly suitable for a highly dynamic environment (Andres and Byrnes 1991), e.g. like the Wadden Sea – the shallow sedimentary coastal zone of the south-eastern North Sea. In this tidal system, where unconsolidated sediments are dominating, shoreline changes and sediment shiftings occur fast. The island of Sylt is a dynamic sandy barrier island in the northern Wadden Sea. It is formed by consistent wave wash and strong currents but is also vulnerable to severe erosion.

A long-term series of aerial photographs (1928–2003) of Sylt enables to observe many coastal processes that can be regarded as representative for the Wadden Sea.

The aim of this study is to

- monitor the shoreline and detect affected sections,
- reconstruct the development of the shoreline,
- analyse if coastal processes can be derivate from aerial photographs and if so, try to quantify them and survey if they have changed over the years,
- survey the efficiency of coastal protection measures,
- forecast future development.

17.2 Study Area

The shallow sedimentary coastal zone of the south-eastern North Sea is termed Wadden Sea (Fig. 17.1a). Its most remarkable features are the tidal flats, which are exposed during low tide. The extensive tidal flat areas have formed in the shelter of a chain of sandy barrier islands. The island of Sylt is such a barrier island and the study area of this survey is the northern part of Sylt.

Northern Sylt can be divided into three areas that form morphological units: the west shore is erosive, the north shore accumulative and the east shore is inactive or fixed (Fig. 17.1b).

The west shore of Sylt is formed by broad sandy beaches with a wide-stretched adjoining belt of sand dunes. It is facing the open North Sea and undergoes high erosion rates as it is exposed to strong westerly winds and a consistent westerly wave wash as well as strong currents (Ahrendt and Thiede 2001). The westerly winds are prevailing and make up 50% of the annual frequency and are causing wind-induced

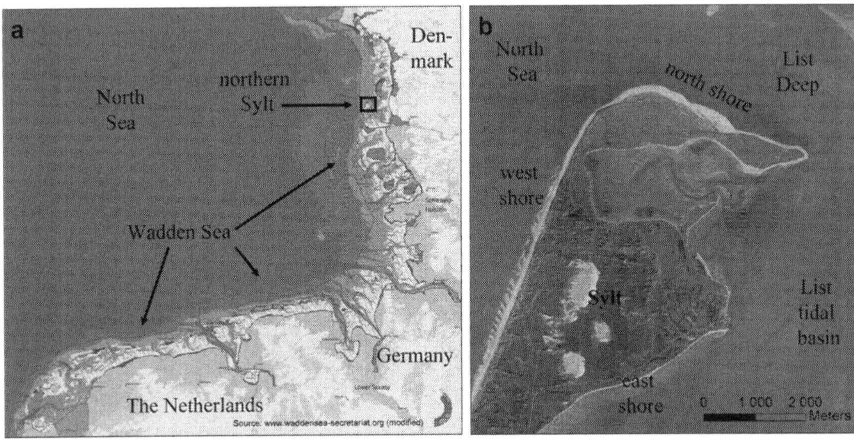

Fig. 17.1 The Wadden Sea is the shallow sedimentary coast of the south-eastern North Sea (**a**). The study area is the northern part of the island of Sylt (**b**), which is located in the northern Wadden Sea

waves (Backhaus et al. 1998). The wave wash loosens the material and the currents transport it away. An annual loss of 1.1 million m^3 sediments is calculated for the entire west shore of Sylt (Ahrendt and Thiede 2001).

The north shore runs in east-west direction and is formed by a 4.3 km long narrow sand spit with sand dunes. The beaches are sandy and wide here. The north shore adjoins the tidal inlet List Deep. List Deep is a 2.8 km wide and 40 m deep channel that connects the List tidal basin with the open North Sea. Strong currents with maximum velocities of 1.3 ms^{-1} can be observed in the inlet, caused by 550 million m^3 water, which are passing it with every tide. Three tidal channels with current velocities of 0.6 ms^{-1} diverge from the List Deep (Backhaus et al. 1998).

One of these tidal channels runs close to the east shore of Sylt. However, most of the east shore adjoins the shallow List tidal basin with its tidal flats, which have formed in the sheltered lee-side of the island. The shallowness goes also along with rather slow current velocities which are often below 0.1 ms^{-1} on the tidal flats (Backhaus et al. 1998). Among its sheltered position, the east shore is further characterised by solid coastal protection constructions, like dykes and petrified coastal sections.

17.3 Materials and Methods

Aerial photographs from 1928 to 2003 showing the northern part of the island of Sylt at low tide were visually analysed and the shoreline digitised with GIS.

The 2003 aerial photographs were provided as a digital Orthophotomosaic (OPM). This means that the photographs were already scanned, georeferenced, processed (e.g. colour adjustment, cutting of overlaps) and put together as a complete

Table 17.1 Overview of the aerial photographs

Year	Date	Colour	Ground pixel size in m	Scale
1928	1928	black-white	0.64	1:24000
1936	27th Aug. 1936	black-white	0.64	1:24000
1958	1st June 1958	black-white	0.55	1:24000
1988	11th May 1988	infrared	0.41	1:10000
1998	22nd Nov. 1998	black-white	0.47	1:15000
2001	07th Dec. 2001	colour	0.85	1:25000
2002	08th Dec. 2002	colour	0.8	1:25000
2003	14th Oct. 2003	colour	0.5	1:25000

mosaic. The photos from 1928, 1936 and 1958 were provided as files while the aerial photographs from 1988, 1998, 2001 and 2002 were allocated as contact prints (Table 17.1). They were scanned with a resolution of 800 dpi and saved as Tagged Image File Format (TIF) documents. The TIF files were imported into GIS using ArcGIS 9.1, ArcMap and georeferenced with it. The OPM of 2003 served as base map to which all other aerial photographs were georeferenced to. The OPM of 2003 was chosen because of its high quality and accuracy, which was checked at numerous control points with a differential GPS (DGPS) device.

However, georeferencing was possible because an appropriate number of bench marks, mainly constructions like buildings, road intersections, harbours etc., which have been existing already since 1928, can be identified in the pictures. The accuracy of georeferencing was improved by DGPS field data. The total spatial error of the photographs (Total RMS Error) is mainly below 5 m.

Before a shoreline could be derived from the photos, at first it had to be defined as there are many different definitions (Boak and Turner 2005). The mean high water line is often used for sandy tidal beaches (Dolan et al. 1980, Thieler and Danforth 1994, Boak and Turner 2005) and was also chosen for this survey because it can be applied best to all different shores in the study area.

In the forerun of the digitising, field surveys with the DGPS were conducted on days with an average high tide water level (mean high tide level at the gauge List from 1995 to 2006: 586 cm). This was done in order to detect spatial pattern at the shore which allow to determine the mean high tide line in the aerial photographs as well as to validate the digitised data.

The western shoreline could be determined because sand bars are located just west of it. The sandbars directly adjoin the shoreline and are only exposed during low tide. The northern shore of Sylt borders directly the narrow and deep tidal inlet Lister Deep which has steep descending slopes. Here, different water levels have only minor horizontal effects and the shoreline could be taken as the observed land-water-transition. The transition between tidal flats and beaches or land is the mean high water line on the eastern shore. It could be found due to sediment colour changing from dark to light.

The shorelines were derived from the georeferenced aerial photographs by visual analysis and manual on-screen digitising in ArcGIS 9.1, ArcMap. This is the most common technique (Boak and Turner 2005).

17.4 Results

The aerial photographs from 1928 just show only the west and north shore of the island of Sylt whereas the pictures from 1936 to 2003 cover the entire northern part of the island. Therefore, the spatial analysis for the west and north shore can go further back in time than for the east shore and the total area.

Figures 17.2 and 17.3 show the spatial development of the northern part of the island until 2003. The total area increased by 0.16 km^2 from 1936 to 1958 (Fig. 17.3a). A distinct loss of area can be observed from 1958 to 1988 which is associated with the loss at the west shore. From 1988 onwards, the size of the area increased again. Overall, the northern part of the island of Sylt gained by 0.25 km^2 from 1936 to 2003 (Fig. 17.3a).

Even more significant shiftings can be observed in the three subareas. A considerable coastal retreat is noticed at the west shore. In total, the west shore lost 0.46 km^2 from 1928 to 2003 (Fig. 17.3a). The spatial retreat is roughly uniform over the entire section and no zones with especially severe erosion were identified (Fig. 17.2).

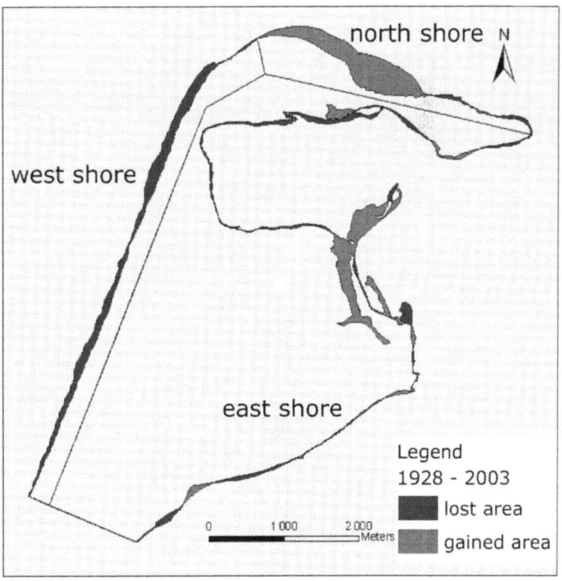

Fig. 17.2 Spatial changes of the northern part of the island of Sylt from 1928/1936 to 2003

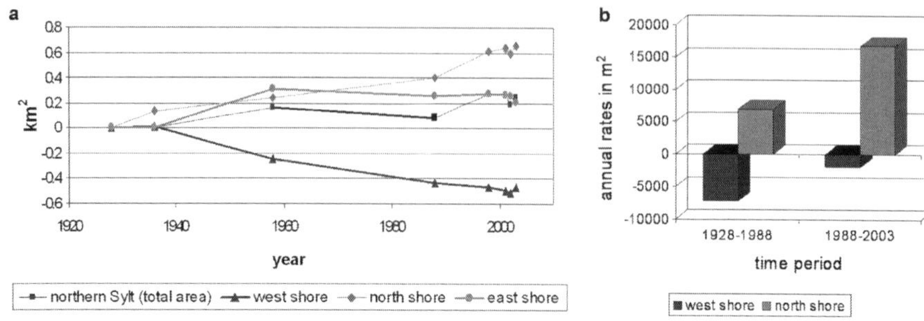

Fig. 17.3 Spatial development of the northern part of the island of Sylt and the three subareas (1928/1936–2003; cumulative frequency curve) (**a**). Annual erosion and accumulation rates at the west and north shore for the time periods 1928–1988 and 1988–2003 (**b**)

The north shore is characterised by a large gain in area. From 1928 to 2003 the north shore area increased by 0.66 km^2 (Fig. 17.3a). The gain in area is not homogeneously distributed along the north shore but a distinct accumulation area can be detected in the western half of the north shore area (Fig. 17.2).

The east shore gained considerably in area from 1936 to 1958 (0.31 km^2). This increase had an effect for the total area and caused the general gain, which is mentioned above. From 1958 to 2003 the east shore lost slightly 0.11 km^2 but there are no distinct erosion or accumulation areas. The net increase of the east shore area is 0.2 km^2 from 1936 to 2003 (Fig. 17.3a).

However, the west and north shore are the most active areas and worth of closer consideration.

17.4.1 Comparison of West Shore and North Shore

From 1928 to 2003 the west shore lost about 463000 m^2 by erosion as the north shore area gained 655000 m^2 by accumulation. This reflects a coastal retreat of about 6200 m^2 per year in the west and an accumulation of 8800 m^2 in the north (Fig. 17.3b). Having a more detailed look at smaller time intervals, a significant change in both erosion rate and annual gain in area can be noticed. From 1928 to 1988 the regarded west shore area decreased by approx. 430000 m^2. This reflects a coastal retreat rate of about 7000 m^2 per year. From 1988 to 2003 the retreat was 30000 m^2 which means an annual coastal retreat of about 2000 m^2.

The north shore area grew in contrast to this. It gained about 410000 m^2 from 1928 to 1988. This means a gain in area of 6800 m^2 per year. Regarding the period from 1988 to 2003 the accumulation was 250000 m^2 and the annual rate increased considerably to 16600 m^2 (Fig. 17.3b).

17.5 Discussion

The developments of the three subareas result together in the development of entire northern Sylt (total area) and affect it proportionate (Fig. 17.3a).

The west shore is a highly erosive coastal section as it is exposed to strong westerly winds, a constant wash of waves and strong currents. Severe coastal retreat is especially observed from 1928 to 1988. But why did the coastal retreat rate decrease from 7000 m^2 (1928–1988) to 2000 m^2 (1988–2003) per year?

At first it has to be mentioned that sand replenishment is carried out on the island of Sylt since 1972 and for this reason the survey period was divided into 2 time intervals. As the aerial photographs from 1988 are closest to that date, the time period was split in 1988 to show the status of the shore "before" and "after" sand replenishment and to check the efficiency of this coastal protection measure.

The artificial supply of sediment almost compensates the eroded material and stabilises the position of the western shoreline. Sand replenishment is conducted almost annually at alternating sections at the western shore (ALR 2007). Altogether 14 million € are assured for sand replenishment on Sylt for the period 2007–2009 (ALR 2007, Landesregierung S-H 2007). Although sand replenishment is expensive and the effect last only for a limited time, in the case of Sylt there is no alternative solution to it (Ahrendt 1993, CPSL 2005). Solid structures such as dykes, tetrapodes, petrified shores etc. turned out to be inefficient at the west shore (Ahrendt 1993) and are not compatible with tourism on the beach which is the most important economic factor for the island. Furthermore, sand replenishment turned out to be the best environmental practice and is recommended to be applied wherever feasible (CPSL 2001, CPSL 2005).

However, even though the coastal retreat is reduced, it can be assumed that the erosion rate is still the same or even increased. This can be due to a rising sea level and the large quantities of unconsolidated sediments which are now available for erosive processes.

The artificial supply of sediment to the depositional system can be regarded as the reason for the increased accumulation rate at the north shore. Otherwise it would be unlikely to achieve such an increased accumulation in such a short time. The sand, which gets replenished at the west shore, gets loosened by waves and removed by long-shore currents. Tidal currents divert a part of the sediment load into the List tidal basin where some of it is deposited at the northern shore of Sylt. Having a look at the 2003 aerial photographs, the accumulation area can be identified as it is not or only sparsely covered with vegetation. This suggests that this area was recently gained.

The accumulation area could establish here despite the close vicinity to the tidal inlet List Deep with its strong currents. This can be explained by the different current patterns at rising and falling tides (Fig. 17.4). When the tides are rising, the incoming water causes high current velocities along the entire cross-section profile of the 2.8 km wide List Deep (Fig. 17.4a). When the tides are falling, the outgoing water gets deflected by the morphology and the shape of Sylt towards the centre and the north of the List Deep (Behrens et al. 1997). This results in slow currents

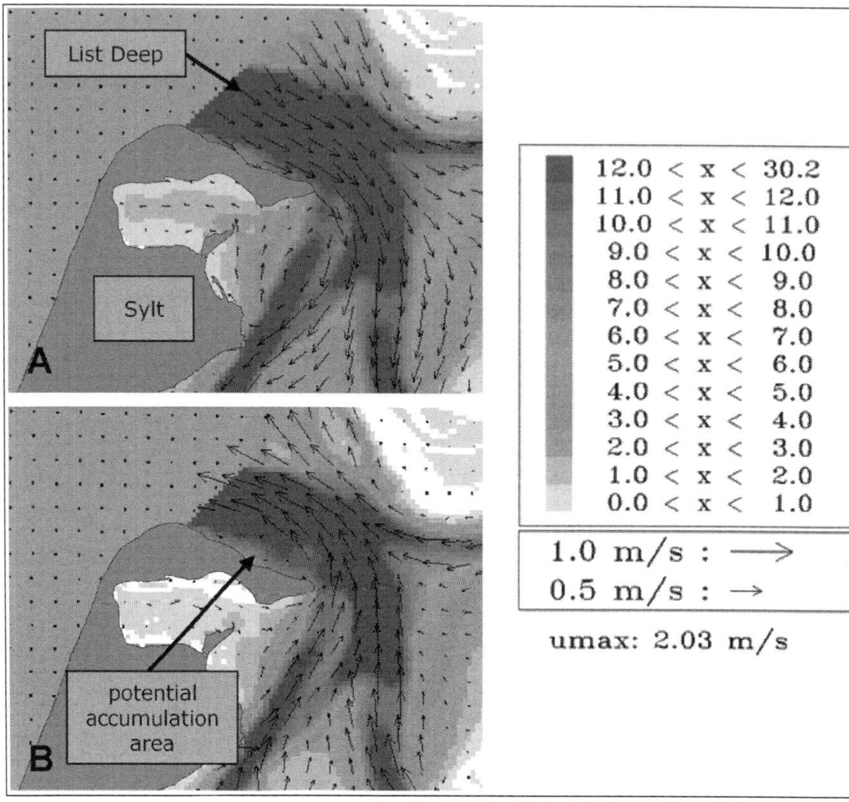

Fig. 17.4 Currents and water levels at northern Sylt at rising (**a**) and falling (**b**) tide

at the northern shore just east of the bulge area where sediments can accumulate (Fig. 17.4b). It can be assumed that these sediments are partly eroded with the following high tide but nevertheless, a positive net accumulation is clearly detected in the long-term view. However, the extent of the potential accumulation area is restricted by high current velocities which border the area and a further gain in area is limited.

Even though sand replenishment represents a large human impact on the natural system and has altered the depositional regime considerably, the question arises what would the development be without it? Computer simulations show that the northern tip of the island of Sylt surely would got detached from the rest of the island within the next 40 to 50 years in case sand replenishments would have been stopped (Ahrendt and Thiede 2001).

The east shore gained much in area from 1936 to 1958. This was a man-made change due to the construction of a dyke in 1937 and the construction of an artificial island in the early 1940s (Wohlenberg 1953). Since 1958, a trend in loss of area can be noticed at the east shore. However, considering the proximity of the tidal channel, that runs southwards close to the east shore, the erosion rates are rather low. This is due to the sheltered position of the east shore as well as the solid coastal protection constructions in this area.

17.5.1 Further Development of Northern Sylt

According to the recommendations of the CPSL (2005) sand replenishment should be applied at the outer coastlines of barrier islands wherever feasible. For these reasons, it is most likely that sand replenishment will be continued at the western shore and no major changes are expected here. The analysis of aerial photographs showed that it successfully balances coastal erosion and stabilises shorelines. Since a sea level rise of about 25 cm is expected for the Wadden Sea by 2050 (CPSL 2001), it is likely that the erosion rate and consequently the amount of sand needed for replenishment and the costs will increase considerably (Landesregierung S-H 2001).

Most changes can be expected in the north shore area but as mentioned above the extent of the potential accumulation area is restricted by high current velocities and a further gain in area is limited. Under current conditions the potential accumulation area would be filled up within the next 10–12 years. As currents and morphology are interacting, it is likely that the accumulation rates will decrease as the actual accumulation area approaches the zone of high current velocities. From this it follows that a cessation of the net accumulation at the northern shore of Sylt can be expected in approx. 20 years and that a dynamic equilibrium will establish. It remains to be seen where the sediments get transported then.

No important changes are to be expected at the east shore. There is no hint of distinct erosive or accumulative areas and shoreline changes are rather slow compared to other areas.

This study shows that different processes can be observed within a small coastal area. The combination of remote sensing data with model data on current velocities allows understanding the interactions and coastal processes as well as making forecasts. Furthermore, the analysis of high-resolution aerial photographs with GIS proofed to be very suitable for the survey of morphological changes and developments. It is a powerful tool to detect erosion "hotspots". This method can be especially recommended for long-term and local surveys at a smaller scale.

Acknowledgements Christian Hass (AWI), the Regional Office of the Schleswig-Holstein Wadden Sea National Park, Toenning (NPA) and the Regional State Office for the Rural Areas, Husum (ALR) are gratefully acknowledged for providing aerial photographs. My thanks to G. Gayer from the GKSS for supplying the current model and special thanks to Trevor Vincent for proof-reading.

References

Ahrendt, K. (1993). Sedimentdynamik an der Westkueste Sylts (Deutsche Bucht/Nordsee). Meyniana 45: 161–179.

Ahrendt, K., Thiede, J. (2001). Naturraeumliche Entwicklung Sylts – Vergangenheit und Zukunft. In: Daschkeit, A., Schottes, P. (Eds.). Klimafolgen fuer Mensch und Kueste am Beispiel der Nordseeinsel Sylt. Springer, Berlin, Heidelberg, New York: 69–112.

ALR (Amt fuer laendliche Raeume, Husum) (2007). Kuestenschutzmaßnahmen Westkueste Sylt 2007. Husum.

Andres, F.J., Byrnes, M.R. (1991). Accuracy of shoreline change rates as determined from maps and aerial photographs. Shore and Beach (January): 17–26.

Backhaus, J., Hartke, D., Huebner, U., Lohse, H., Mueller, A.. (1998). Hydrographie und Klima im Lister Tidebecken. In: Gaetje, C., Reise, K. (Eds.). Oekosystem Wattenmeer – Austausch-, Transport- und Stoffumwandlungsprozesse. Springer, Berlin, Heidelberg, New York: 39–54.

Behrens, A., Gayer, G., Guenther, H., Rosenthal, W. (1997). Atlas der Stroemungen und Wasserstaende in der Sylt-Romo-Bucht. Geesthacht, GKSS Forschungszentrum.

Boak, E.H., Turner, I.L. (2005). Shoreline definition and detection: a review. Journal of Coastal Research 21(4): 688–703.

CPSL (Coastal Protection and Sea Level Rise) (2001). Final Report of the Trilateral Working Group on Coastal Protection and Sea Level Rise. Wadden Sea Ecosystem No. 13. Common Wadden Sea Secretariat, Wilhelmshaven, Germany.

CPSL (Coastal Protection and Sea Level Rise) (2005). Coastal Protection and Sea Level Rise – Solutions for sustainable coastal protection in the Wadden Sea region. Wadden Sea Ecosystem No. 21. Common Wadden Sea Secretariat, Trilateral Working Group on Coastal Protection and Sea Level Rise (CPSL), Wilhelmshaven, Germany.

Dolan, R., Hayden, B.P., May, P., May, S. (1980). The reliability of shoreline change measurements from aerial photographs. Shore and Beach (October): 22–29.

Hapke, C.J. (2005). Estimation of regional material yield from coastal landslides based on historical digital terrain modelling. Earth Surface Processes and Landforms 30: 679–697.

Landesregierung, S-H. (- Ministerium fuer laendliche Raeume, Landesplanung, Landwirtschaft und Tourismus des Landes Schleswig-Holstein) (2001). Generalplan Kuestenschutz - Integriertes Kuestenschutzmanagement in Schleswig-Holstein. Kiel.

Landesregierung, S-H. (- Minister fuer Landwirtschaft, Umwelt und laendliche Raeume) (2007). Haushaltsmittel fuer den Kuestenschutz. Drucksache 16/1203, Schleswig-Holsteinischer Landtag, Kiel.

Li, R., Keong, C.W., Ramcharan, E., Kjerfve, B., Willis, D. (1998). A coastal GIS for shoreline monitoring and management – case study in Malaysia. Surveying and Land Information Systems 58(3): 157–166.

Moore, L.J., Griggs, G.B. (2002). Long-term cliff retreat and erosion hotspots along the central shores of the Monterey Bay National Marine Sanctuary. Marine Geology 181: 265–283.

Rankey, E.C., Morgan, C. (2002). Quantified rates of geomorphic change on a modern carbonate tidal flat, Bahamas. Geology 30(7): 583–586.

Thieler, E.R., Danforth, W.W. (1994). Historical shoreline mapping (I): improving techniques and reducing positioning errors. Journal of Coastal Research 10(3): 549–563.

Wohlenberg, E. (1953). Sinkstoff, Sediment und Anwachs am Hindenburgdamm. Die Kueste 2(2): 33–94.

Chapter 18
GIS in the Trilateral Monitoring and Assessment Program (TMAP)

Gerold Lueerssen

Abstract Since 1978, The Netherlands, Denmark and Germany have been working together on the protection and conservation of the Wadden Sea covering management, monitoring and research. The aim of the Trilateral Monitoring and Assessment Program (TMAP) is to provide a scientific assessment of the status and development of the Wadden Sea ecosystem and and to assess the status of implementation of the trilateral Targets of the Wadden Sea Plan. On basis of expert workshops, integrated data handling and inter-calibration exercises GIS data on Wadden Sea habitats, like salt marshes, blue mussel beds and dunes could be successfully integrated into the de-centrally organized trilateral data handling of the TMAP. Results are published in the Quality Status Report (QSR) Wadden Sea.

Keywords Trilateral Cooperation · Wadden Sea · Trilateral Monitoring and Assessment Program · TMAP · Monitoring · QSR · GIS · Habitats · Wadden Sea Plan

18.1 Introduction

The Wadden Sea Area is a world wide unique wetland of 13,500 km^2 along the southern North Sea coast from the Danish Esbjerg to the Dutch Den Helder. Around fifty in- and uninhabited islands are located within, dikes and estuaries are forming the boundary to the mainland, the North Sea is delimiter to the open sea (Fig. 18.1).

The Wadden Sea is part of various international conventions like Ramsar, Bonn and OSPAR convention, is also designated as Particularly Sensitive Sea Area (IMO) and is to a great extent protected under the EU Birds Directive, Habitats Directive and Water Framework Directive. National management plans like national park laws and planning instruments are complementing the protection status.

G. Lueerssen (✉)
Common Wadden Sea Secretariat, Virchowstrasse 1, 26382 Wilhelmshaven, Germany
e-mail: luerssen@waddensea-secretariat.org

D.R. Green (ed.), *Coastal and Marine Geospatial Technologies*, Coastal Systems and Continental Margins 13, DOI 10.1007/978-1-4020-9720-1_18,
© Springer Science+Business Media B.V. 2010

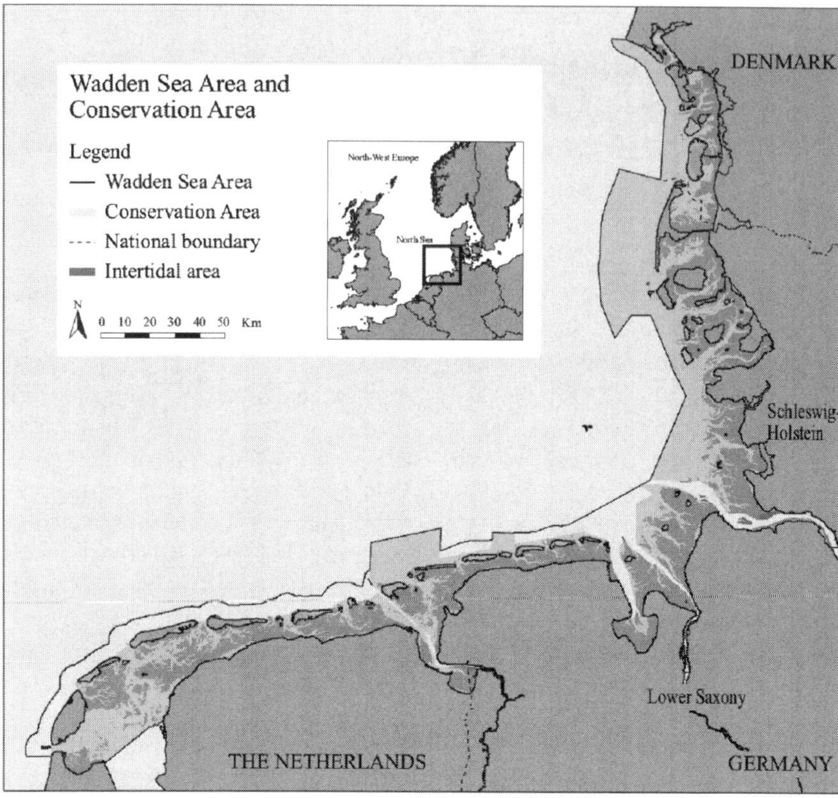

Fig. 18.1 The Wadden Sea area (Trilateral cooperation area) (See also Plate 25 on Page 406 in Color Plate Section)

Since 1978 the Netherlands, Denmark and Germany have been cooperating on the protection and conservation of the Wadden Sea covering management, monitoring and research, as well as political matters. Basis of the Cooperation are the every four years hold Trilateral Governmental Conferences (TGC). The Common Wadden Sea Secretariat is the coordinating body for all trilateral activities. The 8th TGC in 1997 (CWSS 1997) adopted the Wadden Sea Plan as management instrument to implement the so-called Targets and the Trilateral Monitoring and Assessment Program (TMAP) to evaluate the status of the Targets. These ecological Targets stand for natural dynamics, absence of disturbance, absence of pollution for habitats, water and sediment, birds, mammals and cultural landscapes.

18.2 The TMAP

The TMAP plays a central role in the Trilateral Cooperation. The program (including concomitant research) provides the scientific information for a periodic assessment of the status, the evaluation of the ecological Targets and to identify general developments in the Wadden Sea ecosystem. The data of the harmonized national monitoring programs are stored in the trilateral data management system, which supports experts with data for assessment and reporting. Every five years results are published in the Wadden Sea Quality Status Report (QSR), which gave an overview on the status of the Wadden Sea ecosystem and recommendations for management and further monitoring. The latest QSR has been published in May 2005 (Essink et al. 2005). A process to optimize and to adapt the TMAP with regard to the EU Water Framework Directive and Natura 2000 is presently ongoing.

18.3 TMAP – Monitoring Parameters

The TMAP covers a broad range of monitoring parameters from physiological processes over population development to changes in landscape and morphology. The selection of the parameters for the TMAP was based on selected topical focal points, so-called "Issues of Concern": Climate change, input of nutrients and pollutants, commercial fisheries, recreational activities and agricultural practice. The 28 parameters groups of the TMAP (Table 18.1), the Common Package, as well as the sampling strategies stem from these five issues of concern and the requirements of the trilateral Targets of the Wadden Sea Plan.

An elementary component of the TMAP is the common, de-centralized data handling, which makes monitoring data available for trilateral assessment. For this purpose, so-called TMAP data units have been installed in each country which are in charge of storing the monitoring data of the Common Package parameters. The downloading of the actual data is restricted to selected expert groups responsible for assessing and publishing data.

Table 18.1 Monitoring parameter groups of the TMAP

Chemical parameters	Biological parameters	Habitat parameters
Nutrients	Phytoplankton	Blue mussel beds
Metals in sediment	Macroalgae	Salt marshes
TBT in sediment	Eelgrass	Beaches and dunes
Contaminants in blue mussels,	Macrozoobenthos	
Flounders and bird eggs	Breeding birds	*Human use parameters*
	Migratory birds	Mussel/Cockle/Shrimp fishery
General parameters	Beached bird surveys	Recreational activities
Geomorphology	Common seals	Agricultural utilization
Flooding		Coastal protection measures
Land use		
Weather conditions		

18.4 TMAP and GIS Data

Monitoring data of the habitat related TMAP parameter groups (salt marshes, blue mussel beds and beaches and dunes) have been stored in a central geographical information system (GIS). GIS has been chosen, because it was already in use in the countries, because of the possibility of detailed spatial analysis and assessment, the incorporation of other monitoring data with spatial reference and for presentation of spatial data.

The compilation of trilateral common and harmonized GIS data sets started with expert workshops including the participation of data handling experts in the year 2000. Inventories on methods used to obtain the data, on aims of the national monitoring programs and on the GIS data itself were the basis of establishing trilateral methods and classification for the compiling of GIS data in a common trilateral database. The monitoring experts had to "translate" nationally used classifications of their habitat GIS data into a trilateral developed and agreed classification system. This approach avoided additional monitoring efforts and improved the re-use of national data. Practical inter-calibration exercises were needed and carried out to verify the trilateral results and to improve the quality of the trilateral data.

Currently data sets on salt marshes, blue mussel banks and dunes are available. The national GIS data has been collected at the secretariat, transferred to the trilaterally used projection Universal Transverse Mercator (UTM32) with the World Geodetic System 1984 (WGS84) and offered to the experts and users via the TMAP data units.

18.5 GIS Used for Salt Marsh Monitoring

There are almost 400 km^2 of salt marshes are present on the main- and islands of the Wadden Sea. Salt marshes are an interface between land and sea and are strongly influenced by geomorphological, physical and biological processes. They provide a habitat for a wide range of different species (Fig. 18.2).

The Targets from the trilateral Wadden Sea Plan applied about salt marshes are:

- An increased area of natural salt marshes.
- An increased natural morphology and dynamics, including natural drainage patterns of artificial salt marshes, under the condition that the present surface is not reduced.
- An improved natural vegetation structure, including the pioneer zone, of artificial salt marshes.
- Favourable conditions for migrating and breeding birds.

Monitoring of salt marshes is part of the TMAP Common Package. It has been carried out in all countries, but by using different methods, frequencies and classification schemes which led to a heterogeneous data basis, which was not harmonized and therefore not comparable.

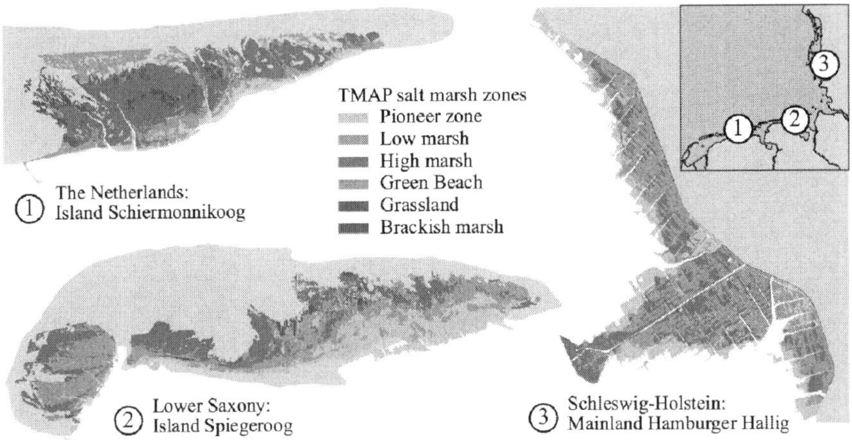

Fig. 18.2 Salt marsh zones in the Netherlands and Germany. The used geographical and attribute data is directly comparable (Essink et al. 2005) (See also Plate 26 on Page 406 in Color Plate Section)

The harmonisation of national salt marsh data started with an inventory of the methods used and the collection of nationally GIS data. It showed that very detailed vegetation and biotype data based on aerial pictures from 1995 to 2001 with a high spatial resolution for a scale of 1:5,000 and 1:10,000 including the needed ground truth verifications were available. In a first step the experts agreed on six salt marsh zones, which could be defined by aggregating the collected vegetation and biotope type data. Furthermore, a classification of salt marsh grazing (intensive, moderate and extensive) and draining (never, more than 10 years ago and within the last 10 years) could be carried out and included in GIS. A compromise was able to be reached on the available grazing and drainage data with even more heterogeneous sources in the countries. The main reasons of success in establishing a trilateral salt marsh GIS database has been the usage of already available data and that common results could be reached without any additional monitoring efforts.

In a second step the experts further divided the six salt marsh zones up into a trilateral vegetation type classification of 25 vegetation types. The GIS data on vegetation typology, in combination with the data of the other salt marsh parameters can provide evaluation and assessment information for the trilateral salt marsh management to reach the trilateral Targets on salt marshes. The GIS database can also deliver the data for describing the "favourable conservation status" of the Habitats Directive and the "good ecological status" of the Water Framework Directive.

In future GIS data on salt marshes will also be used in combination with data from the trilateral bird counts. A detailed analysis of changes in the salt mash vegetation types allows to find out dependencies with bird count numbers. Results can be used for better support of the salt marsh management (Fig. 18.3).

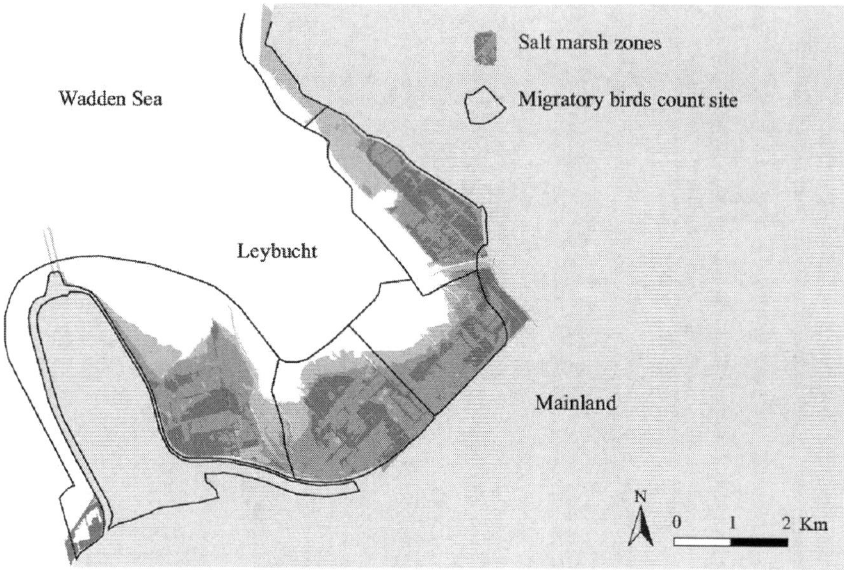

Fig. 18.3 Salt marsh zones (NLPV 1997) combined with migratory bird count sites (NLWKN 2004) in the Leybucht Lower Saxony, Germany (See also Plate 27 on Page 407 in Color Plate Section)

18.6 GIS Used for Monitoring of Intertidal Blue Mussel Beds

Beds of the blue mussel (*Mytilus edulis*) are important biogenic structures in the Wadden Sea ecosystem, serving as habitat and as food source for a number of species. The area of blue mussel beds can vary to a great extent even within a year. The high influence of winters, spat fall and the general dynamic of the Wadden Sea allows monitoring only in limits given by these factors.

The Targets from the trilateral Wadden Sea Plan applied to intertidal blue mussel beds are:

- A natural dynamic situation in the Tidal Area.
- An increased area of geomorphologically and biologically undisturbed tidal flats and subtidal areas.
- An increased area of, and a more natural distribution and development of natural mussel beds, Sabellaria reefs and Zostera fields.

Each country in the Wadden Sea Area carries out monitoring on blue mussel beds in close relation to the aims of the national management plans. Methods with assessment of aerial photographs and ground-surveys were used to determine the location, size and shape of intertidal mussel beds. Because the used aerial pictures were coloured or black and white and had different resolutions respective scales, the by digitizing obtained GIS data were not directly comparable. Furthermore, because of different definitions of mussel beds and the at this time not known comparability of the methods the data was not usable on trilateral level.

The harmonisation of GIS data on blue mussel beds was carried out in five trilateral monitoring workshops by blue mussel bed and data handling experts in 2000–2004. By developing a blue mussel bed measuring protocol and developing a common trilateral definition of intertidal mussel beds a basis for a trilateral GIS database was established.

The available trilateral GIS data on intertidal blue mussel beds has been collected and compared. In inter-calibration exercises national experts assessed the blue mussel bed sizes from the heterogeneous aerial photographs used by their colleagues in the other countries. The same experts additionally carried out a common exercise of ground truth (with GPS) by a detailed monitoring the similar blue mussel bed to find out differences in nationally used methods.

The trilateral experts agreed that the national used methods to get size and location of blue mussel beds are comparable on trilateral level. The results obtained in the workshops and in the exercises were within the normal variability of observed changes of blue mussel bed sizes in the Wadden Sea and therefore usable.

The difference in the monitoring methods does not allow for getting common information on the age of a mussel bed, the type of location (stable bed or not) and the sediment structure of the mussel bed basis.

For the QSR 2004 a detailed overview on mussel bed sizes could be given. By additionally using the GIS data on tidal flats also the percentage of coverage of tidal areas by blue mussel beds could be calculated for sub areas, e.g. the 12 QSR sub areas (Fig. 18.4).

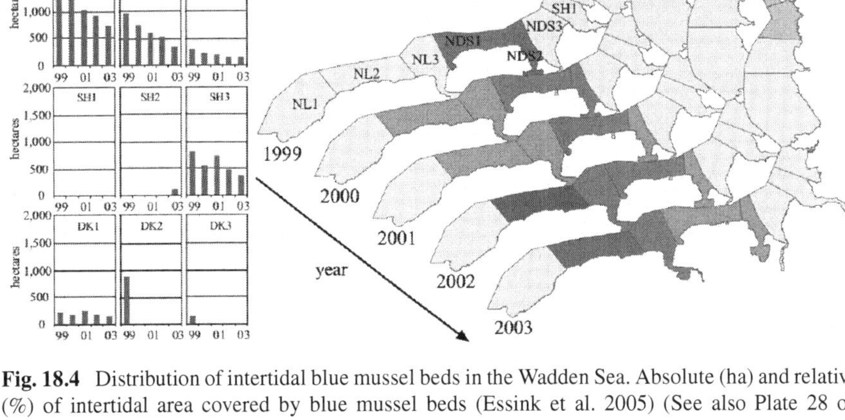

Fig. 18.4 Distribution of intertidal blue mussel beds in the Wadden Sea. Absolute (ha) and relative (%) of intertidal area covered by blue mussel beds (Essink et al. 2005) (See also Plate 28 on Page 407 in Color Plate Section)

18.7 GIS and TMAP Data Handling System

The TMAP data handling system has been developed and implemented to provide experts and managers with the necessary harmonized monitoring data in high quality for trilateral and national reporting. However, the national monitoring programs will not in all cases produce directly comparable monitoring data and therefore data for immediate, fast and efficient assessment. This can be put down to the problems of differently used national data models, the composition of available TMAP data and the different distribution and processing state at the national TMAP data sources.

As response to the national situations, the de-central concept of the TMAP data units has been chosen, which comprises identical and harmonized data models, an integrated data catalogue, which allows an overview on the availability, composition and processing state of TMAP data and an Internet service, which enables interactive data access and transfer.

The open approach enables data storage within administrational boundaries and provides an easier and independent dealing with national data on the trilateral level in relation to data ownership, data quality and data originator closeness.

The TMAP data handling system offers harmonized monitoring data of high quality, which are comparable, reliable, fast accessible and as up-to-date as possible. The four TMAP data units are located in each of the four regions of the Wadden Sea (Netherlands, Lower Saxony, Schleswig-Holstein/Hamburg and Denmark). Via the TMAP database, the Java applet, and an application server users can select data by means of the catalogue, and registered users can immediately download the resulting TMAP data in the trilateral exchange format.

Harmonized spatial (GIS) data on salt marshes and data on blue mussel beds are offered as ArcView shape files including the harmonized attribute data. The centrally compiled and harmonized GIS covers have been separated according to the monitoring boundaries of the countries and are provided and offered as GIS data via the data units. An additional common ArcView project file allows to easily fit together the national shape files on trilateral level. The project file provides a common description and presentation of the stored polygon data. A complete integration of GIS into the data unit (map server) has not been carried out because of high implementation and maintenance efforts at four data unit sites and the limited resources of the trilateral data handling.

18.8 Conclusions

For the first time trilateral harmonized GIS data on salt marshes and blue mussel beds are available for the entire Wadden Sea. The main reason for success were several expert workshops, which integrated also the data handling and carried out inter-calibration exercises. TMAP data units offer GIS data as shape files, physically distributed to the data units of the countries, but easily re-joinable in a trilateral GIS database. The available results improve the assessment quality for the trilateral Targets of the Wadden Sea Management Plan. The data can also support other TMAP parameters, like bird parameters, with valuable background information. The GIS data can also fulfil the requirements of other international reporting obligations, like the EU Habitats Directive or the Water Framework Directive.

References

CWSS – Common Wadden Sea Secretariat (ed.) (1997): Trilateral Wadden Sea Plan 1997. Ministerial Declaration of the Eighth Trilateral Governmental Conference on the Protection of the Wadden Sea, Stade, Annex 1., Wilhelmshaven, pp. 13–83.

Essink, K, Dettmann, C., Farke, H., Laursen, K., Lüerßen, G., Marencic, H., and W. Wiersinga (eds.) (2005): QSR 2004 – Wadden Sea Quality Status Report 2004. Wadden Sea Ecosystem No. 19, Common Wadden Sea Secretariat, Wilhelmshaven.

NLPV – Data from the GIS database of the Niedersachsen Wadden Sea National Park Administration (1997), Wilhelmshaven, Germany.

NLWKN – GIS Data from Lower Saxony Water Management, Coastal Defence and Nature Conservation Agency (NLWKN), former NLÖ, Governmental Bird Conservation Station, Hannover-Hildesheim, Germany.

Chapter 19
Integrating GIS with Hydrodynamic Model for Wastewater Disposal and Management: Pearl River Estuary

Sandy M.N. Ng, Onyx W.H. Wai, Z.H. Xu, and Y.S. Li

Abstract The Pearl River Estuary (PRE) has been experiencing serious water quality deterioration due to the tremendous economic and industrial developments in the nearby regions during the recent years. The integration of a hydrodynamic model, a geographic information system (GIS), and a database of collected field data can provide efficient tools for evaluating potential environmental impact and managing wastewater disposal in PRE. Essential capabilities include: (1) Display of field data in forms of spatial layer, time series, and profile, (2) Data interpolation combining spatial and temporal techniques, (3) Model mesh grid configuration, (4) Model input editing, and (5) Model output interpretation and display. The integrated system is conveniently designed to apply for environmental impact assessment (EIA).

Keywords Hydrodynamic modeling · Wastewater management · Water quality · Environmental impact assessment

19.1 Introduction

When managing wastewater disposal in coastal regions, an effectual approach is to obtain a qualitative and objective assessment of the existing hydrodynamic and environmental condition, and apply numerical modelling for predicting future consequences. Both field measurements and model simulations generate huge quantities of geographically-referenced data, which are undoubtedly difficult to handle without the assistance of computer information technology. GIS, with its capability to store, organize, manipulate, analyze, retrieve, display, and output geographically-referenced information, provides the essential needs to unify different kinds of data into one management platform (Demer, 2000). By linking a GIS with a hydrodynamic numerical model for wastewater discharge, the integrated system will

S.M.N. Ng (✉)
Centre for Water Research, University of Western Australia, Perth, WA, Australia
e-mail: ng@cwr.uwa.edu.au

D.R. Green (ed.), *Coastal and Marine Geospatial Technologies*, Coastal Systems and Continental Margins 13, DOI 10.1007/978-1-4020-9720-1_19,

© Springer Science+Business Media B.V. 2010

be a useful tool not only for predicting and evaluating potential environment impacts due to coastal wastewater disposal activities but also for assessing the accuracy of the numerical model by comparing field-collected data and model-generated results.

Although GIS is demonstrated to be an invaluable asset in hydrodynamics, there are still relatively few such applications, particularly for integrations with models which facilitate three-dimensional simulations. Perhaps one of the reasons behind this is the difficulty to establish a rigorous model representing the ever-changing nature of marine environment. Marine environmental data exhibit variation not only in both location and depth, but also through time (Lucas, 2000). So far most of the current hydrodynamic-GIS models are limited to two-dimensional capabilities. An example is the IDOR 2D which is a closely-coupled depth-averaged model for simulation of pollutant transport (Tsanis and Boyle, 2001). Since hydrodynamics is intrinsically three-dimensional and dynamic, such models will oversimplify the real situations and generate results with limited accuracy. While the fundamental concept of GIS is mapping of information in two-dimensional ground, the challenge is how to unfold this two dimensionality for applications into the three and four-dimensional world.

Another cause which impedes greater use of GIS in hydrodynamics is the difficulty to collect marine and coastal data. In general, there are two approaches to collect marine-related data. One method is to perform actual field surveys and measurements, from deploying shipboard instrument, by the use of satellite remove sensing, or by means of automatic data loggers at fixed monitor stations. However, such means of data collection, requiring expertise in both equipment and skills, are inevitably costly and time-consuming. A second approach is to gather data from existing sources. This may include data derived from published or archived materials, such as maps and graphs, on-site observations, and satellite images, provided by public and commercial domains. Although acquisition of these data is usually cheaper, very often, due the lack of comprehensive data documentation and metadata, the reliability of these data is in question (Barlett, 2000).

In this chapter, the development of a GIS for wastewater management focuses in addressing the issues for the three-dimensional, dynamic nature of hydrodynamics, and for the spatially and temporally sparse distribution of field collected data. This study consists of three main parts: (1) retrieval and display of field data, (2) integration with hydrodynamic model, and (3) application for EIA.

19.2 Study Area

The Pearl River Estuary (PRE), located in the southern part of China, is a triangular-shaped estuary with a surface area of more than 2000 km^2(see Fig. 19.1). The estuary receives inflow of freshwater from four major river outlets at the northwest side: Hengmen, Hongqimen, Jiaomen, and Humen and drains into the South China Sea. This results in a total annual average volume of freshwater discharge of 1.742×10^8 m^3 and an average annual discharge of 5300 m^3/s. The estuary has

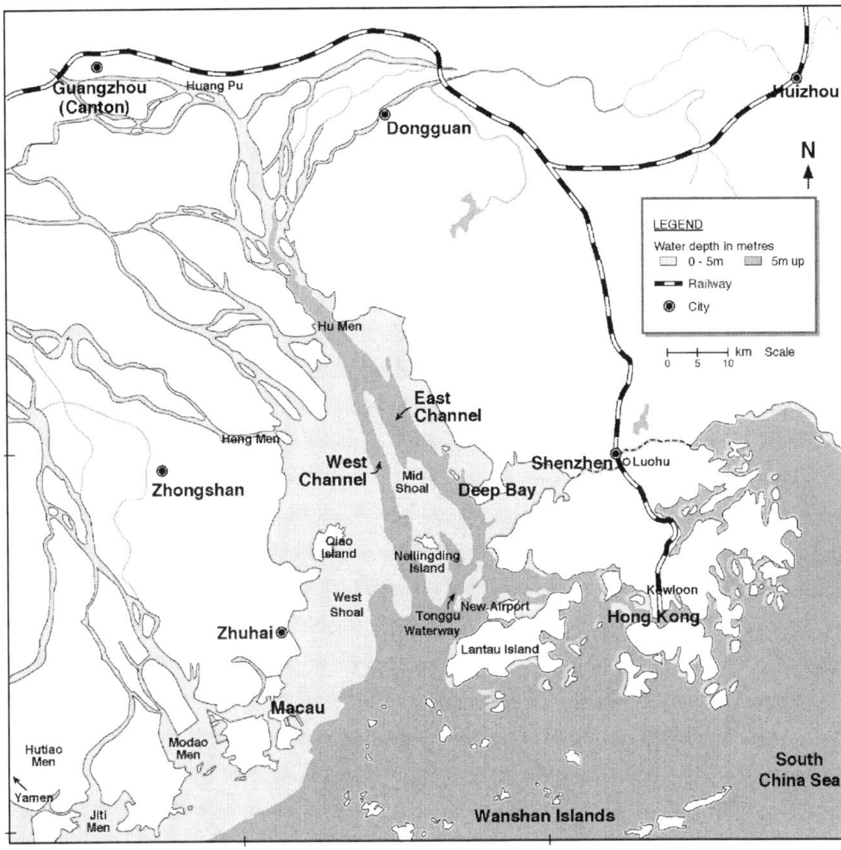

Fig. 19.1 Pearl river estuary (After Ng (2004))

width varying from 5 km at the northern end to about 35 km at the southern end, and a longitudinal length of approximately 70 km. The water depth increases from north to south, and decreases from east to west in the southern end. With the exception of two deeper shipping channels (one at the east flowing through Hong Kong towards Victoria Harbour and the Lamma Channel and the other at the west through the Lantau Channel) with water depth varying from 20 to 30 m, most of the estuary is quite shallow with a water depth between 2 and 10 m (Chen and Heinke, 2001).

In the recent decades, rapid economic and industrial developments in the Pearl River Delta (PRD) region has produced excessive release of pollutants and thus led to severe deterioration of water quality in the PRE system. While Hong Kong is located adjacent to the PRE, the degrading water quality in the PRE also poses a threat to the marine waters of Hong Kong. Due to its overwhelming demand for environmental concerns, the PRE including the regions of Hong Kong waters is defined as the area of interest for this study.

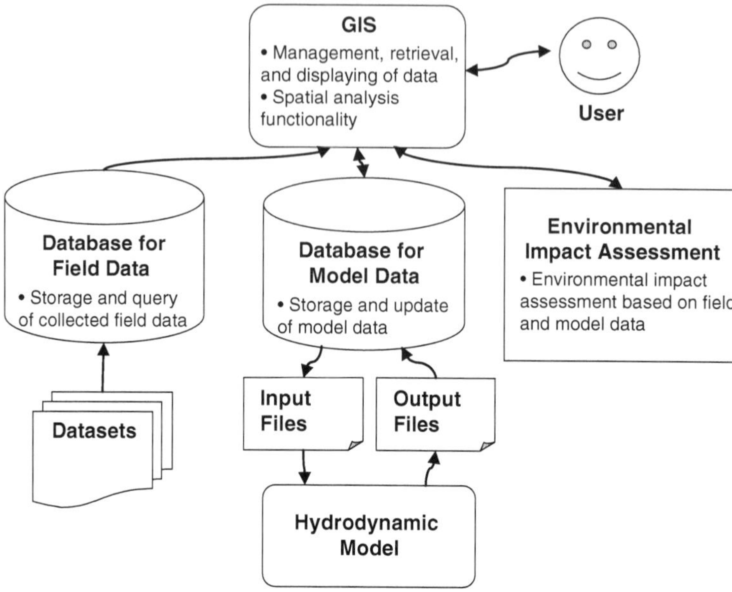

Fig. 19.2 Conceptual diagram of the GIS-hydrodynamic integrated system (After Ng (2004))

19.3 GIS Project

Figure 19.2 is a conceptual diagram of the GIS-hydrodynamic model integrated system. The integrated system links together three key system components: (1) a database for storing field-collected data, (2) a database for storing model input and output, and (3) a set of tools for EIA application, to a single operation platform for efficient retrieval, analysis, and interpretation of information. The GIS application is developed using ESRI ArcView through programming in the software's built-in Visual Basic for Application (VBA) environment and manipulation of ArcObjects. The system exploits the basic functionality of ArcView as well as one additional extension, the Spatial Analyst. The target outcome of the GIS application is a customized interface which enables users with minimal backgrounds in the use of GIS software to conduct GIS-related hydrodynamic analysis.

19.4 Retrieval and Display of Field Data

Hydrodynamic and water quality data of the PRE have been acquired from a number of sources and compiled into a database system using the MS Access software. The collected data encompasses various physical, chemical, and biological parameters, including temperature, current velocity, pH, salinity, turbidity, dissolved oxygen, nutrients (nitrogen and phosphorus compounds), metals, organics and coliform bacteria. The developed database system is further linked with a GIS for retrieval and display of information.

19.4.1 Data Display

There are three basic ways to display raw measurement data. GIS software provides the essential ability to display data in form of a *spatial layer* which is somewhat a snapshot of a specified parameter at a particular depth and instant in time. In order to seek solutions to capture the continuous-changing, three-dimensional characteristic of marine environment, *time series* and vertical *profile* plots are also incorporated into the system (see Fig. 19.3). The graph displays of these two types of display are developed using *Microsoft Office Chart Component*. The output is an x–y scatter plot with *time* given in the x-axis and the *specified hydrodynamic or water quality parameter* in the y-axis for *time series*; and with the *specified hydrodynamic or water quality parameters* given in the x-axis, and *distance from water surface* for y-axis for *profile*.

The retrieval of data in creating a spatial layer, time series, and profile display is processed using Structured Query Language (SQL) query command. The user can select a type of data display and specify a list of input parameters including data source, data set, measurement table, the hydrodynamic or water quality parameter, the measurement depth level (e.g. surface, middle, or bottom) or depth range (i.e. from a specific distance from water surface to another), and the measurement period. Based on the specified criteria, the GIS will command the MS Access database system to extract the data for creating the data display.

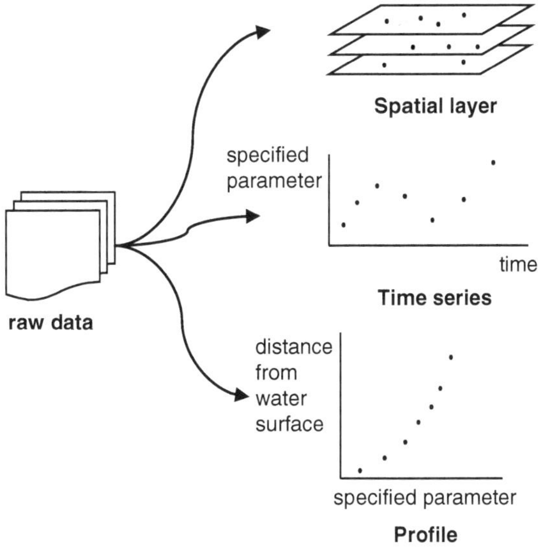

Fig. 19.3 Data display (After Ng (2004))

19.4.2 Data Interpolation

The system makes use of both spatial and one-dimensional interpolation techniques for estimating values at points in space and time in which no measurement was taken. For *spatial interpolation*, there are four basic methods to create continuous predictions from sampled point values: (1) *Inverse Distance Weighted (IDW)*, (2) *Natural neighbor*, (3) *Spline*, and (4) *Kriging*. The selection of which of the interpolation methods to apply depends on the type and characteristics of the specific sampling dataset. While *IDW* is a deterministic technique which is best for creating variables that diminish influence with distance from sampled location, *natural neighbors* is a good general most suitable where sample data points are unevenly distributed. *Spline* will be suitable for interpolating tidal variation as it works best for generating smooth, gradually changing surface. Finally, since *kriging* accounts for distance or directional bias to data, it will be best-suited for exploring pollutants dispersed from a single point source.

In order to derive values at unsampled point in time and depth along a vertical water column (i.e. *temporal and profile interpolation*) for specified fixed station, one-dimensional interpolation tools are developed based a range of mathematic functions, including *polynomial* up to six degrees, *spline (linear and natural cubic)*, *logarithmic*, *exponential*, *power*, and *trigonometric* up to twenty-five frequency components. Except for the *spline* method, data are fitted using the *Least-Square* method. To enable user to examine the goodness-of-fit of the selected interpolation function, the evaluation of the *root mean square R* of the deviations is also included. Additionally, the program also provides guideline in the selection of interpolation method for hydrodynamic parameter which conforms particularly well with certain mathematical functions according to hydrodynamic and sediment transport theories. For example, trigonometric function will be suggested for fitting tidal data since tidal level generally varies in a harmonic manner, while logarithmic function will be recommended for current velocity as it is known that velocity distribution in boundary layer follows the log-law.

Tempo-spatial and *profile-spatial* interpolation functions are also included to account for the fact that the sampling time or depths among the survey stations are often not simultaneous with one another. As shown in Fig. 19.4 for *tempo-spatial* interpolation, fitted curve is first obtained from the time series at each sampling stations. Based on these curves, values at certain specified time instants can be estimated to create a set of point feature spatial layers, which can further be converted into rasters using spatial interpolation methods. And the same interpolation concept applies for *profile-spatial* interpolation.

19.4.3 Dynamic Display

As the hydrodynamic characteristic of marine environment is endlessly changing, dynamic display is an important tool for examining and analyzing temporal variation

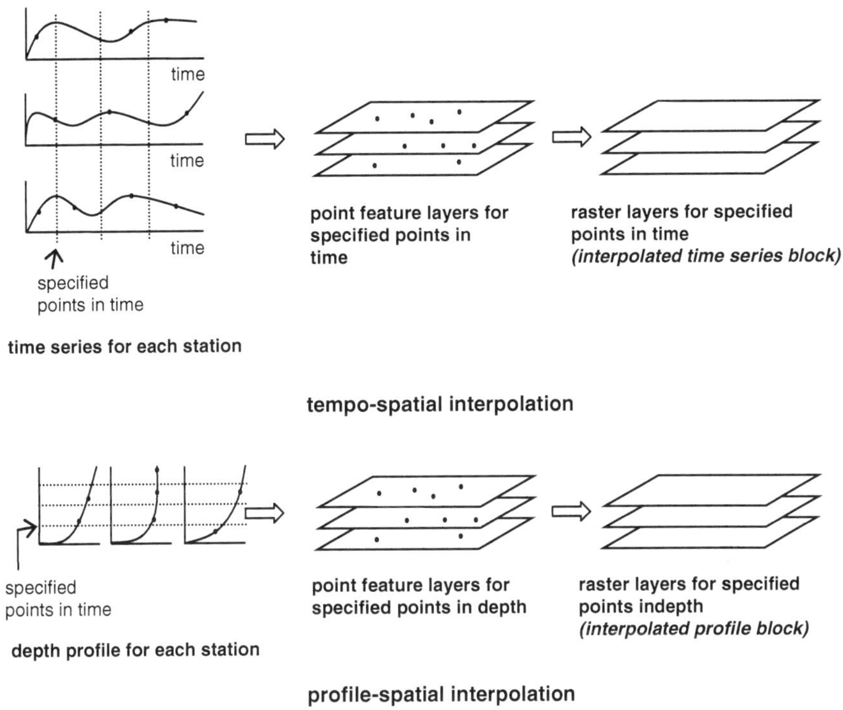

Fig. 19.4 Tempo-spatial and profile-spatial interpolation (After Ng (2004))

of coastal data. A dynamic simulation of the spatial layers varying with time can be executed by selecting an *interpolated time series block*. The dynamic display also allows the user to pick any point within the study area to create a *time series* plot with a cursor along the temporal curve which moves simultaneously with the spatial display. Additional features include playback controls such as pause, stop, and forward buttons, play speed adjustment, and slider to indicate the progress of simulation.

19.5 Integration with Hydrodynamic Model

The GIS is integrated with a three-dimensional parallel finite element hydrodynamic model developed by Wai and Lu (2000). This model was further modified by Jiang (2003) to include cohesive sediment and heavy metals transport simulation. The GIS-hydrodynamic model integration process is carried out using the *interface integration* method (Shamsi, 2002), in which the transfer of information is automated using pre-processor and post-processor components. While the pre-processor is for analyzing and exporting GIS data to model input files, the post-processor is for importing of model output and displaying it in the GIS interface. Moreover, the GIS has been customized to allow for generation and editing of the model mesh grid.

19.5.1 Pre-processing and Mesh Grid Configuration

The data input requirements for the hydrodynamic model include configuration of mesh grid, bathymetry of the study area, inflow and outflow for source elements, suspended and bed sediment parameters, tidal flat drying and flooding control variables, and initial conditions and boundary conditions of tidal elevation, salinity, current velocity, sediment concentration, and metal (or water quality) concentration. For efficient retrieval of the input data through the GIS interface, a MS Access database is created to store the input data. The data in the tables from the database are linked directly to pivot tables in the user input dialog boxes, which enables users to retrieve and update the input values as needed.

The configuration of model mesh grid is referenced based on the geographic coordinates of nodal points and the assignment of nodal points for each mesh element. Programming routines have been developed for generating the setup of model mesh grid and for carrying out refinements. There are basically two ways in which the user can the reposition the locations of existing nodal points. One is by modifying the values of the geographic coordinates in the nodal coordinate table, and the other is by moving nodal points through the GIS display.

Additional data displaying functions are also incorporated for examining water depth, initial conditions, and open boundary conditions in GIS. Water depth, initial conditions, and open boundary conditions are stored in a set of tables. To enhance the understanding of information, the user may visualize the values through the GIS interface. The displaying of input data in form of a spatial layer is available for water depths and initial boundary condition. Figure 19.5 shows a comprehensive representation of the water depth, which gradates from light to dark as it varies from shallow to deep. For open boundary conditions, the user can select a nodal boundary point to create a *time series* plot or vertical *profile* plot of the specified input parameter.

One important application of the collected data is for setting model initial boundary conditions. To derive values of initial concentration at nodal points of a configured mesh grid, a raster layer of concentration must first be created from the collected dataset. The raster layer of concentration can then be overlaid onto the mesh grid, and the values at the grid nodal points are extracted from the raster layer and stored into the model database.

19.5.2 Post-processing

The data displaying methods for model data are fundamentally the same as those for field data (i.e. spatial layer, time series, and profile). The retrieval of output data from

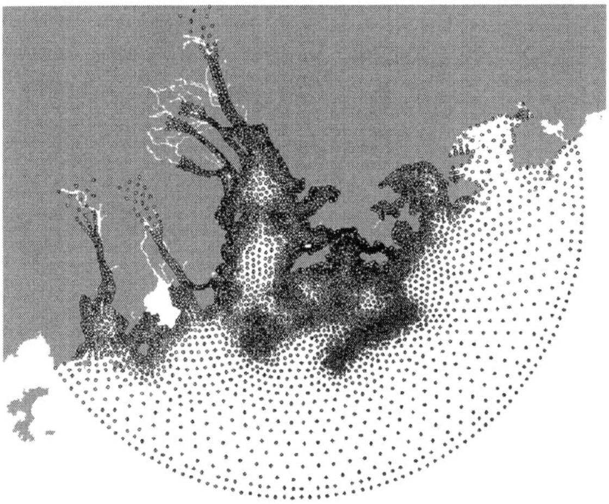

Fig. 19.5 Display of water depth from model input in GIS (After Ng (2004)) (See also Plate 29 on Page 408 in Color Plate Section)

model database is also carried out based on SQL query. To create a spatial layer, the user must specify the *hydrodynamic or water quality parameter*, the *time instant of simulation*, and the *water depth layer*. For display of a time series plot, the user must specify the *hydrodynamic or water quality parameter, the nodal number*, and the *water depth layer*. Likewise, the required input parameters for creating a profile are the *hydrodynamic or water quality parameter, the nodal number*, and the *time instant of simulation*. Spatial interpolation may also be used to create continuous predictions of model outputs, and the generation of *interpolated time series block* or *interpolated profile block* is also available. Moreover, an *interpolated time series block* can further be applied for dynamic display.

19.6 Application for EIA

In examining spatial trend in water quality data, one of the most significant concerns is to identify the locations subjected to most severe environmental impact. An EIA module is being developed to quantify the extent of wastewater impact. The water quality parameters of interest when assessing the impact of wastewater disposal include dissolved oxygen (DO), nutrients (nitrogen and phosphorus), unionised ammonia, E. coli, pH, salinity, temperature, suspended solids, metals, and chlorophyll-a. The environmental standards applied for this module are based on the outline for water quality objectives (WQOs) for marine waters of Hong Kong from the Environmental Protection Department (EPD) of Hong Kong. A procedure has been written to normalize each of these water quality parameters to a value with 0 representing zero concentration and 1 for the maximum acceptable concentration. A weighting scheme has been adopted to prioritize the importance of each water quality parameters (i.e. which of the parameters contribute the most to environmental impact and which ones the least?). And an environmental impact index can be computed using the following mathematical expression:

$$INDEX = 1/n \sum_{i=1}^{n} w_i x_{nom,i} \quad i = 1,2,...,n$$

where n is the total number of water quality parameters considered, w_i is the weight assigned to the i-th parameter (a value between 0 and 1), $x_{nom,i}$ is the nominal value of the i-th parameter, and *INDEX* is the impact index. This impact index formulation is applied using the map algebra technique in GIS. A set of raster layers of the considered water quality parameter can be combined to produce a layer of impact index, which will provide an overall picture of the areas potentially at risk. Furthermore, this module allows for the flexibility to adjust the maximum concentrations as well the weighting factors for exploring wastewater disposal impacts under different environmental assessment scheme.

19.7 Conclusions

This chapter has presented a GIS, which bridges together data collected from field and those generated by hydrodynamic model to a single geographic platform for efficient retrieval, display, manipulation, and analysis of information. The integration not only allows for comparison of information through various data display functions such as spatial layer, time series, and profile, but also extends the use of field data for model inputs. The integrated system is further applied for EIA. While only one application of GIS for EIA involving the use of GIS map algebra technique has been discussed, there are numerous other ways for examining environmental conditions to be explored for future study. These include assessing the relative importance of water quality parameters, investigating for temporal patterns and trends, and simulating scenarios to evaluate environmental impacts under different situations.

Acknowledgments The authors would like to acknowledge the financial supports provided for this work by the Hong Kong Research Grants Council (research grant no: PolyU 5143/04E) and the Hong Kong Polytechnic University (research grant no: G-YD93).

References

Barlett, D.J. (2000). Working on Frontiers of Science: Applying GIS to the Coastal Zone, in: Wright, D.J. and Barlett, D.J. (eds), *Marine and Coastal Geographical Information Systems,* Taylor & Francis, Philadelphia, pp. 11–24.

Chen, J.C., and Heinke, G.W. (2001). *Pearl River Estuary Pollution Project Summary Report,* Hong Kong University of Science and Technology, Hong Kong.

Demer, M.N. (2000). *Fundamentals of Geographic Information Systems*, John Wiley & Sons, New York.

Jiang, Y. (2003). "Three-dimensional Numerical Modelling of Sediment and Heavy Metal Transport in Surface Waters", PhD Dissertation The Hong Kong Polytechnic University, Hong Kong.

Lucas, A. (2000). Representation of Variability in Marine Environmental Data, in: Wright, D.J. and Barlett, D.J. (eds), *Marine and Coastal Geographical Information Systems,* Taylor & Francis, Philadelphia, pp. 53–74.

Ng, S.M.Y. (2004). *Development of a Geographic Information System (GIS) for Wastewater Disposal Hazard Mitigation in Pearl River Estuary.* Technical Report, The Hong Kong Polytechnic University, Hong Kong.

Shamsi, U.M. (2002). *GIS Tools for Water, Wastewater, and Stormwater System,* American Society of Civil Engineers, Reston.

Tsanis, I.K., and Boyle, S. (2001). "A 2D Hydrodynamic/Pollutant Transport GIS Model", *Advances in Engineering Software,* vol. 32, pp. 353–361.

Wai, O., and Lu, Q. (2000). "An Efficient Parallel Model for Coastal Transport Process Simulation", *Advances in Water Resources,* vol. 23(7), pp. 747–764.

Chapter 20
Geographic Information Systems and Sustainable Water Resources Management in Nigeria

Ifeyinwa Christabel Okeke

Abstract This chapter describes the importance of Geographic Information Systems in sustaining water resources in Nigeria. Water is a central resource and an indispensable ingredient for all human activities. Nigeria as a country is naturally endowed with abundant surface and groundwater resources yet the effective and efficient management of these resources has so far remained elusive. There is an increasing awareness that water resources exist in limited quantities and the available supply varies considerably during the course of a year. As a result, it calls for an urgent need to find ways of saving, reusing and recycling as well as developing geographic information methodologies to improve water resources.

Keywords GIS · Sustainability · Water resources · Management

20.1 Introduction

Nigeria is a rapid developing country with a population of 148 million people (NPC, 2006). She is naturally endowed with abundant surface and groundwater resources. The quality of life depends on water and whether as resource or commodity it is the basis of our agricultural, municipal, industrial, environmental and aesthetic purposes. Thus it is a central resource and an indispensable ingredient for all human activities. The increasing pressure of the ever-growing human population, increased living standards and the concomitant economic activities are exerting tremendous pressure on the water resources making water to be taken for granted, abused and unnecessarily exploited. Coupled with lack of advance planning, the various water sources are becoming unsuitable for beneficial use in a number of areas. Ground water levels are rapidly getting depleted while fresh water sources are being polluted in spite of water demand daily increase. Water is a complex system requiring

I. C. Okeke (✉)
Department of Geography, University of Lagos, Akoka, Lagos, 101014, Nigeria
e-mail: ifychristabelokeke@hotmail.com

D.R. Green (ed.), *Coastal and Marine Geospatial Technologies*, Coastal Systems and Continental Margins 13, DOI 10.1007/978-1-4020-9720-1_20,
© Springer Science+Business Media B.V. 2010

thorough assessment and monitoring techniques thus, access to information is a vital part of the good management. The use of information and communication technologies offers a vast range of possibilities for establishing new approaches to the collection and use of water data. Geographic Information Systems (GIS), with its combination of mapping and database management offers an optimized water management tool. This chapter describes the importance of GIS in sustainability of water resources.

20.2 Study Area (Nigeria)

Nigeria as a country is located on the West Coast of Africa with its shoreline extending about 800 km. It has a landmass of about 92,400 km and is endowed with water resources. It situates between latitude 4°10′ and 13°50′ North and between Longitudes 2°15′ and 14°45′ East of the Greenwich meridian. She is bounded by Niger republic to the North, Benin Republic to the West, Cameroun Republic to the East and Atlantic through the Bight of Benin in the South. Nigeria's climate is tropical with generally high temperatures and characterized by strong latitudinal zones, drier northwards and humid inland to the coasts. The South-West (SW) and North-East (NE) trade winds blows across the country ushering in the rainy season (March-September), little dry season in July–August and the dry season (October–February). A micro-climatic condition called harmattan brings dusty haze from the Sahara and is significant during dry season. Rainfall is the key climatic variable with marked alteration in wet and dry seasons in most areas. As a source of water, it varies from region to region. The mean annual rainfall throughout the country varies from over 3,000 mm in Niger Delta to less than 500 mm in the extreme North-East and North-West. Topographic relief plays a significant role in local climate with Jos and Mambilla Plateau having mean annual rainfall of about 1,500 mm. During the rainy season, water banks and reservoirs over flow thus rain harvesting is carried out through different methods and receptacles to make water available more especially during the dry months. Vegetation is strictly a function of rainfall and is evident by the thick mangrove swamp of the tropical rain forest.

20.2.1 Population Projections (Pp)

Adequate water supply is central to life and civilization. Water accounts for about 65% of the human body and the minimum needed to carry out the functions of living properly is about 65 Litres per day in urban areas and 25 Litres per day in rural areas. The global demand for water is thus rising with increases in demand as population increases. The Nigerian initiative "Water for people, water for life" programme aimed at achieving 100% water supply coverage for all state capitals by the year 2007 was not achieved in spite of the emphasis that water is critical to people's survival, as a major variable in contemporary global politics with domestic implications for family cohesion, education, youth development, health and sanitation, agriculture for food security and industrialization (This Day newspaper, February,

2005). Nigeria is similarly off-track from achieving similar millennium development goals (MDGs) of 75% target by the year 2015 (UNICEF, 2006). The daunting task of many countries not having additional sources of water to develop economically and even those countries that have additional sources of water; the time periods required to implement those projects are likely to be much longer than expected at present. Global water consumption estimates in different parts of Nigeria such as Anambra State SW local government areas show that additional water would be required for domestic, industrial, agricultural and miscellaneous purposes. The power failure of Nigeria also points the need for more water for hydropower generation. Therefore, as the population doubles, living standard and industries increase, competition for water becomes more intense and the complexity of water problems grows and become of increasing concern to both policy makers and private citizens. Luijten (1999) reiterates that the complexity of water problems impact sustainable development and quality of life which in turn threatens food security, human health and natural ecosystems. It then becomes evident that we must all think in terms of development for the future by users and communities concerned becoming the guardians of sound water management practices.

20.2.2 Climate Change (Cc)

The change in climate has potentials of affecting humanity as well as in the distribution of water resources. Climate change results in redistribution in time and space of our water resources. This could be as a result of low precipitation in rainy season, rapid increase in human activities and delayed development of water resources project. The downward trend in rainfall, runoff and available water resources in Nigeria with increasing demand for water in agriculture and other sectors remains a challenge while El-Nino events continue to demonstrate the water shortage impacts in all regions of Africa. The Intergovernmental Plan on Climate Change (IPCC) report on temperature increase, precipitation change, increased variability and sea level rise all impact directly on available water resources, increasing stress on the already scarce water supplies (WMO/UNEP, 1992). It is predicted that after 2025, climate change could also make conditions worse if precipitation in major food producing regions and evaporation rates increases. Therefore timely recognition of this threat together with adequate planning and management of improved water systems will allow reduction of drastic effects in the 21st century. Continuous update of data from maps, GIS and satellites should be carried out while long-term effects of climate change should be inculcated in decision-making.

20.3 Methodology

The chapter employs Geographic Information Systems model as a tool for sustainability of water resources in Nigeria. Figure 20.1 shows a sketch model for GIS and sustainable water resources. Accepting the challenge of a good water resources management is a difficult task that requires powerful tools. Inappropriate data and

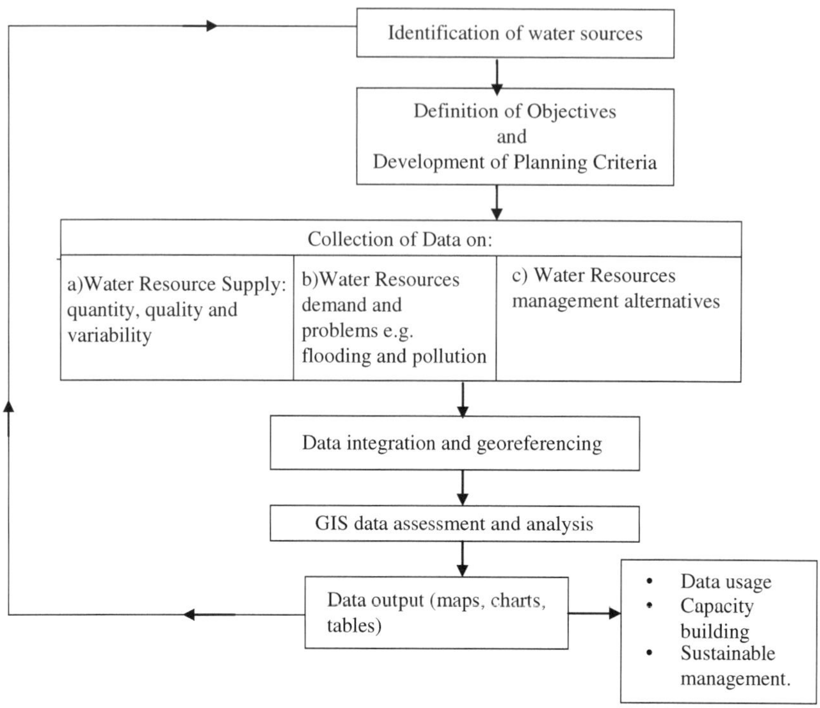

Fig. 20.1 A GIS and water resources model for sustainability
(Source: compiled by author)

misunderstanding of water management processes leads to inefficiency and misuse of water resources. Therefore to manage water properly, there is need for access to a relevant, comprehensive, current and accurate data. In arid semi-arid regions of Nigeria, this challenge is greater due to harsh climatic conditions such as water scarcity that often characterize the region. Geographic Information Systems (GIS) aids data collection, analysis and visualization and assists water experts to analyze complex situations and to supply decision makers with precious information. The beauty of the tool is the combination of location information (i.e. latitude/longitude) with descriptive tabular information that describes the feature being plotted. Therefore, the utilization of ground water or surface water as a water source could employ Geographic Information System (GIS) as a tool using the data collected through source water assessments which helps to protect and manage water resources.

The model operates through location of the zone concerned, identification of the objective for sustainable management of water resources in the area. Collection of statistical data on water supply and demand, sketching of base map and recording of data on the base map using the (GPS) global Positioning System to geo-reference. GIS is be used to explore the databases, undertake data analysis and output the result in any desired form (maps, graphs, tables). Adequate reliable hydrological data is a prerequisite for Water Resource Assessment (WRA), because it covers a wide range of fields. These information and data involves technical (water collection, treatment, distribution and evacuation), social and economic aspects. Sustainable water management facilitates the development of constructive applications of new geo-information technologies. Water shortages facing some Nigerian cities reveal the true value of this critical resource. Several rivers and lakes undergo a marked reduction in flow leading to increased conflicts over water resources use in rural, urban and agricultural areas. Other sources of water such as wells are under threat by over-exploitation. Groundwater stores about 95% of the freshwater yet they are becoming threatened by unflared exploitation and contamination from many sources. Thus a Geographic Information System aids information extraction and also helps to protect the source of water supply.

20.3.1 Geographic Information System and Water Assessment

Longley et al. (2001), defined GIS as a "Tool for performing operations on geographic data that are too tedious or expensive or inaccurate if performed by hand" It is also a mechanized method of keeping inventory of geographically distributed features and facilities through data input, management (storage and retrieval), manipulation and analysis as well as different outputs. GIS technology handles data for coastal area management, large databases, assesses water hazards, and provides efficient database and storage, retrieval, analysis, updating and visualization facilities. Water suppliers can use GIS in conjunction with the source water assessments to help protect their water supplies by keeping updated inventory of water quality, contamination, background and infrastructural information such as geology on location of wells, houses, and pipelines as well as producing intelligent maps and graphics that allow one to view data concerning water supplies. The schematic of the GIS tool in water assessment is shown in Fig. 20.2.

20.3.2 Sustainable Development

Mankind faces the problem of ensuring a rational use of water resources to avoid depletion and pollution. Thus, there needs to be a plan for the management of water that will protect the long-term future of the environment while encouraging sustainable development (Oyebande, 1989). Sustainable development being the development that meets the needs of the present without compromising the ability of future generations to meet their own needs (WCED, 1987) is a necessity to promote, restore and maintain water resources utilization because, it implies an extended criterion for planning, designing, operating and maintenance. Sustainable development of water resources is an essential future for stable human development should be the concern of water related project planners, engineers and scientists because it permeates all sectors of the society. The Dublin statement, "Scarcity and misuse of fresh water pose a serious and growing threat to sustainable development and protection

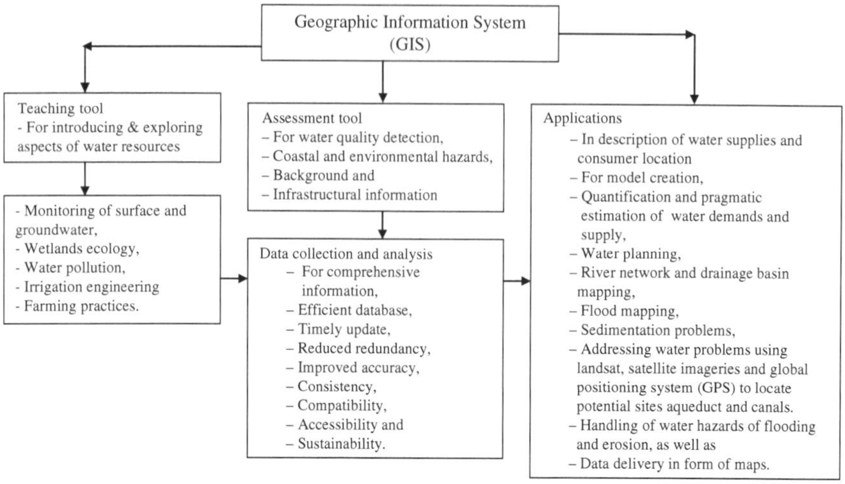

Fig. 20.2 GIS, a water assessment tool
(Source: compiled by author)

of the environment, human health and welfare, food security, industrial development and ecosystems on which they depend are all at risk unless water and land resources are managed more effectively in the present decade and beyond than they have been in thee past" (ICWE, 1992). Education is a strategic tool for sound environmental management which could be practiced through geographic information (Okeke, 2004). Our lives impact the environment, no matter where we live and the best way to protect the environment is to learn about them (Okeke, 2003). This could be done through the use of hydrological research in a changing environment through the use of public information systems and effective technology such as Geographic Information System (GIS) is required for adequate management of water resources. Good governance in water management along the integrated water resources management (IWRM) principles is necessary for water productivity increase. Thus sound and effective management of water resources has a key function in sustainable development.

20.3.3 The Planning Challenge

The Nigerian master planning efforts worked towards sustainable development in assessing the demand of water and the water resources available according to needs, priority and potentiality. Oyebande (2004) therefore, suggests that the Federal Ministry of Water Resources (FMWR) with the overall leadership responsibility of the water sector should establish clear policies and guidelines for its own role and those of other operational agencies and stakeholders (including the private sector) so as to improve water supply services. Various water resources projects have to be planned and managed to ensure the availability of water in sufficient quality and quantity at the right location and time. Planning is necessary because of intensified pressure on scarce water resources, which causes challenges on water resources. The acute shortage of freshwater experienced in Nigeria presently points towards the need to plan for optimum utilization, coherent policies and programmes for the development and management of water resources. The government of Nigeria has a national water policy whose aim is to provide water to all inhabitants of Nigeria by the year 2020, increase the present inadequate level of services to 120 Litres per capital per day, 60 Litres per capital to urban, peri-urban and rural areas by the year 2020, increase the capacity of local, state and federal government to assist communities to obtain water supply facilities that could be maintained.

20.4 Recommendations

Water is a precious resource that should be assessed, conserved and protected for efficiency, equitability and sustainability to benefit mankind. A range of coping strategies is thus necessary if sustainable development is to be realistic. As in the African Water Vision, the following recommendations are necessary:

- Water is a complex system and access to information is a vital part of the good management. More attention should therefore be paid to information and communication technologies, which offers a vast range of possibilities for establishing new approaches to the collection and use of water data.
- There should be sustainable access to safe and adequate water supply and sanitation because water sustains life. Climate information and skillful forecasts is needed to manage water better.
- The urgent need for quality data on flow, quality, collection, storage and analysis of water must be a priority in national planning. To properly handle the data for water resources management, a geospatial information technology such as Geographic Information System (GIS) would be inevitable.

Water scientists should develop and maintain information on water resources availability, assessment, monitoring as well as understanding hydrological regimes affecting water resources.

20.5 Conclusion

Water is life as well as the basis of all living ecosystem and habitats that pervades human lives and embraces different cultures. Sustainable development of water resources therefore depends on the effective functioning of natural and human water delivery systems which is severely limited by inadequate water supply to an environment, rising human population, industrial growth and unwise management of water practices.

Hydrological research in a changing environment through the use of public and geographic information systems (GIS) is thus required for adequate management of water resources, capacity building and policy formulation. Thorough water assessment also needs to be carried out timely.. Finally, effective and efficient water management cannot be attained without the availability and use of comprehensive, accurate and easily accessible geospatial information. This provides researchers, resource managers and decision makers with a tool for effective and efficient storage as well as manipulation of data for both scientific management and policy-oriented information.

References

ICWE, US Agencies, Interagency Task Force (1992), "Protection of the Quality and Supply of Freshwater Resources". Country Report, USA to the International Conference on Water and the Environment: Development Issues for the 21st century, Dublin.

Longley, P. et al. (2001). Geographic Information System and Sciences. John Wiley and Sons, England.

Luijten, J.C (1999). A tool for community-based water resources management in hillside watersheds. Ph.D dissertation. University of Florida, Gainesville, FL. http://etd.fcla.edu/etd/uf/1999/amp 7392.

NPC (2006). National Population Census figures for 2006. National Population Commission of Nigeria.

Okeke, I.C. (2003). Coastal Challenges and the Challenges of Coastal Education in Nigeria. In fifth International Symposium on GIS and Computer Cartography for Coastal Zone Management conference proceedings. CoastGIS'03 Ducale, Genova, Italy, 16th to 18th October, P. 102. http://www.gisig.it, Full proceedings on CD ROM.

Okeke, I.C. (2004). Women, Geoinformation and Sustainable Environmental Management. In Nwilo, P.C, Uluocha, N.O and Saka, D.N (Eds.) Application of Geoinformation to the Development and Management of Nigeria's Resources (GEOSON proceedings), pp.145–152. JAS publishers.

Oyebande, L (1989). Water Resources *Journal of the Nigerian Association of Hydrogeologists* vol.1 No 2, October. Nigerian Association of Hydrogeologists pp.221.

Oyebande, L (2004). Fostering Implementation: Know How for Action. African *Pre-Conference* on Water for Food and Ecosystems: Make it Happen. Addis Ababa, Ethiopia. November 4–6.

This Day Newspaper, "Water for People, Water for Life Programme" Friday Feb/18, 2005, vol 11, No 3589 pp.42.

UNICEF, (2006). Water, the challenge. Retrieved on 23 September, 2007. Web site: http://www.childinfo.org

WCED, (1987). *Our Common Future*, Oxford University Press, Oxford. (WCED –World Commission on Environment and Development).

WMO/UNEP, (1992). IPCC supplement: scientific Assessment of climate Change. (Intergovernmental Panel on Climate change).

Chapter 21
Spatial Analysis of Traffic and Risks in the Coastal Zone

Ronald Pelot and Lauren Plummer

Abstract Management of the coastal zone for the purposes of conservation and sustainability is a well-described concept in the literature. This chapter describes the importance of modelling various forms of maritime traffic as an intrinsic element of this process. Effective use of GIS in this context implies extensive data acquisition and cleaning, vessel path simulation, traffic density maps, and incident (spills or accidents) distribution when applicable. Risks can be determined by evaluating detrimental occurrences relative to exposure levels. Effective visualization and quantitative analyses can then be fruitfully applied to various coastal zone management spheres, such as oceans use planning, emergency response planning, and marine protected area delineation.

Keywords Maritime activities · Coastal zone management · Traffic · Risk · Spatial analysis

21.1 Introduction

Over 40 years of international coastal zone management (CZM) initiatives are documented in the literature. Hildebrand (2002) estimates that between 4000 and 5000 books, documents, and articles have been written about coastal zone management with distinct specialty areas. Ricketts (2000) suggests a schematic strategic approach for integrated coastal zone management in Atlantic Canada. As of 2000, 78 coastal zone management initiatives are "on record" from coast-to-coast in Canada (NetCoast, 2003). As of 2002, approximately 700 global coastal zone management initiatives are documented (Hildebrand, 2002; Sorenson, 2002a, b;

R. Pelot (✉)
Department of Industrial Engineering, Dalhousie University, Halifax, B3J 2X4, Nova Scotia, Canada
e-mail:Ronald.Pelot@dal.ca

This updated chapter was published earlier in the Journal of Coastal Conservation 11:201–207 (2007).

D.R. Green (ed.), *Coastal and Marine Geospatial Technologies*, Coastal Systems and Continental Margins 13, DOI 10.1007/978-1-4020-9720-1_21,
© Springer Science+Business Media B.V. 2010

NetCoast, 2003). Cicin-Sain and Knecht (1998) discuss various coastal and ocean activities that threaten the coastal zone, and consequently require a comprehensive management plan.

It is estimated that 90% of world trade travels by sea and in 2001, seaborne trade approached 6 billion tons (Behnam and Faust, 2003). Increased use of Canadian coastal waters for commercial fishing and shellfish poses another potential threat and requirement for better coastal zone management information. In the past decade, employment within maritime industries such as aquaculture as well as the production of offshore oil and gas has increased exponentially, while supporting a substantial portion of the nation's economy (DFO, 2005). Tourism and recreational practices have also greatly increased in recent years, engendering further threats to maritime coasts and environment (DFO, 2005). Approximately 10 million Canadians participate in recreational boating each year (Pelot et al., 2004) and similarly, increased tourism (current and projected) via the cruise line industry in previously uncharted waters present unique considerations. The introduction of "bigger and better" cruise line passenger vessels also presents the opportunity for new, increased demand and/or potential negative impacts on the coastal zone, that may not have existed before, and therefore, demands a new thinking and approach of the coastal zone (Pelot et al., 2005).

In recent years Canada has become an international leader in the development of policies addressing the issue of integrated coastal zone management. In 1996, Canada became the first nation to construct and implement the "Oceans Act" legislation, in which broad ocean management is the primary focus (DFO, 2005). In response to this Oceans Act and with the ratification of the United Nations Convention on the Law of the Seas in 2003 (DFAIT, 2003), Canada has since implemented "Canada's Oceans Strategy" (Government of Canada, 2002a). As an accompanying document to "Canada's Ocean Strategy", the "Policy and Operational Framework for Integrated Management of Estuarine, Coastal and Marine Environments in Canada" (Government of Canada, 2002b) outlines the nation's suggested plan and approach for the management of marine and related areas. In addition, the "Oceans Action Plan" provides supplementary instruction concerning Canada's movement towards developing a functioning framework with which to operationalize sustainable development within practices that govern maritime waters (DFO, 2005).

Despite these novel developments in maritime regulation, enforcement and integrated management policies however, opportunities for the retrieval of information focused on the effects of maritime traffic and activity on coastal areas are rare. Reliable quantitative modelling of maritime traffic addresses several aspects of Oceans management. The traffic volumes and types, by location and time period, indicate potential conflicts in uses of the waterspace, which can be managed to some extent through good governance. Coastal development is also intrinsically linked to traffic, whether through port development, fish plant location, or marinas and wharves for example. The traffic brings benefits through economic activity and recreational enjoyment, but conversely exposes the ocean and coastline to various risks associated with accidents and spills.

According to Islam and Tanaka (2004), the negative effects of numerous anthropogenic activities on oceanic and coastal zones have been increasing at an alarming rate, and this trend is predicted to continue. These authors present many of the major pollutants integral to the current degradation of coasts worldwide. Many of these are inherently related to shipping and other commercial activities, such as oil drilling and dredging practiced within the marine environment (Islam and Tanaka, 2004). Kris Christen illustrates the extent of this damage on coasts along the United States of America. Coastal threats such as the introduction of invasive species by the release of ballast water, a common practice among cargo tankers and large container ships, are resulting in significant alterations in coastal habitats. These alterations are leading to adverse effects on all aspects of this environment (Christen, 2004). In general, shipping as well as recreational boating activities within the marine environment present various other potentially harmful threats to the coast. As an example, the extent to which the aquatic environment and its inhabitants are affected by the acoustic output of large and small vessels is wide-ranging. This type of aquatic pollution is often overlooked, however it is significantly detrimental to the behavioural patterns of marine species (Zacharias and Gregr, 2005). Incidents related to cargo shipping have also been significant contributors to coastal zone damage. It has been estimated that almost 6000 tons of oils are released or accidentally spilled into the oceans annually as a result of shipping processes (IISD, 1995).

The need for improved coastal management planning and decision-making requires a holistic understanding and knowledge of vulnerable coastal areas and the predictive occurrence of threats to these areas (Christen, 2004; Zacharias and Gregr, 2005). The International Maritime Organization (IMO, 1978) has determined particularly sensitive sea area (PSSAs) based on the potential for damage to these areas and their ability to counteract or survive them. Currently, Geographic Information Systems (GIS) presents the ability to spatially display coastal areas that are protected, or in need of further management and protective legislation based on environmental impact assessments (Douven et al., 2003). Although the incorporation of GIS to facilitate decision-making procedures associated with coastal zone management has been extensively practiced (Douven et al., 2003), work focused on coupling the GIS data concerned with environmentally assessed zones with the related impacts of traffic in these threatened areas has been relatively rare. As defined by Zacharias and Gregr (2005, p.88), the vulnerability of coastal areas is: "the probability that a feature will be exposed to a stressor to which it is sensitive". The only way in which vulnerability can be accurately predicted is to integrate all stressors involved in the potential degradation of the study area. To date, GIS mapping has addressed the issue of spatially depicting sensitive areas, as well as to identify the degree of potential damage from certain stressors such as oil spills, for example. This mapping would be greatly improved and rendered more complete if traffic patterns through these identified sensitive areas, were also incorporated (Pelot, 2001).

21.2 Traffic Modelling

There are many different forms of maritime activity. As each activity has its own associated patterns and behaviours, it is necessary to simulate each activity's traffic separately. For example, the characteristics of large commercial shipping vessel traffic (such as a tanker, bulk, or cargo ship) are very different compared to those of smaller vessels (such as a kayak, barge or fishing vessel), and accordingly should be modelled differently. This chapter discusses these various activity and incident types categorized into several groups, with their traffic modelled in distinct GIS layers, given appropriate consideration for land avoidance, route selection, and behavioural patterns.

21.2.1 Integration of Disparate Data Sets

The main challenge in the collection and formatting of data sets for spatial traffic analysis is not because the activity types are different per se, but rather because the availability, precision, and reliability of these sources are extremely variable. To provide a perspective on the traffic generation algorithms, it is useful to note that the ideal situation would be to have detailed continuous position information on all vessels plying the waters. With the advent of Global Positioning Systems (GPS), this type of data is often collected for ships, as well as through "black-box" monitoring programs for fishing compliance enforcement. In this case, centralized aggregation of the data and/or access are administrative hurdles, but many of the technical processing difficulties would disappear.

In fact, almost none of the data are available in this preferred form. A summary of the main classes of data is presented in Table 21.1, followed by a description of the processes developed to prepare them for spatial analysis. Custom software,

Table 21.1 Classes of available maritime traffic data

Class	Available information	Example
A	Continuous time-stamped position data; vessel info	Canadian West Coast shipping transits; Automatic Identification System (AIS)
B	Specific origin & destination, and intermediate waypoints; travel dates; vessel info	Canadian East Coast shipping transits
C	Specific origin & destination; travel dates; vessel info	Ferries; Cruise ships
D	Specific origin; general destination; travel dates; vessel info	Commercial fishing by NAFO zones
E	Specific origin; general destination; frequency of trips; vessel info	Lobster fishing; ecotours
F	General origin; general destination; frequency of trips; classes of vessels	Recreational boating

the Maritime Activity & Risk Investigation System or MARIS (Pelot, 2001) was developed to perform the various traffic generation and querying tasks essential for spatial traffic and risk analysis.

21.2.2 Contiguous Track Generation with Land-Avoidance

Given trips with specific, but limited, position information (Classes B & C), the origin, destination, and possibly sporadic waypoints can be connected to form continuous vessels tracks. This process engendered several challenges:

- Connecting the known points by straight lines (or following the Great Circle Route when applicable) is the most easily justified assumption, but often leads to unrealistic paths;
- The feasibility of these paths is improved by constraining where the vessels *cannot* go, including a land-avoidance algorithm; the land-avoidance relies on a node-network with the shortest path calculated through an efficient implementation of the Floyd-Warshall Shortest Pairs algorithm (Hilliard and Pelot, 2002).
- The quality of the paths is further improved by enforcing where the vessels *should* go, such as restricting movements to shipping lanes where applicable;
- Sometimes segments of trips are missing, and a filtering system was developed to ensure contiguity of the trajectory (Wootton and Pelot, 2002). For example, a self-reporting system might record a ship that states it is at position X heading towards position Y, and the next recorded event is that it is anchored at port Z. Despite the lack of an explicit record, the trajectory from Y to Z must be included in the path simulation.

21.2.3 Absence of Waypoints

For some types of traffic, specific information about the vessel's destination is unavailable (Classes D & E). Examples include some fishing fleets that are only required to report according to the Northwest Atlantic Fisheries Organization Subdivision (NAFO) where their catch was made. To accommodate this situation, a dispersion algorithm (Pelot et al., 2002) was developed as part of a track simulation process. The concept of this process is to circumscribe a feasible vessel destination area, such as a NAFO subdivision or fishing grounds, and then within it randomly select a destination point, which can then be connected to the vessel's origin, using the land-avoidance algorithm. Embellishments to this procedure include multiple feasible regions for a given trip, non-uniform probabilities of selection if some areas are more likely destinations than others, and multiple movements within the feasible region(s). In this latter case, statistics are gathered on the distribution of vessel turn angles (by class of vessel) as well as inter-turn distances, which can then be sampled to realistically simulate large-scale vessel movements.

In some cases, specific travel dates and times are recorded (Classes A to D). In other instances, only general frequency of trips is known (for example, number of trips per day, month or year). In this case, trips are simulated to represent the correct average transit rate.

21.2.4 Estimated Traffic

Some types of maritime traffic, recreational boating in particular, are not amenable to recording individual vessels trips (Class F). Two avenues have been taken to address this dearth of information: direct observation through sampling (via satellite images, ground-level radar, aerial observation, aerial radar/photography or ground-level observation), and surveys (of the general public or experts). Some work has been completed in this area (Pegler et al., 2003; Pelot et al., 2000) and research is ongoing to improve the estimates derived via these various mechanisms.

The trips for this traffic are generated in the same manner as described in the previous section, with the added uncertainties about origin, frequency and type of vessel. However, despite the lower quality of this data and the intensive efforts required to collect and upgrade them, they serve as a reasonable foundation for most CZM evaluations.

21.2.5 Dedensifying Traffic Data

In some instances, continuous traffic data are denser than required, and consume inordinate amounts of computer storage space. The principal sources of such data are radar output from vessel traffic services (VTS), and on-board GPS units set to a high sampling rate. In this case, de-densification of the paths is required to remove superfluous intermediate points. In general, if there are multiple points on a straight line segment between two vertices where the ship turned, then these intermediate points can be eliminated with no loss of important information. However, some heuristics are required to determine what comprises a straight section, and what degree of turning is significant relative to the overall scale of a vessel's movements (Pelot and Wu, 2007).

21.3 Traffic Analysis and Visualization

Once the vessel tracks have been generated, various evaluations can be carried out. The basic ones are path-independent such as number of trips, number of visits to a given port, or number of transits through a designated or planned Marine Protected Area. Others are indirectly related to the paths, requiring trajectories to calculate metrics such as average distance from shore or distance travelled in Canadian waters. However, to examine traffic concentration or other aspects of spatial distribution, two procedures are available: grouping adjacent trajectories to establish routes, where possible, accentuating dense traffic areas (Lin, 2008); or, subdividing the study area and calculating traffic counts in each subdivision. Such partitioning can be accomplished through gridding (Fig. 21.1), or by applying administrative boundaries or other arbitrary delineation.

Options for scaling and positioning the grid may be invoked, but calculations must not be performed indiscriminately as erroneous results may ensue. The two key difficulties are firstly the well-known modifiable areal unit problem (MAUP) whereby statistical results may be sensitive to the grid size, and secondly the compatibility of the grid size with the given data accuracy. If there are few way-points in a shipping dataset and the track generation algorithm has induced coarse

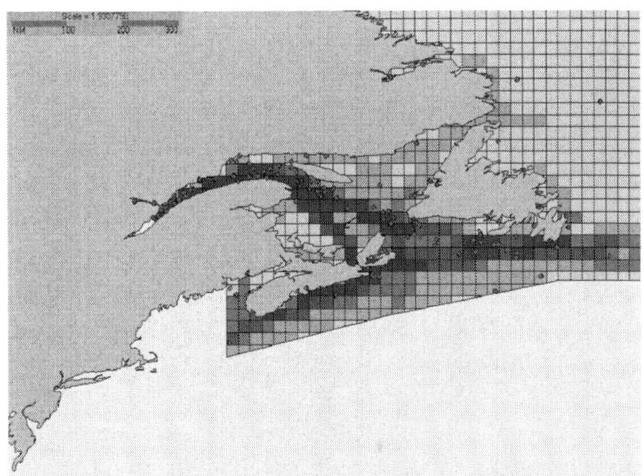

Fig. 21.1 Shipping traffic density distribution through gridding (See also Plate 30 on Page 408 in Color Plate Section)

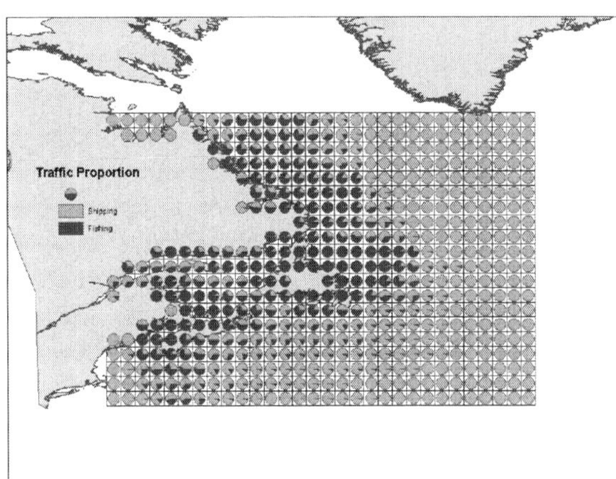

Fig. 21.2 Fishing versus shipping traffic proportion distribution in Atlantic Canada (See also Plate 31 on Page 408 in Color Plate Section)

generalizations in the paths, there is no point parsing the resulting set of tracks with a very fine mesh.

Gridding is also an effective means for merging the different data sets. As an example, Fig. 21.2 shows the proportion of shipping versus fishing traffic over the course of a typical year around Canada's East Coast (Pelot and Hilliard, 2004). Trends can also be detected in traffic type, traffic volume, and/or traffic location.

21.4 Value of Spatial-Based Analysis

Traffic data alone can be brought to bear on many decisions. Since the MARIS databases encompass a wide range of activities from small recreational boats up to ships, the mix and concentration of traffic can be examined for improved coastal zone management (Shahrabi and Pelot, 2003), oceans management, and traffic management. One consideration in the latter case involves the placement of aids to navigation, which can be reviewed periodically based on need and types of users of waterways (Pelot et al., 2002).

As a decision aid, the GIS views afford managers a wealth of information for better understanding the characteristics of maritime traffic in different areas. In some cases, such data displays, whether raw incident data, simulated traffic data, or calculated risk distributions, are sufficient for providing guidance. For other types of situations, explicit mathematical modelling yields more comprehensive results for complex multi-criteria decisions. Finally, maps of traffic, incidents or derived variables, serve as a very effective means for imparting information, knowledge or decisions to senior management or to the public. Alternatively, it can be used to educate the public or mariners about the perils of boating, hazardous areas, incident hot spots, congested areas, or other spatial attributes of the maritime environment.

In contrast, any maritime management decision or program where location is a factor can benefit from spatial analysis. The location element can be the result of data analysis such as identifying incident clusters (Shahrabi and Pelot, 2003), the explicit goal of the decision such as resource deployment location, or the de facto nature of some decisions which are constrained by administrative boundaries. For example, many fishing licenses are allocated for particular areas, so it makes sense to examine activity and incidents for those fleets within their respective operating zones (Tavera-Torres, 2001). Alternatively, various boundaries for an anticipated MPA (Marine Protected Area) can be overlaid on traffic maps for impact assessment (Pelot and Wootton, 2004).

Furthermore, GIS can also be used to calculate indirect spatial information, such as the distance from shore of maritime incidents (regardless of the specific location), or neighbourhood effects, a feature of spatial statistics where surrounding features can be included when examining relationships. In general, the availability and quality of data dictates the degree to which reliable spatial analyses can be completed, but these techniques make a significant contribution to the understanding and management of the marine environment.

21.4.1 Incident Analysis

A much wider assortment of spatial analysis techniques exists to operate on point data such as incident locations (discharges or accidents), compared with the line data in the preceding section. There are several methods for identifying clusters of incidents according to various criteria. To adjust for the random nature of incidents, as well as the ambiguity which may surround the georeferencing of their occurrence, Kernel Density analysis (Silverman, 1986) permits smoothing of the incident patterns to produce gradient maps (darker shades for higher incident concentrations) as shown in Fig. 21.3 for fishing accidents. Such pictorials are also very effective for temporal comparisons, whether monthly, seasonal, or annual (Shahrabi, 2003).

Another valuable perspective is to analyze incidence probabilities as they relate to traffic densities. Incidents can include the risk and occurrence of accidents such as collisions, groundings, oil spills, fires or explosions, capsizing and sinking. Previous applications of GIS modelling for coastal zone management have the ability to depict the level of impact on coasts from various incident occurrences, however the location and probability of the occurrence of such incidents is highly related to the location of the specific vessel types that may be involved. More sophisticated spatial modelling applies Dual Kernel density interpolation, a risk-adjusted method whereby the point incidents are fit with statistical distributions, as with Single Kernel density analysis but in this case the spread in each locale is conformed to a secondary variable, in this case the local traffic density. The outcome of this procedure is a relatively continuous mapping of incident rate variations across the study area, as Fig. 21.4 shows for fishing vessels.

21.5 Conclusion

The rate at which maritime activities, pollution, exploitation and climate change are negatively affecting coastal areas worldwide is shocking. There is a great need for the implementation and enforcement of legislation responsible for the protection of these coastal zones. In the past three decades, extensive work has been done to establish a successful framework from which decisions and regulations to facilitate coastal zone management may be adopted. Canada has become a global leader in this domain, becoming "the first country in the world to adopt comprehensive oceans management legislation" (DFAIT, 2004), referring to the Oceans Act of 1996.

In a relatively novel approach to coastal zone management, GIS has become a useful tool, with its ability to spatially depict vulnerable coastal areas, as well as to

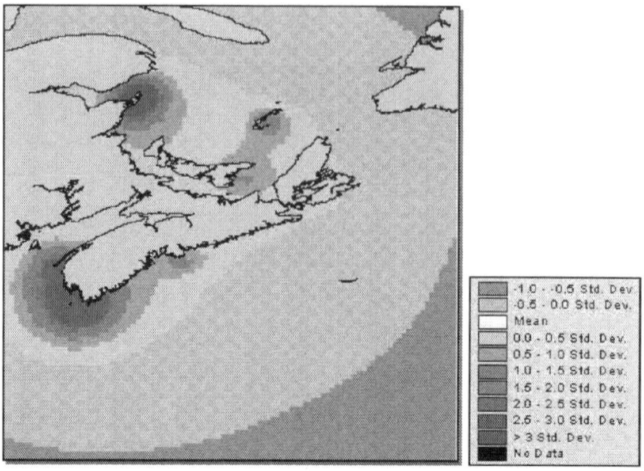

Fig. 21.3 Single Kernel density analysis of fishing incidents (See also Plate 32 on Page 409 in Color Plate Section)

Fig. 21.4 Fishing incident rate distribution using Dual Kernel density interpolation (See also Plate 33 on Page 409 in Color Plate Section)

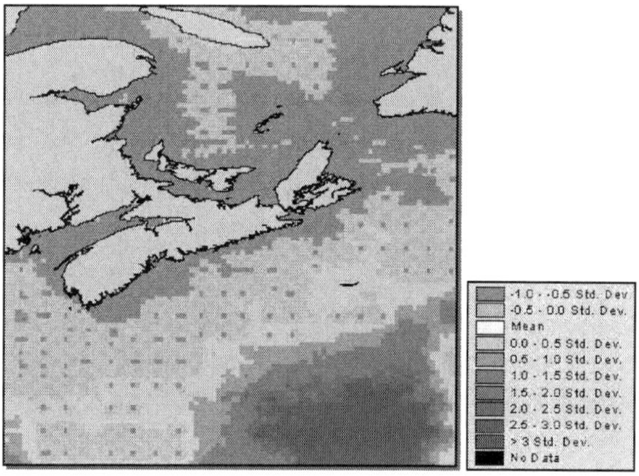

predict and illustrate the level of impacts from various potential incidents, such as oil spills, on these susceptible areas. An issue that has not yet been fully addressed however, is the information available regarding traffic patterns and densities and its relations to the degradation of coastlines. Spatially representing marine traffic patterns, incidents and risks, facilitates the processes of decision making as well as policy implementation and enforcement concerned with coastal zone management.

Acknowledgments This project is supported by the Canadian Coast Guard (CCG), the GEOIDE National Centre of Excellence (NCE), and the Natural Sciences and Engineering Research Council (NSERC). This research relies on excellent technical assistance by members of the MARIN (Maritime Activity & Risk Investigation Network) research group in the Department of Industrial Engineering, Dalhousie University.

References

Behnam, A., Faust, P. (2003). "Twilight of flag state control", *Ocean Yearbook*, vol. 17, pp. 167–192.

Christen, K. (2004). "Americans receive wake-up call on oceans", *Environmental Science and Technology*, vol. 38, no. 13, pp. 242A–244A.

Cicin-Sain, B., Knecht, R.W. (1998). *Integrated coastal and ocean management: Concepts and practices*, Island Press, Washington, D.C.

DFAIT (Department of Foreign Affairs and International Trade) (2003). *Canada Ratifies UNCLOS*, [Online]. Available from: <http://www.lac-bac.gc.ca/webarchives/20061107161547/http://w01.international.gc.ca/minpub/Publication.aspx?isRedirect=True&publication_id=380488&Mode=print> [2 July, 2009].

DFAIT (Department of Foreign Affairs and International Trade) (2004). *Canada and the Oceans*, [Online]. Available from: <bac.gc.ca/webarchives/20060213013641/http://www.dfait-maeci.gc.ca/sustain/EnvironIssu/canOcean/oceans-en.asp> [2 July 2009].

DFO (Department of Fisheries and Oceans Canada) (2005). *Canada's Ocean Action Plan: for Present and Future Generations*, [Online]. Available from: <http://www.dfo-mpo.gc.ca/oceans-habitat/oceans/oap-pao/pdf/oap e.pdf> [2 July, 2009].

Douven, W.J.A.M., Buurman, J.J., Kiswara, W. (2003). "Spatial information for coastal zone management: the example of the Banten Bay seagrass ecosytem, Indonesia", *Ocean and Coastal Management*, vol. 46, pp. 615–634.

Government of Canada (2002a). *Canada's Ocean Strategy*, [Online]. Available from: <http://www.cos-soc.gc.ca/dir/cos-soc_e.asp> [23 June 2005].

Government of Canada (2002b). *Policy and Operational Framework for Integrated Management of Estuarine, Coastal and Marine Environments in Canada*, [Online]. Available from: <http://www.cos-soc.gc.ca/doc/im-gi/index_e.asp> [23 June 2005].

Hildebrand, L. (2002). *Integrated Coastal Management: Lessons Learned and Challenges Ahead.* Discussion document for Managing Shared Water/Coastal Zone Canada 2002 International Conference, Hamilton, Ontario, Canada, pp. 1–12.

Hilliard, R.C., Pelot, R.P. (2002). *Development of a Path Generation Algorithm for Simulating Maritime Traffic*, MARIN Report #2002-07, Dalhousie University, Department of Industrial Engineering, Halifax, Nova Scotia, Canada.

IISD (1995). *Projet de société: Canada and Agenda 21*, [Online]. Available from: <http://iisd.ca/worldsd/canada/projet/c17.htm> [24 June 2005].

IMO (1978). *International Maritime Organization: Special Areas and Particularly Sensitive Sea Areas,* [Online]. Available from: <http://www.imo.org/Environment/mainframe.asp?topic_id=760 > [May 2005].

Islam, M.S., Tanaka, M (2004). "Impacts of Pollution on Coastal and Marine Ecosystems Including Coastal and Marine Fisheries and Approach for Management: a Review and Synthesis", *Marine Pollution Bulletin*, vol. 48, pp. 624–649.

Lin, D. (2008). *Spatial Analysis of Ship Routes for Maritime Security and Safe Navigation,* PhD of Engineering Thesis, Dalhousie University, Department of Industrial Engineering, Halifax, Nova Scotia, Canada.

NetCoast (2003). *ICM efforts in Canada, USA, other countries, and international efforts,* [Online]. Available from: http://www.netcoast.nl/journal/icmefforts_canada.htm [November 2004].

Shields, R., Neville, P., Pelot, R., Wootton, D. (2002). *Development of a Dispersion Algorithm for Simulating Fishing Vessel Traffic*, MARIN Report #2002-06, Dalhousie University, Department of Industrial Engineering, Halifax, Nova Scotia, Canada.

Pegler, K.H., Coleman, D.J., Pelot, R.P., Keller, P.C (2007) "An Enhanced Spatio-spectral Template for Automatic Small Recreational Vessel Detection", *Photogrammetric Engineering & Remote Sensing*. Vol. 73, no. 1, pp.79–88.

Pelot, R.P., Kendrick, P. Shields, R., Deveaux, J.P., (2000), *Recreational/Tourism Marine Activity Assessment in the Bay of Fundy*. MARIN Report #2000-05, Dalhousie University, Department of Industrial Engineering, Halifax, Nova Scotia, Canada.

Pelot, R.P. (2001). *Maritime Activity and Risk Investigation System (MARIS): A GIS Tool for Maritime Traffic and Incident Analysis,* MARIN Report #2001-01, Dalhousie University, Department of Industrial Engineering, Halifax, Nova Scotia, Canada.

Pelot, R.P., Hilliard, R.C., Dickie, S. (2002). *Canadian Navigational Aids Client Distribution.* MARIN Report #2002-05, Dalhousie University, Department of Industrial Engineering, Halifax, Nova Scotia, Canada.

Pelot, R.P., Hilliard, R.C. (2004). *Relative Traffic Densities and Accident Likelihood for Fishing and Shipping Vessels in Atlantic Canada,* MARIN Report #2004-06, Dalhousie University, Department of Industrial Engineering, Halifax, Nova Scotia, Canada.

Pelot, R.P., Wootton D. (2004). *Maritime traffic Distribution in Atlantic Canada to support an evaluation of a Sensitive Sea Area proposal,* MARIN Report #2004-05, Dalhousie University, Department of Industrial Engineering, Halifax, Nova Scotia, Canada.

Pelot, R.P., Delbridge C., Wu, Yan, Hilliard, R., Wootton, D. (2006). *Location-Based Risk Analysis of Recreational Boating Activity,* MARIN Report#2005-02, Dalhousie University, Department of Industrial Engineering, Halifax, Nova Scotia, Canada.

Pelot, R., Hilliard, C., Mukesh, M.D., Deacoff, C., Wootton, D., Moyst, H. (2006). *Cruise Ship Activity and Risk Analysis for Improved SAR Response Planning,* MARIN Report #2006-03, Dalhousie University, Department of Industrial Engineering, Halifax, Nova Scotia, Canada.

Pelot, R.P., Wu, Y. (2007) "Classification of Recreational Boat Types Based on Trajectory Patterns." *Pattern Recognition Letters,* vol. 28, no. 15, pp.1987–1994.

Ricketts, P.J. (2000). "Integrated coastal zone management in Atlantic Canada: Looking toward the third millennium", *Ocean Yearbook,* vol. 14, pp. 291–359.

Shahrabi, J. (2003). *Spatial and Temporal Analyses of Maritime Fishing and Shipping Traffic and Incidents,* PhD of Engineering Thesis, Dalhousie University, Department of Industrial Engineering, Halifax, Nova Scotia, Canada.

Shahrabi, J., Pelot, R.P. (2003). *Geographic Information Systems in Coastal Zone Management, with Applications of Spatial Analysis to Marine Traffic in Canada: Proceedings of the International Management Conference,* Sharif University, Iran.

Silverman, BW (1986) *Density estimation for statistics and data analysis.* Chapman & Hall: London.

Sorenson, J. April 2 draft (2002a). *Baseline 2000 background report: The status of integrated coastal zone management as an international practice,* 1st iteration, University of Massachusetts, Boston.

Sorenson, J. August 26 draft (2002b). *Baseline 2000 background report: The status of integrated coastal zone management as an international practice,* 2nd iteration, University of Massachusetts, Boston.

Tavera-Torres, C. (2001). *A Quantitative Analysis of Fishing Vessel Incidents in the Canadian Atlantic Region,* Master of Engineering Thesis, Dalhousie University, Department of Industrial Engineering, Halifax, Nova Scotia, Canada.

Uremocich, C., Pelot, R.P. (2002). *Implementation of Eastern Canada ECAREG Shipping Data: Preparation & Procedures,* MARIN Report #2002-01, Dalhousie University, Department of Industrial Engineering, Halifax, Nova Scotia, Canada.

Zacharias, M.A., Gregr, E.J. (2005). "Sensitivity and Vulnerability in Marine Environments: an Approach to Identifying Vulnerable Marine Areas", *Conservation Biology,* vol. 19, no. 1, pp. 86–97.

Part III
Spatial Planning

Chapter 22
Predictive Models to Inform Spatial Planning for Scottish Marine Fish Farms

C. Greathead, M. Gubbins, I.M. Davies, and D. Tulett

Abstract Spatial planning for coastal aquaculture ensures that the locations of farms are suitable to all relevant stakeholders. The Scottish Government has published "Locational Guidelines", where lochs and voes around Scotland are divided into three Categories according to their sensitivity to further aquaculture development, to aid this process. Predictive modeling is used to estimate the degree of nutrient enhancement and benthic impact arising from existing aquaculture development. These are reviewed every three months and are then published on the Marine Scotland-Science (MSS) website. The models used for the assessment are described.

Keywords Fish farming · Locational guidelines · ECE · Benthic impact · Categorisation

22.1 Introduction

An important aspect of spatial planning for coastal aquaculture is that the locations of farms are appropriate and acceptable to all stakeholders. To help guide industry and regulators the Scottish Government have published "Locational Guidelines for the Authorisation of Marine Fish Farms in Scottish Waters" (Gillibrand et al., 2002). In this document, lochs, voes, bays and sounds around Scotland were divided into three Categories according to their sensitivity to further aquaculture development.

Predictive modeling was used to estimate the degree of nutrient enhancement and benthic impact arising from existing aquaculture development. The Categories are reviewed every three months and are then published on the MSS website at www.frs-scotland.gov.uk. Only the 112 lochs that are listed in the MSS sea loch database are included in the Locational Guidelines.

C. Greathead (✉)
Marine Scotland-Science, Scottish Government, PO Box 101, 375 Victoria Road, Aberdeen, AB11 9DB, UK
e-mail:greatheadc@marlab.ac.uk

D.R. Green (ed.), *Coastal and Marine Geospatial Technologies*, Coastal Systems and Continental Margins 13, DOI 10.1007/978-1-4020-9720-1_22,
© Springer Science+Business Media B.V. 2010

22.2 Methods

The box models used in assessing sea lochs make two simplifying assumptions about the topography, hydrodynamics and water exchange of the sea loch systems.

The equilibrium concentration enhancement (ECE) model (Gillibrand and Turrell, 1997) predicts the enhancement of dissolved inorganic nitrogen (DIN) from fish farming sources. This is the product of the total consented biomass of all the finfish farms in the loch and the nitrogen source rate, divided by the flushing rate of the sea loch. The values from the ECE model are then sorted and indexed by applying a semi-logarithmic scaling to the ECE values from 0 to 5, such that each sea loch can be assigned an index of nutrient enhancement.

The carbon deposition model estimates the area of seabed impacted by the deposition of particulate carbon from fish farms using a modified "Gowen" model (Gowen and Bradbury, 1987). This model has been modified to be more specific to fish farms and sea lochs. The degraded (infaunal trophic index <30) area is calculated by summing the areas under each farm in the loch where carbon deposition rates exceed 0.70 $kgC/m^2/year$ and converted to a percentage of the loch surface area. Each sea loch is then assigned an index from 0 to 5 for benthic impact (Fig. 22.1).

The two indices (0–5) are combined to give a single index for each sea loch, scaled from 0 to 10. This provides an indication of the relative sensitivity of a sea loch system to further fish farming development and was divided into three categories.

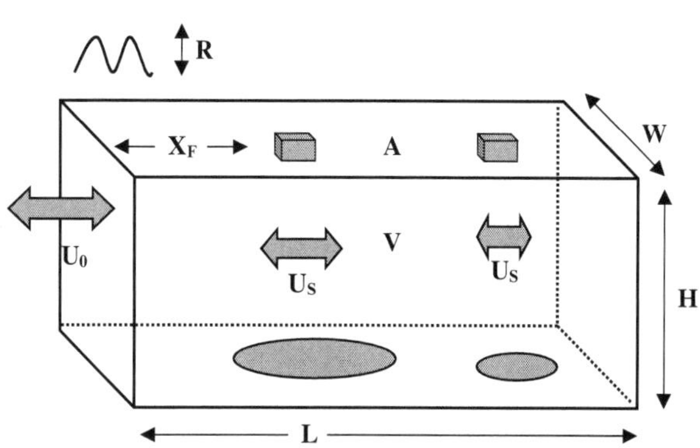

Fig. 22.1 Schematic diagram representing the simplifications used by the carbon deposition model. The sea loch is represented by a rectangular basin with identical physical characteristics to the real system. The tidal current amplitude decreases from U_0 at the mouth to zero at the head, with values at each site, U_S, calculated accordingly. Fish farms are represented by the filled rectangular block, with the areas of impacted seabed denoted by the filled ellipses. All other parameters and variables are described in the text (See also Plate 34 on Page 409 in Color Plate Section)

Category 1: areas where the most precautionary approach to further fish farming development should be adopted (Indices 7–10).
Category 2: areas where the new development or expansion of existing sites would not result in the area being re-categorised as Category 1 (Indices 5–6).
Category 3: areas where there appears to be better prospects of satisfying nutrient loading and benthic impact requirements (Indices 0–4).

A quarterly review of the "Locational Guidelines" is produced and published on the Marine Scotland-Science website (www.frs-scotland.gov.uk).

The maps are produced using an ArcMap project that exports the maps straight to pdf files. All the information is sourced from a single shapefile that is updated from a database.

22.3 Discussion

Both models attempt to assess the cumulative impact of all consented finfish farms in each sea loch system. The "Locational Guidelines" are based on relative levels of impact and are designed to be precautionary due to the simplifications and assumptions used in the models that are necessary for such complicated systems. They were not designed to describe location-specific levels of biological impact of fish farm wastes on the marine environment.

To ensure that the impact of the fish farming industry on the biodiversity, landscape, cultural heritage and recreational uses of the sea and coast is minimised and significant impacts avoided, the Scottish Government is investigating the relocation of fish farms to more dispersive areas. This "National Relocation Initiative" should allow farms to optimise their production and reduce their environmental impact on, for example, wild salmon interests or nationally protected areas.

In addition, Area Management Agreements and Groups bring together fish farmers and other interested groups in an area to coordinate farm activities such as the timing of stocking, fallowing and sea lice treatments, and to improve overall utilisation of coastal resources.

Future improvements to the models will include adjustments to allow the ECE values to reflect changes in internal mixing and dispersiveness in relation to the distance of the farms from the loch mouth.

References

Gillibrand, P.A. and Turrell, W.R. (1997), The use of simple models in the regulation of the impact of fish farms on water quality in Scottish sea lochs. *Aquaculture*, vol. 159, pp. 33–46.

Gillibrand, P.A., Gubbins, M.J., Greathead, C. and Davies, I.M. (2002), Scottish Executive locational guidelines for fish farming: predicted levels of nutrient enhancement and benthic impact. *Scottish Fisheries Research Report*, No. 63, 2002.

Gowen, R.J. and Bradbury, N.B. (1987), The ecological impact of salmonid farming in coastal waters: a review. *Oceanography and marine biology: an annual review*, vol. 25, pp. 563–575.

Chapter 23
Towards A GIS-Based Methodology for Marine Protected Area Zoning

Sarah A. Loos and Rosaline R. Canessa

Abstract In response to concerns of ad hoc approaches, site selection algorithms, such as Marxan, are being used to systematically and scientifically recommend sites for marine reserves. Similar challenges have been faced when siting zones (which aim to protect key habitats while permitting various uses) within broader Marine Protected Areas (MPA). This research explores Marxan as a decision support tool for MPA zoning. It aims to answer two questions: *Can the use of Marxan be streamlined, thereby removing some of the guesswork associated with its use?* and *How can zoning configurations be developed to incorporate large amounts of data and stakeholder opinions while being transparent, repeatable, and scientific?* This chapter reports on experimentation with various Marxan settings using the Southern Strait of Georgia, British Columbia, Canada as a study area as well as interviews conducted with zoning practitioners, in the context of developing Marxan as a decision support tool for MPA zoning.

Keywords GIS · Marine protected areas · Zoning · Marxan · Decision support

23.1 Introduction

According to the widely quoted IUCN definition, a marine protected area (MPA) is defined as:

> Any area of intertidal or subtidal terrain, together with its overlying water and associated flora, fauna, historical and cultural features, which has been reserved by law or other effective means to protect part or all of the enclosed environment (Kelleher and Kenchington 1992).

MPAs can be divided into several classes including those of complete protection (no-take/no-entry areas), and those which accommodate complete protection alongside a variety of uses such as recreation and fishing. The latter category, called

---

S.A. Loos (✉)
Department of Geography, University of Victoria, PO Box 3060 Victoria, BC, V8W 3R4, Canada
e-mail: sarahloos@gmail.com

D.R. Green (ed.), *Coastal and Marine Geospatial Technologies*, Coastal Systems and Continental Margins 13, DOI 10.1007/978-1-4020-9720-1_23,
© Springer Science+Business Media B.V. 2010

multiple use MPAs, will be discussed here. To avoid conflict and to preserve ecologically important areas within multiple use MPAs, different uses and levels of protection must be separated into zones. Zoning in the marine environment is becoming increasingly common as marine areas face increasing demands. Zoning is a complicated task, as it must balance different uses and values, and involves a variety of biophysical and socio-economic information. Zoning is also an inherently spatial problem. Arbitrary, ad hoc zoning methods have been shown to result in inefficient and ineffective MPA designs (Stewart et al. 2003; Villa et al. 2002). There is a need, therefore, for consistent and rigorous methods for zoning. Thus, there are calls for spatial decision support tools to assist in developing zoning plans.

One such tool is Marxan which has been developed to work in conjunction with Geographic Information Systems (GIS) to design ecological reserve systems (Ball and Possingham 2000; Possingham et al. 2000). Marxan runs a heuristic algorithm, called simulated annealing, which has been found to be a superior algorithm for solving complex site selection problems (McDonnell et al. 2002; Stewart et al. 2003).

Using GIS, the study area is divided into cells known as planning units, which can be of any shape or size, depending on the study area and data richness (Game and Grantham 2008). Input data layers typically depict spatial distribution of ecological features. There are a number of parameters assigned by the user:

- the number of runs of the algorithm and the number of iterations within each run;
- targets indicate how much area of each feature should be included within the solution;
- Boundary Length Modifier (BLM) controls the degree of planning unit clustering;
- planning unit cost represents the undesireability of a particular unit; and
- Species Penalty Factor (SPF) controls the importance of meeting each feature's targets.

The objective function of the algorithm is to minimize costs while meeting targets. Results from this iterative algorithm can generate two types of spatial outputs: (i) a number of alternative siting solutions; and (ii) an aggregate of all solutions, called a summed solution, showing the "popularity" of individual planning units and how important they are for meeting targets.

There are several challenges in applying Marxan to MPA zoning. Marxan was developed to site reserves with a single objective, e.g., to maximise biodiversity. The complexity of the decision problem increases considerably when developing a system of zones in a multiple use MPA which entail multiple objectives, such as maximising biodiversity, minimising impacts of fisheries and maximising marine recreation opportunities. Such an MPA might include a number of reserve sites, as well as other types of zones allowing different types of uses. In addition, Marxan is, unfortunately, a difficult program to use, and there is a great deal of flexibility and uncertainty in setting variables for the algorithm such that it is an exploratory rather than deterministic tool. In response to these challenges a Marxan Good Practices

Handbook (Ardron et al. 2008) was developed. This manual draws on the experience of the greater Marxan community and it offers users a simplified and streamlined approach to experimenting with settings. The handbook has not removed the need to experiment with settings; it does, however, offer guidelines for determining appropriate input parameters.

The goal of the research is to contribute to the development of scientific and rigorous GIS-based decision support tools for zoning by extrapolating the use of Marxan from marine reserve siting to multiple use MPA zoning by:

- testing for sensitivity under various settings; and
- modelling and integrating solutions for different types of zones.

This chapter reports on a survey of planning experts and discusses the results with respect to the challenges and opportunities of developing Marxan as a decision support tool for MPA zoning. It also presents the results of Marxan testing and subsequent zoning methods.

23.2 Methods

The research was initiated with a Functional Requirement Study (FRS) (Fig. 23.1). This included a literature review, consultation with MPA planners through semi-structured questionnaires and interviews and definition of decision parameters. A comprehensive spatial data set was also assembled. The main focus of the research comprised developing zoning models and testing the sensitivity of Marxan in developing zoning solutions. This included addressing:

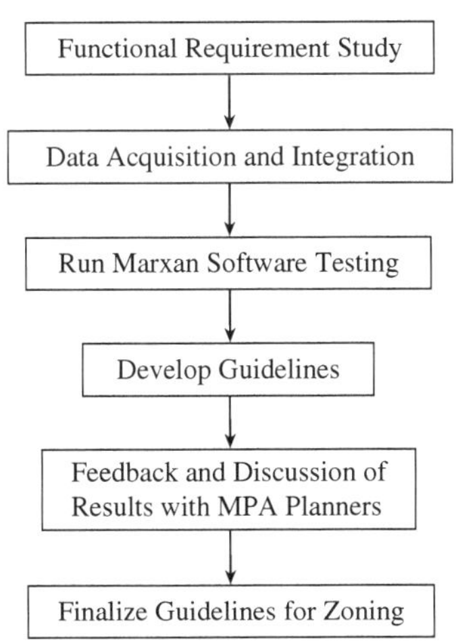

Fig. 23.1 Research design

- What effect do the size and shape of planning units have on the results of simulated annealing?
- What, if any, is the "cost" of data gaps?
- What is the relationship between the number of runs and the number of planning units? What is the relationship between the number of runs to the complexity / number of data layers?
- What effects do different Boundary Length Modifiers have on the ability of solutions to meet goals under different data and spatial conditions?
- What effect does the number of simulated annealing iterations have on the efficiency of the solution and ability to meet targets?

A literature review revealed a paucity of available published information on specific approaches to MPA zoning. Therefore, to acquire an understanding of past and future zoning efforts, MPA zoning practitioners and experts were interviewed during the winter of 2004/2005. Participants were chosen based on zoning knowledge and experience. Representatives from Canadian MPAs as well as a range of participants from the USA and Australia were selected. In total eleven people participated.

Participants were sent a questionnaire and asked to respond in writing. The answers provided the basis for a subsequent telephone interview that expanded and clarified the written answers. Of the eleven respondents, five had already participated in zoning, and six were preparing for the zoning process.

Zoning experts were asked questions about their zoning experiences, or anticipated experiences if zoning had not yet occurred. Questions were related to both the process and the outcome of zoning efforts. In addition, attitudes pertaining to the suitability of GIS-based site selection tools for zoning were probed. The results of the questionnaires set the stage for software testing.

Several Marxan settings and configurations were tested; planning unit size and shape, data gaps, study area configuration, boundary length modifier (BLM), planning unit cost, and species penalty factor.

Each of the settings listed above were tested in two different study area configurations: (i) Open Ocean, within which planning units were uninterrupted by land; and (ii) Inlets and Passages, characterized by narrow channels and islands. The purpose of this testing was to determine how Marxan reacts to features (i.e. land in this application) that interrupt the planning unit grid.

Following the software testing three methods for developing zones were developed and tested. Discussion with zoning experts and a review of literature indicated that three basic zones are commonly used, and that model was followed in this research. Two zones were modelled: a "conservation" zone, and a "recreation" zone, while a third "multiple-use" zone would include any areas remaining after the other two zones were developed and refined.

For each zoning method, the conservation zone was developed first, followed by the recreation zone. The same conservation zone was used for the three methods. In the first method, each zone was developed separately, then joined. In the second method, areas of importance for the conservation zone were excluded when developing the recreation zone reflecting a conservation priority in the MPA, then the two zones were joined. In the third method, areas of importance for the conservation zone were assigned a cost of 3 when developing the recreation zone, then the two zones were joined. The higher cost made it less likely that conservation units would be included in the recreation zone, unless they were very important for meeting recreation targets. For each of the three methods, an overlay was used to join the two zones, and areas of overlap were flagged and rated according to their importance for each zone.

23.3 Results

Results can be divided into two sections: Marxan testing and zoning development. The first, Marxan testing, resulted in some guidelines that, previous to the release of the Marxan Good Practices Handbook (Ardron et al. 2008), had not been published. It should be noted that some results may be a direct result of the configuration of the study area, particularly the occurrence of numerous islands, inlets and narrow channels as well as areas of open water, and will likely be different if tested in other locations.

It was discovered that smaller grids are preferable as they provide more possible configurations of planning units. Square planning grids are also preferable, as they cluster slightly more efficiently than hexagons.

Data gaps were found to adversely affect the clustering ability of Marxan, though with higher targets this was less apparent. This indicates that higher targets should be used with sparse data.

Of the study area configurations tested, the Open Ocean scenario reacted more to BLM than the Inlets and Passages scenario. This result confirmed that Open Ocean and Inlets and Passages areas should be separated for Marxan analyses. If run together the software tends to choose open areas because clustering is easier.

The species penalty factor (SPF) testing conducted in this research was preliminary, and more testing is needed to examine the relationship between SPF and other factors. It was discovered, however, that with increasing SPF the overall area selected in a solution increases, the majority of which is composed of non-target species planning units. The Marxan Good Practices Handbook indicates that care should be taken to not choose an overly large SPF, and that one should use values that barely meet targets (Ardron et al. 2008).

The inter-related nature of many settings, such as BLM and cost, also became apparent through testing. For example, a change in one requires a change in the other to maintain similar levels of clustering. In addition, the subjective nature of Marxan inputs means that the need for experimentation cannot be completely eliminated. Because the values used for Marxan settings are inter-related, and depend on the study area and zoning objectives, it is almost impossible to provide guidelines that would be applicable across different applications. Experimentation is a necessary component of Marxan use.

The second set of results, from the zoning development, was developed partly based on the Marxan testing as well as expert interviews. Three zoning methods

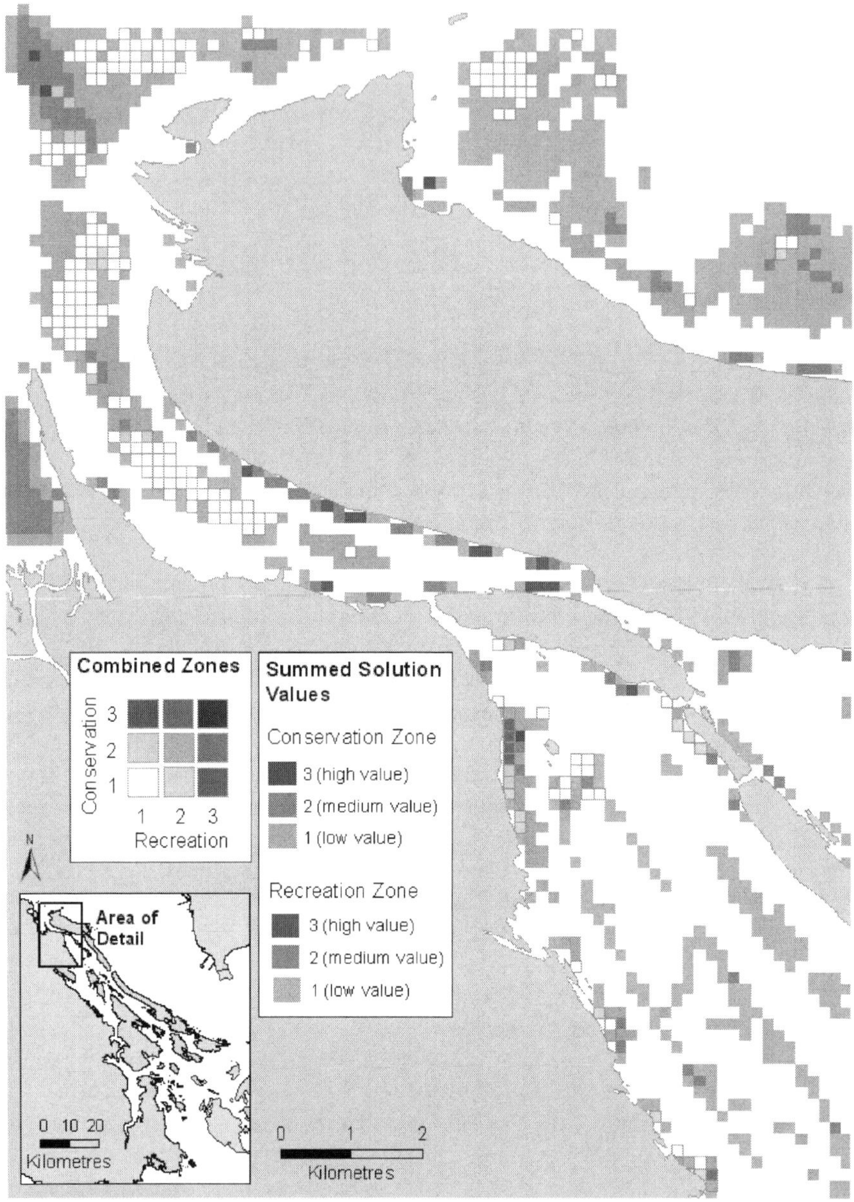

Fig. 23.2 Area of detail from method three combined zones (See also Plate 35 on Page 410 in the Color Plate Section)

were tested, and the suitability of using a particular method depends on the nature of the problem being addressed. The first method, described above, gives stakeholders an opportunity to decide what areas of overlap should be attributed to the conservation or recreation zone, rather than applying overall rules before the zones are combined. However, if priority of the conservation zone is a given, the third method would be useful because it incorporates this priority into the development of the recreation zone. This creates fewer areas of high value conflict with the conservation zone when the two zones are combined. Figure 23.2 shows the results of an output using the third method. The overlap areas, outlined in green, are planning units that fall within both zones and would need to be manually (through expert advice or stakeholder input) assigned to a zone. The relative values of overlapping units for each zone are provided through colour-coding, which can be used as a guide when sorting them into zones.

23.4 Discussion

During this research several challenges associated with the use of Marxan for marine zoning came to light. These, along with suggestions for its role in the zoning process, are discussed below.

Terms such as efficient, adaptable, systematic, effective, scientific, and transparent have been used to describe ideal zoning tools and processes (Lewis et al. 2003; Villa et al. 2002). Through this research it became apparent that Marxan offers some of these characteristics. While it has the power to incorporate large amounts of information, and to quickly adapt to changing requirements, Marxan is deficient in other areas. Because there are many interrelated settings with few specific guidelines, "tweaking" plays a large role in Marxan use (Ardron 2005; Fernandez 2005) making Marxan an exploratory rather than solution-oriented tool. While the algorithm itself is scientific and the map outputs are numerical, the guesswork and manipulation of settings dilutes the systematic and objective potential of Marxan. One can argue that in this aspect Marxan is a modest improvement over traditional ad hoc methods. In addition, true transparency is almost impossible with the software. While the settings used to develop zoning can be shared, with stakeholders for example, it would be very difficult for those unfamiliar with Marxan to evaluate the appropriateness and motivation for using a given set of settings.

In interviews with zoning practitioners it became apparent that there is a divide between the required inputs of Marxan and the way zoning has been traditionally approached. Terms such as BLM, species penalty factors, and even targets in some cases, are not concepts that are applied in non-Marxan-based zoning development. Probably the biggest challenge for MPA managers and zoning practitioners when using Marxan is translating zoning goals into specific software settings. For example, while most interview participants indicated that zones covered as much space as was feasible, Marxan requires specific targets, in percentages, for every data layer. Figure 23.3 shows a list of zoning user needs compared with the range of functions available in Marxan. Several requirements for zoning identified by questionnaire participants and zoning literature are included in Marxan settings. These include the avoidance of fragmentation, the incorporation of existing uses and public opinion, and the inclusion of certain amounts (targets) of particular features in solutions.

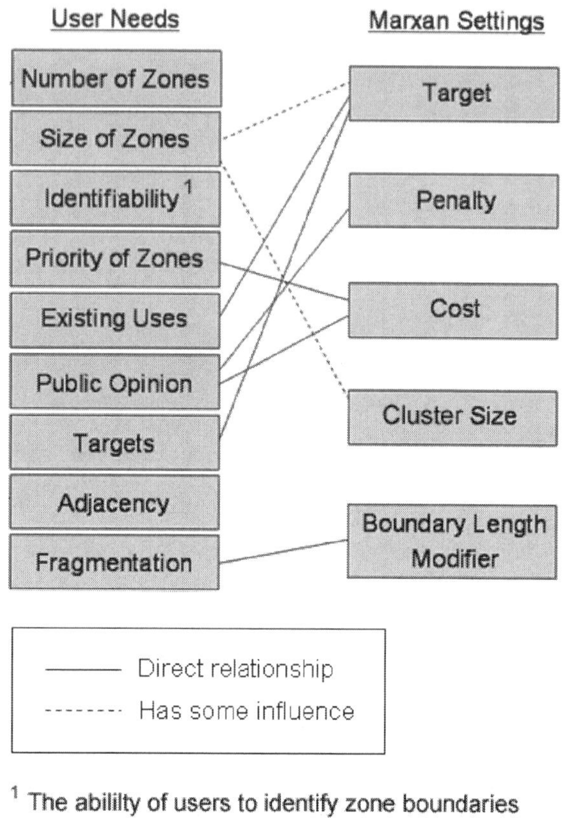

Fig. 23.3 Functionality of Marxan compared with user needs for zoning (See also Plate 36 on Page 411 in the Color Plate Section)

Other areas, such as the number, identifiability, and relative placement of zones, are not supported by the present functionality of Marxan. The size of zones is somewhat controllable through Marxan, though if the summed solution output is used the size is ultimately determined by the person who refines the zone boundaries.

Marxan provides a powerful, flexible method for integrating and analysing large amounts of spatial data, however, when it comes to developing actual protected areas or zones, Marxan outputs are not suitable for immediate use. The "best" solution from a set of Marxan runs is of limited usefulness, as it gives no indication of the relative importance of planning units. In addition, since every solution contains a random element, there is no guarantee that the areas selected for the best solution are actually of highest value. It is far better, therefore, to use the summed solution output (see Fig. 23.2), which serves to identify hotspots, and also provides an indication of the relative value of each planning unit. These outputs are a starting point, and bear little resemblance to zones. Refinement and boundary development are the next steps required, and must be done manually.

Because Marxan does not have the ability to develop multiple zones simultaneously, zoning must be done in steps as outlined earlier in this chapter. Zones are developed separately, joined, and the areas of overlap examined manually. While more time-consuming than a more automated tool, this is an excellent method for including stakeholders in the zoning process. Maps such as Fig. 23.2 can be used by stakeholders to refine zones. This uses Marxan outputs (thereby taking advantage of the algorithms and data processing powers of Marxan) and allows input from stakeholders (thereby facilitating consensus development and buy-in). A give-and-take approach, where units are swapped in and out of the solutions, could be used. The ability of stakeholders to visualize the relative values of planning units for each zone would facilitate such trade-off analyses. This visualization of conflicts between zones is a key contribution to the usefulness of Marxan for zoning.

23.5 Conclusion

There was a perception by some questionnaire participants that the systematic approach offered by GIS and Marxan would lend transparency and scientific rigour to the zoning process. There is, however, as discussed above, a large amount of subjectivity that forms the use of the software. There is a danger, as with any solution developed by a computer, of perceiving the results as impartial, scientific and perfect. Simply recording settings is not sufficient to make the process transparent and there is a need, as one interview respondent stated, to have the "bias of the zoning process identified up front." It became clear through this research that it is the underlying purpose of zones, whether for conservation or for fisheries development, that determines Marxan settings and outputs. This is important, as an emphasis on conservation goals will produce a very different result from a process that emphasizes economic values. Unless specifically stated it would not be possible for a non-Marxan user to identify this bias.

Despite the limitations listed above, the actual outputs provided by Marxan have great potential to support zoning development, particularly involving stakeholders, particularly when used in an iterative exploratory decision support context rather than a solution-finding context. The summed solution output not only identifies hotspots for each zone, but it also indicates the relative importance of planning units for each zone. This creates a starting point and provides alternatives that have the potential to help stakeholders reach consensus. Through an iterative process of give-and-take stakeholders and MPA planners could use summed solution outputs to develop and refine zone boundaries.

Three methods for combining zones using the summed solution were developed in this research. They identify areas of overlap between zones, and the relative importance of these overlap areas for each zone. It is anticipated that these methods could help stakeholders and zoning practitioners develop multiple zones simultaneously. Practical testing of these methods is needed to determine their usefulness.

As zoning continues to gain recognition as an effective management tool for multiple use areas the tools and theories supporting zoning will continue to advance. The present version of Marxan has served as an excellent starting point, but it is time to move forward. The developers of Marxan are in the process of extending Marxan into MarZone to enable modeling multiple zones, multiple costs and multiple objectives. In the mean time, a certain amount of *caveat emptor* should accompany the use of Marxan should it be applied to marine protected area zoning.

References

Ardron, J (2005) Personal Communication. Marine Analyst, Living Oceans Society. January 12, 2005

Ardron JA, Possingham HP, and Klein CJ eds (2008) Marxan Good Practices Handbook. External review version; 17 May, 2008. Pacific Marine Analysis and Research Association, Vancouver, BC, Canada. 155 pages. www.pacmara.org

Ball IR and Possingham HP (2000) Marxan (V1.8.2): Marine Reserve Design Using Spatially Explicit Annealing, a Manual

Fernandez L (2005) Personal Communication. Manager, Representative Areas Program, Great Barrier Reef Marine Park Authority. February 4, 2005

Game ET and Grantham HS (2008) Marxan User Manual: For Marxan version 1.8.10. University of Queensland, St. Lucia, Queensland, Australia, and Pacific Marine Analysis and Research Association, Vancouver, British Columbia, Canada

Kelleher G and Kenchington R (1992) Guidelines for Establishing Marine Protected Areas: A Marine Conservation and Development Report. IUCN: Gland, Switzerland

Lewis A, Slegers S, Lowe D, Muller L, Fernandes L and Day J (2003) Use of Spatial Analysis and GIS techniques to Re-Zone the Great Barrier Reef Marine Park. In Woodroffe CD and Furness RA (eds) Coastal GIS 2003: an Integrated Approach to Australian Coastal Issues (Wollongong Papers on Maritime Policy No. 14). Centre for Maritime Policy, Wollongong, NSW

McDonnell MD, Possingham HP, Ball IR, and Cousins EA (2002) Mathematical Methods for Spatially Cohesive Reserve Design. Environmental Modeling and Assessment 7:107–114

Possingham HP, Ball IR, and Andelman S (2000) Mathematical methods for identifying representative reserve networks. In Ferson S and Burgman M (eds) Quantitative methods for conservation biology. Springer-Verlag, New York, pp 291–305

Stewart RR, Noyce T, and Possingham HP (2003) Opportunity cost of ad hoc marine reserve design decisions: an example from South Australia. Marine Ecology Progress Series 253:25–38

Villa F, Tunesi L, and Agardy T (2002) Zoning Marine Protected Areas through Spatial Multiple-Criteria Analysis: the Case of the Asinara Island National Marine Reserve of Italy. Conservation Biology 16:2 pp 515–526

Chapter 24
Functional Connected Areas in Regional Planning: Sea and Land Uses Interaction

Leon Gosar, Manca Plazar, Tanja Preseren, and Franc Steinman

Abstract The regulations and the competence related to the sea often define various, decision-making levels and institutions. To a large extent spatial planning indirectly dictated and still dictates the use of the sea in the narrow coastal belt (since these are the so-called functionally connected lands). The discussion about the sea related issues will also have to focus on the consequences brought by the authorized uses and the applicable legislation on the individual (spatially defined) areas. The modelling approach used for the entire Slovenian coastline (Fig. 24.1) and extent of regional spatial development plan is presented. The methodology presents a turn from qualitative assessments to the suitable conceptual and quantitative solutions which could lead to the comprehensive development of whole region.

Keywords Marine area · Water management · Integrated coastal management · GIS · Spatial development planning

24.1 Introduction

In the past, the subjects related to the sea were predominantly overshadowed by other water management contents, which regulated especially the inland waters, including the affluents of the coastal sea. Also in the world, increased attention has been paid to the problems related to sea management where, only recently, functionality connected marine and land areas are dealt with comprehensively. The European Community has already undertaken the recording of the sea management goals as well as the preparation of the regulation for integral management in the coastal area, although numerous subjects are already covered by the recently adopted Directive establishing a framework for Community action in the field of water policy.

L. Gosar (✉)
Faculty of Civil and Geodetic Engineering, Chair of Fluid Mechanics, Jamova 2 SI-1000, Ljubljana, University of Ljubljana, Slovenia
e-mail: leon.gosar@fgg.uni-lj.si

D.R. Green (ed.), *Coastal and Marine Geospatial Technologies*, Coastal Systems and Continental Margins 13, DOI 10.1007/978-1-4020-9720-1_24,
© Springer Science+Business Media B.V. 2010

The coastal sea and the near-coastal area have many specific characteristics, which are not comparable to the characteristics of the inland water regime, therefore they require special consideration both within the regulations governing the water problems, as well as within the strategy of water management and development planning (spatial planning, regional economy planning, etc.).

The regulations and the competence related to the sea (given indirectly or directly) define various decision-making levels: state, regional and municipal one, taking into account various development sector interests. The coast (or more precisely the costal line) is usually adopted as a dividing line between the two legislations regulating the management with the land area and the marine area. Joining the land and the marine (water) areas into a uniform method of planning the use of the physical space – by means of suitable expert bases to allow the use of the sea – most often remains the key instrument also for the introduction of the water management contents into spatial planning. The use of the physical space, land and sea has to be adapted to the natural resources or to the water regimes of the sea, coast and near-coastal areas. Such an approach is also the basic guidance included in all international conventions, which at the same time require the implementation of analyses of cost-efficiency of water-related activities. Only these two together, i.e. harmonised natural resources and sound development goals, can serve as a support in the decision-making process on how to achieve sustainable water management (Fig. 24.1).

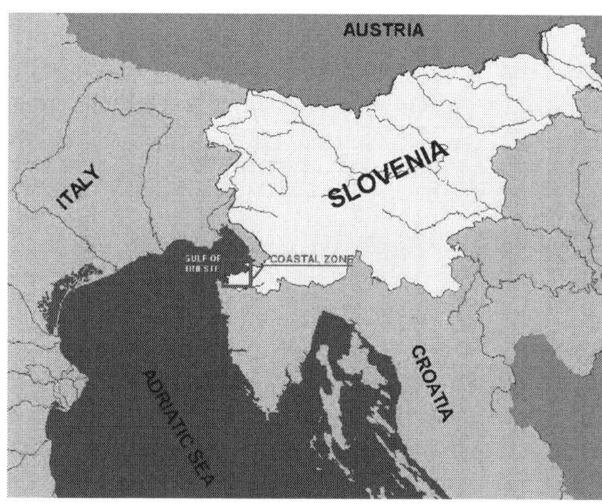

Fig. 24.1 Coastal zone in Slovenia (See also Plate 37 on Page 411 in Color Plate Section)

24.2 Sea and Land Uses

To a large extent, land use planning indirectly dictated and still dictates the use of the sea in the narrow coastal belt (since these are the so-called functionally connected lands). The discussion about the sea related issues will have to focus also on the consequences brought by the authorized uses and the applicable legislation on the individual (spatially defined) areas. Regarding the recognised notion of the use of the sea and the water rights, it is first necessary to record and present the actual state of the use of the sea, the rights and the obligations associated with the obtained water rights arising from the national regulations and regulations of the local communities, which relate to an individual area of the authorised use or to the performance of activities.

The preparation of suitable expert bases for water (sea) and waterside area (near-costal area) management can provide efficient transposition of the water protection policy (in all its forms) and the principles of sustainable water management to all levels of spatial planning and decision-making process on the activities which affect the environment. Therefore, the use of the physical space, land and sea has to be adapted the marine environment in terms of quantity and quality, which is the basic guidance included in all international conventions; but, at the same time, we should not neglect the analyses of cost-efficiency of the possible activities as a support in the decision-making process.

24.2.1 Legal Regime Term

In order to be able to assess the volume of the rights, obligations and limitations, it will be necessary to analyse, for the purpose of the uses of the sea, the legal regimes of the areas related to these uses, as they are defined in the legislation. The volume of the existing rights, obligations and limitations has to be a starting point for future planning and authorisation of the sea use.

A legal regime is considered to be a string of legal rules, as introduced by the regulation, which lays down the method of exercising the granted right of use and the obligations related to it in a clearly defined area. These rules can be set only when there exists a legal basis and criteria, on the basis of which a legal regime can be unambiguously determined. A legal regime, as a rule, interferes with the property rights of, for example, persons of the private law, as well as of an individual user (e.g. a tourist).

At the same time, the legal regime of lands, be it a marine or near-coastal land, which primarily presents the limitations on the use of the sea or water lands, exerts influence also on the other (e.g. neighbouring) uses of the sea or the costal area, due to functional integrity of the water regime of the sea as a fluid or the sea as a receiver of all waters, etc. The legal regime of the sea use has to be harmonised with the legal regime of the public good, which already applies to the sea (as a natural resource), the goal of which is to allow the general use to anybody to the same extent and under the same conditions.

We can note that there are many different sector legislations which intervene in the field of waters and which introduce a number of areas with limitations, i.e. areas with a legal regime. The recording of all areas, the protection of which depends on the regulated water regime, is required also by the Directive establishing a framework for Community action in the field of water policy, which demands that the water management basis and the planning record the contents of the adopted legal regimes and the regulation, on the basis of which the legal regime was established, in order to take into account also their water rights.

24.2.2 Water Right and Legal Acts As an Integrative Tool – Reality Check Problem

The issue of integration and integrative management has already been a part of environmental science for some time. At the same time the issue of decision support tools has been developed alongside informatics. The question arise: how could we define a good decision support tool in the light of integrative management? And one of the answers is, that it is able to function in the real world (it survives the reality check). The hardest reality check which could be recognized in the real world is probably the litigation procedure. In the case of conflicts (conflict of opinion, conflict of interest, etc.) the decision made by the mutually recognized arbiter stands for the generally recognized reality check. By this we have clearly shown the importance of the close connection to legal acts in any of the applications considered.

General legal acts as well as individual (or specific) legal acts as individual water rights generally are, are representing for the legal reality, which is representative also for a certain area (i.e. marine area). Reverse connectivity of this reality to the actual physical (also chemical, biological, and other) environment is another link that has to be maintained by means of established mechanisms like official supervision and the presence of (inspectors), monitoring and other recognized services which recognises a reliable link between the legal reality and actual developments in the real world.

Water rights (individual and general) are therefore an essential element for any model that serves as a decision support tool in the field of water management and spatial planning, which is also the fact recognized by many researchers working on the matter. The integration of this dimension in the complex models of water management is still an effort to be made. It is beyond the scope of the present chapter to decompose integration procedures by means of the integration of different aspects (economic aspects, technical aspects, and legal aspects), with essential developments in this field.

24.3 Functional Connected Areas – Land and Sea Interaction

An example of the functional connection of the use of real estates and the use of the sea can be presented by the circumstances which have to be taken into account due to the legislative regulation of the EU Directive concerning the quality of bathing water. The public interest (suitable water in bathing areas) is a condition, under which such use of the sea can dictate the limitation and determination of the conditions for the use of the land real estates and the activities on them. However, the criteria have to be set, with which the limitations are established within a minimum volume (the principle of proportionality), which still guarantees a suitable quality of the bathing waters.

Alongside laws on waters and water management, even sectorial laws applied to water can enforce additional areas with legal regimes. The water legislature enforces the well-known water resource protection areas adjacent to water bodies, riparian areas (rivers and streams) and coastal areas (standing water), maintenance strips along water infrastructure (dykes etc.), flood retention areas, areas for reservoirs etc. The Water Framework Directive in requesting that protected areas, whose existence is conditioned by the presence of water or water management systems, are listed in water management guidelines, while areas, which should be protected in the future, have to be harmonised with water management plans.

Legal regimes that stem from uses of the sea or the continent with adjacent uses functionally tied to the sea (Fig. 24.2) are conditioned by particular sectors and their regulations:

- protected area – nature preservation,
- protected area – preservation of cultural heritage,
- aquatory of ports, anchorages, shipping routes – reservation for shipping,
- beaches and bathing waters (managed or declared) – recreation and leisure,
- fishing reserves and mariculture aquatories – fisheries.

For the contents shown in Fig. 24.2, we have to emphasise again that enforced legal regimes determine conduct, even of coincidental users (local population, tourists, etc.). From the remaining parts of the coastline, which are (still) outside areas with applied legal regimes (whereby free access to the sea is possible), one can see that such areas on the Slovenian coast are very few.

Because of the conflicting demands for physical space and the limited possibilities, an integral spatial conception of the Coast should be provided that will coordinate the interests of development activities with spatial possibilities and protective requirements. Simultaneously, the conditions for the development of a quality tourist programme should be established and a permanent public access to the coast and to bathing sites should be provided.

The Slovenian coast is located in a water deficient area (Fig. 24.1). Water deficiency is a restricting factor in the development of activities using and discharging large quantities of water, therefore the development of such activities should be thoroughly studied from the environmental, spatial, technological and economic aspects.

As a result, only the activities that cannot be performed anywhere else because they require the presence of the sea, shall be developed at the sea and in the coastal area, if such activities do not reduce the quality of water but rather increase the quality of use, whereby they do not hinder public access to the sea and the coast. No developments that may restrict the open view of the sea and threaten the conservation of nature and cultural heritage shall be carried out in the coastal and inshore

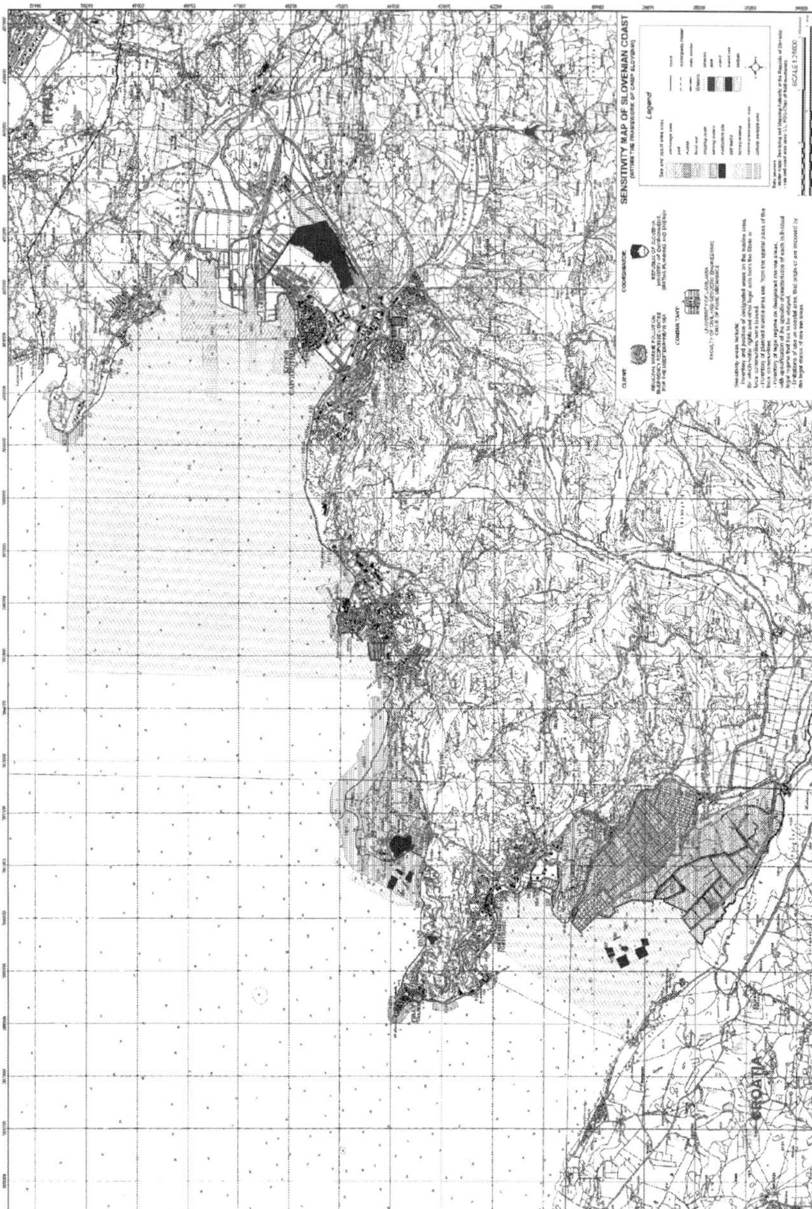

Fig. 24.2 Existing legal regimes as decision support tool for managing sensitivity areas for national contingency plan (See also Plate 38 on Page 412 in Color Plate Section)

area and at the inshore sea. The coastline shall not be reduced, but it can be extended if this is acceptable considering the vulnerability of the zone.

The conception and the programme of measures for the development of recreational and tourist area along the entire length of the Slovenian coast will preserve and emphasize the outstanding landscape features of the area concerned and ensure the protection of valuable natural features and the natural and cultural heritage. The implemented programme will contribute to increased living quality in the area, to the sustainable economic (tourist) development, as well as to the conservation of nature and cultural heritage of the sensitive costal area.

The objective of the project was to prepare the conception of spatial planning (Fig. 24.3), which will be subject to a more detailed planning, and a detailed programme of implementing measures. A detailed spatial planning conception of the coastal area will present clear guidelines on spatial development of the entire Slovenian coast. The designing of alternative conceptions will provide a higher-quality selection of the proposed conception.

At the same time, the vital objective of the project was to coordinate the development and protective interests through the process of spatial planning and management of the coastal area between the key partners (the state, communities/municipalities and other partners), establishing in this manner an integral approach and improving the coordination between different institutions and the subjects involved at the national and the local levels.

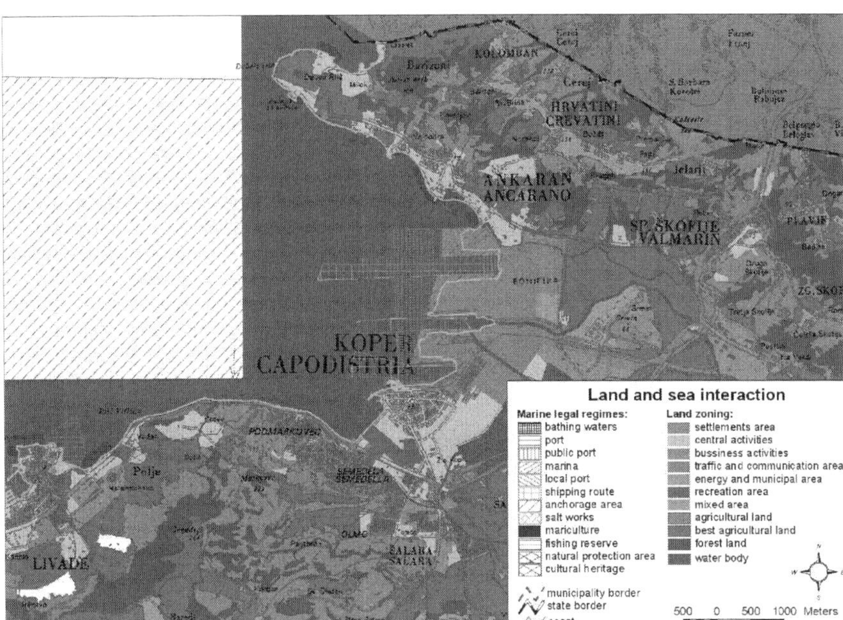

Fig. 24.3 Spatial representation of land and sea interaction as functional connected areas (See also Plate 39 on Page 413 in Color Plate Section)

Preparation of the programme of implementing measures will enhance the implementation of the adopted decisions after being properly presented to the general public. The promotion of the project in the public will also help to increase awareness of the urgency of protection and sustainable management of the coastal area.

The presented consequences of the introduction of the legal regimes require consideration about the future procedures of granting the special use rights or areas defined with a legal regime. The presented approach shows that we already have efficient tools, with which it is possible to define (or design) applicable boundary conditions, in order to be able to plan or allow the uses of the sea and coast area in the future, while taking into account the functional connection of the sea use, real estate use and the implementation of the current and planned activities.

The collected spatial data provide an integrated and hierarchically structured insight into the legislative framework, governing the relations in the area under consideration at the national, regional and the municipal level, including the binding expert groundwork and arrangements, which will facilitate a detailed analysis of eventual disparities between the particular legislative and executive levels. A range of spatial planning groundwork, studies and projects and a framework presentation of the key conflicts in terms of different levels of spatial management (national, regional, municipal and local) have been prepared.

24.4 Conclusions

The contribution of the discussion is especially in the fact that it directs towards the path leading from qualitative assessments (e.g. how the sea is problematic) to the suitable conceptual and quantitative solutions in the field of water management and spatial planning, which will be predominantly based on a suitable expert and legal groundwork, which is a condition for a successful implementation of measures. Quantitative (and integrated) analysis of the coast area of the Republic of Slovenia represents a condition for any decision-making process on the use of the marine and land environment and the activities functionally related to it. The use of the standard solutions of processing of spatial and attribute data will allow the combination and complementation with other contents and expert bases which enter the processes of development planning.

The presented approach combining in close interaction existing (or foreseen) legal regimes show that the sea surface is not an empty space, therefore enforcement of integral planning of water, land and natural resources (the sea) is essential. This does not mean that expanding planning of existing land use on the continent to the sea would be enough. Abroad, corporations are merging the connected subjects of public and ordinary law, thus creating formal and informal ties between states, self-government (municipalities), associations of water users, the public, affected individuals etc. according to principles of partnerships.

Water management entails management (administration) and use, thus it will be necessary even in the field of water management to formalise use as well (water itself, sediments, property – land, buildings, infrastructure). One of the first steps of the state when dealing with its own property according to principles of good management is the establishment of a comprehensive inventory of sea uses, water rights on the sea and functionally connected land. An important aspect is the implementation of principles of sustainable development and the principle of integral dealing with the sea and continent when planning uses for the coastal area. Functional ties between activities on the sea and those on the continent (e.g. manipulation surfaces for mariculture) or vice versa (e.g. water surfaces for bathing) demand harmonised planning of both and should include, according to the scope of the decision process also adequate models (hydraulic, water quality, sediment transport, models of anthropogenic activities, etc.)

A spin-off result of the analysis where the established consequences of poorly thought out enforced legal regimes, which demand serious recollection about the future granting of special rights or limiting areas with legal regimes. The article provides information on the development of the decision support tool that has in its core spatial representation of the legal acts. At the first stage already identification of direct conflicts (actual, potential) is a clear result of the effort. At the same time the created database acts as a reference point for any considerations about the subject. But the identification of conflicts is not the end stage of the developments as we move towards the more advanced approaches of the optimization of the sustainable marine and land use.

Bibliography

Banovec, P., Gosar, L., and Steinman, F. (1999). Coastal Zone Management Plan in Slovenia – Marine Area Uses, User *Proceedings* from 14. ESRI European User Conference, Munich, Germany.

Capobianco, M. (1999). EU Demonstration Programme on Integrated Coastal Zone Management. *Role and Use of Technologies in Relation to ICZM*, Technomare S.p.A., Venezia, Italy.

Gosar, L. (2000). Sea Area Uses in Water Management. *Master of Science Thesis*, University of Ljubljana, Faculty of Civil and Geodetic Engineering, Ljubljana, Slovenia.

Gosar, L., Plazar, M., Prešeren, T., and Steinman, F. (2005) Functional Connected Areas in Regional Planning : Sea and Land Uses Interaction. *6th International Symposium on GIS and Computer Cartography7 for Coastal Zone Management : Defining and Building a Marine and Coastal Spatial Data Infrastructure,* Aberdeen, Scotland.

Gosar, L., Prešeren, T., Plazar, M., and Poljšak, A. (2007) Detailed Conception of Coastal Strip Spatial Arrangements. *MEDCOAST 07: The Eight International Conference on the Mediterranean Coastal Environment*, Alexandria, Egypt.

Steinman, F., Banovec, P., and Gosar, L. (2006) Functional Connected Areas : Sea and Land Uses Interaction. *Advances in Geosciences : Vol. 4 : Hydrological Science,* Hackensack, N.J.: World Scientific.

Chapter 25
The Development of Coastal Information Systems: The Role of Networks in Bringing Spatial Analysis into Planning and Management

Tim Stojanovic

Abstract Developers and users of GIS face many challenges in producing successful IT tools. This chapter focuses on the domain of applied technology, and the issue of how to successfully transfer information to users in the coastal zone, where there are complex marine and terrestrial jurisdictions and responsibilities. The worldwide increase in the volume of spatial and environmental data has led governments and organisations to consider the approach of information management. The chapter reports the findings of research in the UK, which has detailed problems such as: "information overload"; commercial, political and environmental sensitivity of data, lack of information policy or good data handling practices; and the difficulties and failures in establishing distributed, inter-organisational information systems.

The solutions presented include metadata, interoperability, harmonisation (standards), semantic webs, information policy, information mapping and partnerships. The chapter concludes by focusing at the regional level, and describes the experience Severn Estuary Partnership, UK, in establishing a GIS Forum to bring together technicians, users and owners of GIS data within a network, and building on this, to work together to provide a framework, to resolve the current lack of co-ordination and harmonisation of GIS data to aid spatial planning on the Severn Estuary. The chapter also presents research findings about the information flows and data transfers between coastal stakeholders involved in coastal partnerships around the UK. Improved understanding of these issues will aid IT developers and coastal planners and managers in maintaining networks of relationships that enable them to have a good understanding of the coast and to support their decision-making.

Keywords Information systems · Integrated coastal management · Metadata · Coastal networks · Information transfer

---

T. Stojanovic (✉)
Marine and Coastal Environment Research Group, School of Earth and Ocean Sciences, Cardiff University, Cardiff CF10 3YE, Wales UK
e-mail: stojanovic@cf.ac.uk

D.R. Green (ed.), *Coastal and Marine Geospatial Technologies*, Coastal Systems and Continental Margins 13, DOI 10.1007/978-1-4020-9720-1_25,
© Springer Science+Business Media B.V. 2010

25.1 Coastal Information and Data

Managing geographic information is not the primary task that most coastal practitioners are setting out to achieve. Yet to acquit their various responsibilities and aspirations for the coast, practitioners are increasingly forced to deal with information overload from a glut of articles, statistics, surveys and research projects, reports, academic studies, telephone calls, satellite images and GI Tools. It is assessed that ten terra bytes of relevant environmental data is collected each day (Gunther, 1998). At the coast, this situation is complicated by the fact that information changes as quickly as the tide flows, and a division of responsibilities and functions between land and sea which is so complex, that in the UK it involves at least 28 government agencies, departments or statutory bodies. There are over 260 formats of Earth Observation data alone relevant to the coast (Harris, 1999).

25.2 Getting GI to the User

In the UK and Europe, a series of reports at the national level have focused on reasons why geographic information sharing is hampered at the coast [IACMST (1994, 1995, 2005); Smith System Engineering et al. (1996); Millard and Sayers (2000); Franklin and Harries (2002); Centre for Earth Observation (1995); Doody et al. (1998)]. These reports have documented a series of blocks due to the commercial, political and environmental sensitivity of data; lack of information policy or good data handling practices; and the difficulties and failures in establishing distributed, inter-organisational information systems. Despite this developed understanding of the problems, the solutions remain largely unrealised, and failures of IT tools are a relatively common feature in Integrated Coastal Management.

25.3 Solutions from Information Science

One high profile solution is the construction of spatial data infrastructures which will provide diverse communities of users with access to distributed data (Longhorn, 2005). In Europe this is being driven forward by the INSPIRE Directive 2007/03/14 EC. This includes the definition of marine and coastal components (Ferreira et al., 2007). However, this kind of "top down" approach tends to focus on national level and sectoral marine datasets for strategic planning. In contrast, one can consider the great variety of data used for operational management carried out at regional and local levels. The implementation of Integrated Coastal Management involves a variety of local level stakeholders, including coastal planners, engineers, conservationists and scientists from number of organisations co-operating to develop a strategy for a distinct coastal area. Local partnerships of public, private and voluntary sectors are driving "bottom up" GI solutions. The GI to support this approach can be developed from primary tools toward a more integrated approach, but the final goal is envisaged as better understanding through *communities of knowledge.*

A number of methods from Information Science are presented as most fruitful in solving the inter-organisational problems often present when trying to take a collaborative, people-centred approach (Felleman, 1997). Firstly, in order to capture the complexity of decision-making at the coast, successful GIS will need to move beyond the model of a single IT developer utilising a user requirements survey, to a more sophisticated model that is developed in close collaboration with a variety of users. Soft Systems Methodology has demonstrated a number of promising applications in this field (Checkland and Holwell, 1998). Figure 25.1 shows the development of a soft systems model for the management of coastal geohazards . This element of the model was constructed as part of a series of workshops attended by 24 representatives from a range of public organisations. This kind of approach is important in getting users to buy into ownership of a system, and in designing a user friendly front-end.

Secondly, Coastal GIS can sometimes suffer from identity crises which inhibit user uptake, due to the great variety of data layers which may not be clearly related to the functions of management. Information Mapping (Horton, 1988) helps build a conceptual model of information resources that enables coastal practitioners to make sense of complex situations by defining data types and linking datasets to organisations and activities. However, these methods have received little attention in ICM, in contrast to metadata and ontologies (Millard, 2004), which have reached the status of buzzwords.

25.4 Technical Solutions: Metadata

For the UK coastal zone, the generation of metadata commonly occurs on an ad-hoc basis, although some organisations have developed an Information Policy. Some government agencies such as Environment Agency and Centre for Environment Fisheries and Aquaculture Science are active in promoting the use of metadata by their scientists. e-GMS (e-Government Metadata Standard) lays down the elements to be used by government officers when creating metadata for public agency information resources or when designing search systems. Presently, many practitioners find it easier to identify and obtain datasets through internet searches or via personal telephone calls. An array of metadata standards exist, none of which are designed specifically for the nature of marine and coastal datasets, although many provide "extensions" which can be setup to provide for special user needs such as coastal information. The UK GEMINI Standard (v2.1) has been devised by a partnership including the Association for Geographic Information (AGI). This is compatible with ISO 19115 *Geographic Information: Metadata* which been developed by the International Standards Organisation, along with a family of ISO 19000 standards which will deal with issues such as definition of XML schema. An element of EUROSION project focused on the production of a metadata recordset specifically

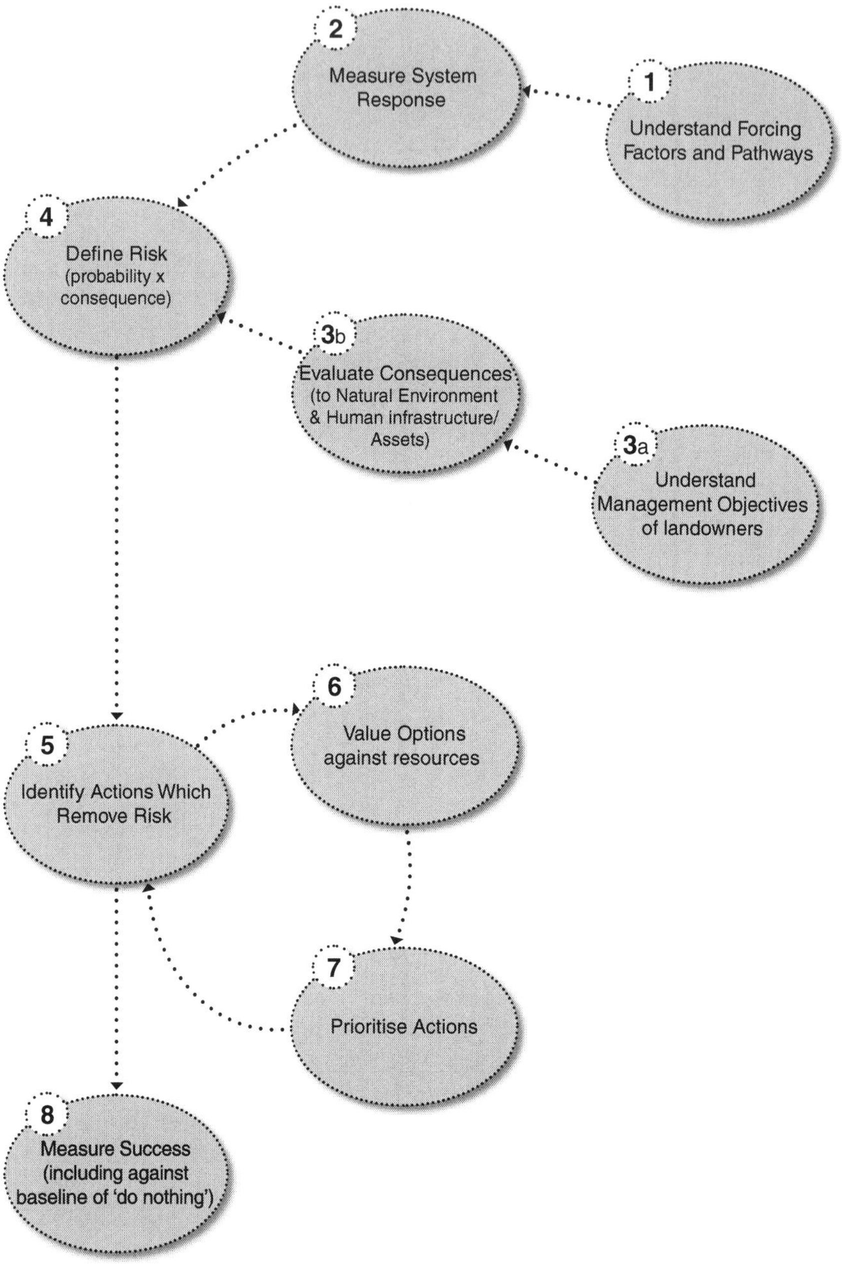

Fig. 25.1 Group model of decision stages in the management of coastal geohazards (See also Plate 40 on Page 414 in the Color Plate Section)

for coastal information (Eurosion, 2004). The Eurosion Metadata Format has been developed as part of the *CoastBase* project and contains a thesaurus which can be used to classify coastal functions.

Whilst metadata is often mentioned as an important solution for the reuse of spatial data, on its own it merely forms a giant suppository or warehouse of data. In the same way that libraries can remain underused because of the lack of time for users to query them, the potential of metadata may remain unlocked unless coupled with more sophisticated networks, such as portals and discovery tools (using web semantics) at national and international levels, and information systems based around users for discrete biogeographic coastal units at the regional and local levels.

25.5 GIS Fora: Co-operation on the Basis of a Coastal Region

The Severn Estuary GIS Forum was established in 2004 and is one of a number of such initiatives facilitated by Coastal Partnerships around the UK (Geowise Ltd., 1998). Membership of the GIS Forum is approximately 35 individuals drawn from wide range of functions (such as nature conservation and contingency planning) including government agencies and local authority officers from the around the Severn Estuary. One of the major products of the forum has been in trying to establish a Memoranda of Understanding on data sharing, in order to overcome blocks to information transfer caused by intellectual property and copyright issues, and to help create bilateral agreements on data sharing. The meetings of the Fora have facilitated co-operation and good practice by providing training on issues such as metadata creation to standards through the use of GI tools; implementation of information policy and the implications of directives on Freedom of Information and the Arhus Convention on Public Participation in Enviromental Decisionmaking; and mapping information sources for the Estuary. Since there is a complex range of new responsibilities in these areas, it is especially helpful to consider how they should be fulfilled in the coastal context. For example, the variety of co-ordinate and reference systems which span the land-sea divide, and which are important in defining the projection of datasets, raise particular requirements for metadata records. The GIS Forum has raised awareness about the range of GI Tools in use around the estuary and has created a metadatabase of over 200 records of spatial data, but the ability of the forum to move beyond technical support and to create and maintain a truly inter-operable approach has proven challenging.

25.6 The Role of Coastal Partnerships

One solution for creating a truly inter-operable approach is to employ technical specialists and GIS officers to develop and harmonise IT tools specifically applied to coastal management for a given region. In the UK, about 60 non-statutory Coastal

Table 25.1 Information transfers for 4 UK coastal partnerships

	Use-type	Rank		FOCI	Tools	volume Per month	Percent
1	Finding out who does what, where and how to contact them	3	√√√	Roles, responsibilities, contacts	Legislation, regulation, contacts databases, ownership/tenure, policy statements, directories, jurisdictions	**570**	21.92
2	Reviewing the characteristics of a place	2	√√√	Descriptions, snapshots, infrastructure (including access), locations	Maps, weather forecasts, landscape assessments, photographs, inventories, GIS	**591**	22.73
3	Understanding the state/level of use/opinions on something	1	√√√	Physical objects or indicators Resources Judgements, values and attitudes Levels of significance Extent of an activity	Stakeholder assessments, public participation, surveillance, regulation, indicators	**689**	26.50
4	Analysing impacts of an activity	4	√√	Modelling, effects of development, sensitivity analysis,	Models, EIA, SIA, planning permission	**250**	9.62
5	Measuring trends	7	√	Ambient monitoring, Habitat morphology	Timeseries, monitoring, indicators	**64**	2.46
6	Researching a process	9	√	Modelling, predicting a process	Surveys, samples, studies	**45**	1.73
7	Getting educated about an issue	5	√√	Laws of nature, management techniques	Tutorials, displays, visualisations, presentations	**144**	5.54
8	Understanding a technical process	10	√	Management techniques, practices, remediation and restoration	Procedure maps, descriptions, manuals	**39**	1.50
9	Developing a scenario	6	√√	Planning, alternatives, projected activities, simulation, choosing a site from alternative locations, risk assessment	DSS, forecasts, GIS	**143**	5.50
10	Responding to an incident	11	√	Remediation options, risk assessment	Communications, manuals	**8**	0.31
11	Resolving a conflict	8	√	Sites of conflict, causes, legal status	Legislation, case law	**57**	2.19
12	Calculate the costs of an activity	12	√	Value of the environment, incentive measures	Land values, resource assessment, cost benefit analysis	**1**	0.04

Source: Stojanovic (2002).

Partnerships exist, and a small number of these have appointed GIS Officers or Information Officers (Stojanovic and Barker, 2008).

The Severn Estuary Partnership (SEP) acts to promote communication and co-operation on an estuary-wide basis and deals with 2 regional governments, 2 countryside agencies, several ports, 3 water companies, 3 Environment Agency regions and 14 local authorities (as well as a wide variety of stakeholders and local communities). Each of these organisations has its own data requirements and GI Tools. A recent (SEP) project has focused on developing a joint GIS for mapping conservation features and related activities, as an approach to unlock the inertia towards estuary-wide GIS. The spatial analysis will support the implementation of a single Scheme of Management on behalf of the Association of Severn Estuary Relevant Authorities, and provide analytical capabilities to assess the extent of significant or cumulative effects of plans, projects and activities on the nature conservation features. This approach is an example of an ICM Partnership setting strategic goals through information policy, and also undertaking the technical work to harmonise coastal GIS. The Severn Estuary Partnership has also recently collaborated with a range of other North West European partnerships in order to produce a guidance document on developing information systems in the coastal zone (Stojanovic, 2007).

Table 25.1 presents combined data on information transfers between partners in 4 UK ICM initiatives, listing the sums of information transfer per month for different functions (Stojanovic, 2002). The picture demonstrated is one of complexity of relevant data sources, both spatial and non-spatial, relying on both formal and informal information systems. Yet spatial analysis carried out within GIS is central to many of the types of assessment required. Embedding this data within the management process is not simply a technical exercise, but involves the development a genuine network of relationships amongst coastal practitioners. As coastal managers face increasing information overload, it is most important that they maintain a network of relationships that will enable them get a good understanding of the coast, its issues, problems and possible solutions. Sometimes, GIS developers have sought to create the virtual or technological networks before these community networks have been put in place. The methods described in this chapter are an attempt to systematically address this gap, and show how the support of ICM partnerships at regional and local levels can provide an opportunity to mobilise geographic information.

Acknowledgments The Author wishes to acknowledge Mr Steve Knowles and Mr Alun Rogers (Severn Estuary Partnership) and Alex Midlen (Essex Estuaries Initiative) for providing information and comment on GIS development.

References

Centre for Earth Observation (1995). Collect User Requirement from Experimental User Community on Coastal Zones. *ACR-CEO-ES*, ACRI Consortium, Cedex, France.

Checkland, P.B. and Holwell, S. (1998). *Information, Systems and Information Systems*. Chichester: Wiley.

Doody, J.P., Pamplin, C.F., Gilbert, C., and Bridge, L. (1998). European Union Demonstration Programme on Integrated Management of Coastal Zones, Thematic Study F- Information

Required for Coastal Zone Management. *3050/STU/9700186, Report* to the European Environment Agency, Copenhagen, Denmark.

Eurosion (2004). *Guidelines for implementing local information systems dedicated to coastal erosion management.* A Report to the Directorate General Environment, European Commission by National Institute for Coastal and Marine Management of the Netherlands (RIKZ), EUCC – The Coastal Union, IGN France International, Autonomous University of Barcelona (UAB), French Geological Survey (BRGM), French Institute of Environment (IFEN), and EADS Systems & Defence Electronics.

Felleman, J. (1997). *Deep Information. The Role of Information Policy in Environmental Sustainability.* US: Ablex.

Ferreira, M., Lucius, I., Longhorn, R., and Millard, K. (2007). Data Harmonisation: The Needs and Aspirations of the EU Coastal Management Community. A report by the MOTIIVE project.

Franklin, F. and Harries, J. (2002). *Delivering Integrated Marine Mapping for the UK,* Report of a workshop held at Church House, London, 11 September, 2002, CEFAS.

Geowise Ltd. (1998). Report of the Information, Research and GIS Topic Group. A Report to the Moray Firth Partnership.

Gunther, O. (1998). *Environmental Information Systems.* Springer: Berlin.

Harris, R. (1999). Earth Observation Data Policy and Europe. Proceedings of the 25th Annual Conference and Exhibition of the Remote Sensing Society. *Earth Observation: From Data to Information.* University of Wales, 8–10 September 1999. pp. 319–332.

Horton, F.W. (1988). *Information Resources Management.* Englewood Cliffs, US: Prentice Hall.

IACMST (1994). Marine Environmental Data. *IACMST Working Group Report No.2,* Inter-Agency Committee on Marine Science and Technology.

IACMST (1995). Operational Marine Environmental Network. *IACMST Working Group Report No.4,* Inter-Agency Committee on Marine Science and Technology.

IACMST (2005). Marine Data and Information Partnership- a proposal. Inter-Agency Committee on Marine Science and Technology.

Longhorn, R. (2005). Chapter 9. Coastal Spatial Data Infrastructure. In: D. Bartlett and J. Smith (Editors), *GIS for Coastal Zone Management.* CRC Press, Boca Raton.

Millard, K. (2004). Ten Questions on Ontologies. *Coast Map News* 7, p29.

Millard, K. and Sayers, P. (2000). Maximising the use and exchange of coastal data. A guide to best practice, *CIRIA Research Report.*

Smith System Engineering Ltd, HR Wallingford, and Southampton Oceanography Centre. (1996). *Annex to the final report* on a study of the information requirements for Coastal Zone Management. *CA212D021B/ 1.0,* Smith System Engineering.

Stojanovic, T.A. (2002). Coastal Strategies in England and Wales: Principles for Managing Information. *Unpublished PhD Dissertation,* Cardiff University, Cardiff.

Stojanovic, T.A. (2007). Guidelines for implementing Local Information Systems at the Coast. Cardiff University and COREPOINT, Cardiff. Available online at: http://corepoint.ucc.ie/Cpages/outputs.htm

Stojanovic, T. and Barker N. (2008). Improving Governance through local Coastal Partnerships in the UK. *Geographical Journal* 174 (4):344–360.

Part IV
Coastal Geomorphology

Chapter 26
GIS Tool for Coastal Morphodynamics Analysis

Joaquim Pais-Barbosa, Fernando Veloso-Gomes, and Francisco Taveira-Pinto

Abstract The aim of this work is to identify morphological and hydrodynamic patterns (e.g. coastal forms, wave breaking zones, rip currents patterns, submerged bars), based on several aerial images. Understanding the patterns, which occurred in the past, may help to increase the knowledge to predict future ones. To accomplish this, a GIS database, an analysis algorithm and a user-friendly GIS tool are currently under development. The GIS database will enable to store all available image survey sets, as well as to store the results of the performed visual and computed analysis. This former analysis consists in a Matlab 7.0 algorithm, which uses mainly the toolbox image processing. For non GIS specialists, a specific tool to visualize, overlap and search is being developed using Visual Basic 6.0, and MapObjects 2.3.

Keywords GIS · Morphodynamics · Morphology · Coastal zone

26.1 Study Area General Description

The study coastal area under analysis is located on the Portuguese northwest coast. The area is bounded by the city of Espinho in the north and to the south by Mira beach. For the purpose of this study, this coastal segment was divided in three stretches; Espinho-Furadouro; Furadouro-Torreira; and Costa Nova-Mira beach (Fig. 26.1).

The sea action is the most energetic and dynamic force acting in the study area and is, therefore, its main modelling agent. The littoral drift currents act mainly in a north-south direction. The area in study has significant coastal erosion problems, which are clearly visible in some areas.

J. Pais-Barbosa (✉)
Faculty of Engineering of Porto University – FEUP, Hydraulics and Water Resources Institute – IHRH, Porto, Portugal
e-mail: jlpb@fe.up.pt

D.R. Green (ed.), *Coastal and Marine Geospatial Technologies*, Coastal Systems and Continental Margins 13, DOI 10.1007/978-1-4020-9720-1_26,
© Springer Science+Business Media B.V. 2010

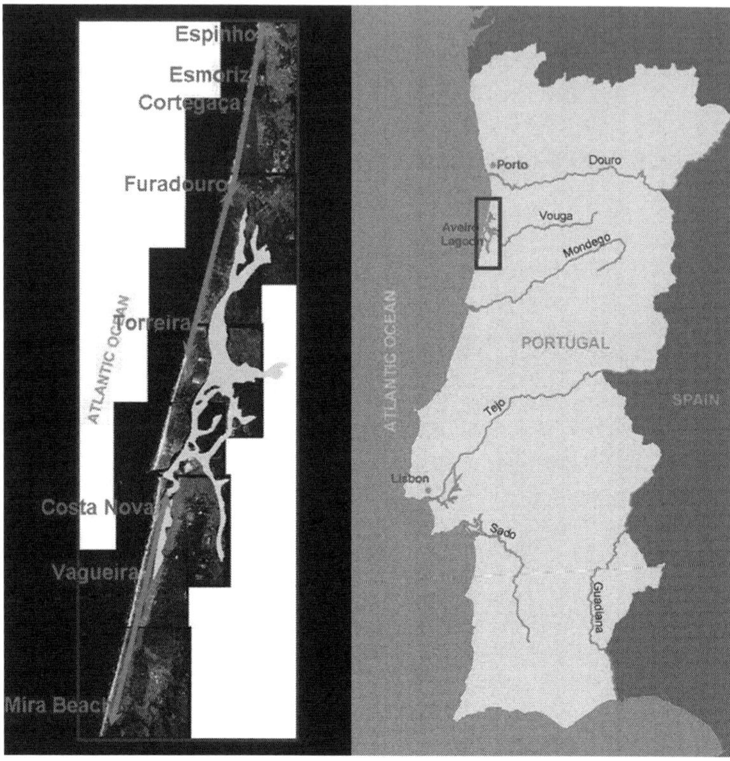

Fig. 26.1 Study area location (See also Plate 41 on Page 415 in Color Plate Section)

This coastal area is plain and has a general orientation NNE-SSW. It is mostly a sandy coast, formed by sandy beaches and sandy dunes with small elevations, with highly dynamic characteristics that change coastal forms constantly.

The wave climate is characterized by medium significant wave heights from 2 to 3 m, with periods ranging from 8 to 12 s and storm significant wave heights exceeding 8 m, with periods reaching 16–18 s. Wave direction exhibits the higher frequencies and intensities in the NW quadrant with 43.8% of occurrence, WNW with 28.0% and NNW with 21.2% (Anon. 1993, 2003).

The tide regime in the western Portuguese coast is of the semidiurnal type, with a period of approximately 12 h 25 m, reaching a range of up to 4.0 m for spring tides. According to the tide range classification suggested in Short, 1991, this shoreline can be classified as meso-tidal (2–4 m). Meso-tidal and macro-tidal (> 4 m) beach systems have been less studied than micro-tidal (0–2 m) ones.

With this research it is expected to improve the knowledge on the identification and typification of hydromorphologic patterns of breaking waves, sea bottom forms, currents and there relation with local wave and tide regime. In addition it should also improve the knowledge on coastal morphology patterns identification in regard to the wave direction, sea bottom forms, wave heights and periods and beach profile. This can lead to a better understanding of the phenomena occurring at the coast, namely coastal erosion.

Nevertheless, the wave-structure/land interactions are rather complex and thus of difficult characterization. This is not only due to the existence of dynamic action of non-linear interactions over different time scales, but also to the great difficulty to obtain long series of georeferenced data. Coastal morphodynamics is changing, either at a time scale of days (e.g. between tides) or of years (e.g. global climate change) (Pais-Barbosa *et al.*, 2003).

26.2 Morphodynamic Classification

Micro-tidal wave-dominated beach systems can be classified using three parameters: breaking wave height (H_b); wave period (T); and grain size (defined by sediment fall velocity, W_s). Gourlay (1968), was the first to combine these parameters into the dimensionless fall velocity by,

$$\Omega = \frac{H_b}{W_s T}$$

Using this dimensionless relation, Gourlay found that when $\Omega < 1$ laboratory beaches tended to be steep and stable; however when $\Omega > 1$–2 they eroded with sediments forming a terrace attached to the beach (Short, 1999).

According to Short (1999), Wright and Short (1984), have adapted Ω for natural beaches. Based both on beach experiments from representative beach types, and numerous observations, these authors found that when $\Omega < 1$ beaches tend to be reflective (steep, bar-less); on the contrary, when $\Omega > 6$, beaches tend to be dissipative (flat, multi-bared); between those ($\Omega = 2$–5) they classified beaches as intermediate (one to two bars) (Short, 1999). A comprehensive range of micro-tidal beach types are illustrated in this study.

In an attempt to classify the meso and macro-tidal beach systems, Short (1991) merged in the same group macro-tidal and meso-tidal beach systems, particularly those where tide range exceeds 3 m for average spring tides; and thus only referring to macro-tidal beaches. One major issue in the study of macro-tidal beaches is to find a common ground for comparisons; these situations do not occur only in high to low sea and swell environments, with varying grain size and sorting, but also where tides range from 3 to 15 m, concerned the previous classification.

According to some studies and field data from the coast of Australia, macro-tidal beaches were divided into three groups, based on gradient, topography, and relative sea-swell energy, (Short, 1991):

- Group 1 – Higher wave beach, planar, uniform slope
- Group 2 – Moderate wave beach, multi bar
- Group 3 – Low wave beach and tidal flat

The knowledge of the morphodynamics of macro-tidal beaches compared to micro-tidal beaches is very restricted, and based on a handful of experiments. There is a need to conduct experiments on these systems so as to improve our understanding of the morphodynamics of the three (or more) suggested group, as well as the critical thresholds which distinguish them from micro-tidal beaches and from tidal flats, (Short, 1991).

26.3 Database Development

26.3.1 Introduction

The coastal area in study has changed deeply over the last century; shoreline shapes and profiles have been constantly modified. An overview of the shoreline development along this coastal stretch can be found, in Veloso-Gomes *et al.* (2004), and in the EUROSION Project reports.

As previously referred, there is not much knowledge of the meso-tidal beaches characteristics. The study described in this chapter aims at improving the knowledge and understanding of meso-tidal morphology and hydrodynamics, focusing on the northwestern coast of Portugal. To accomplish this Geographical Information Technologies (GIT) will be used to develop a Geographical Information System (GIS) database and aerial imagery analysis tools.

Remotely sensed data and images have been widely applied over the years to coastal environments with different degrees of success, largely dependent upon the application. Aerial photography, appear as a great source of information, as a result of the high resolution and spatial detail, and it is a very important support for studies on small areas. In opposite the satellite images, until recently, offered a poor resolution, but with a great spectral and temporal resolution and area coverage (Green and King, 2002).

As mentioned above, the data is being stored in a GIS database, which incorporates several types of vector (topographic surveys) and raster (aerial images) data and alphanumeric information. In addition, the data resulting from the aerial imagery analysis, from both the visual and the GIT tools analysis, are also stored in the database.

With this combination of images, GIS and GIT analysis, it is expected to identify and analyse coastal forms, wave patterns, local currents as well as sea bottom characteristics (submerged bars). The forms, patterns and local currents will be correlated with the hydrodynamic conditions especially the wave regime (period, direction and height) and the tide stage. However, for some surveys the wave parameters are not available making the correlation more difficult.

Following is a brief resume of the database development, as well as the aerial images analysis, that have been carried out.

26.3.2 Data Acquisition and Integration

The data used in this study is diverse, from different sources, and in different supports (digital and hardcopy) (Table 26.1). Besides aerial imagery, vector data is also used, especially to support the georeferencing process.

A preliminary analysis of the data available, show the following major problems:

- Some aerial image surveys do not cover all the three stretches;
- Data with different or non existing coordinate systems;
- Images control points with errors in between 100 and 200 m;
- Parameters of flight and photographic camera are unknotted available or unknown.

The data was divided in two classes, whether it was georeferenced or not. The images in hardcopy format (1996 and 2001) were scanned; this process of scanning can introduce sources of error affecting the subsequent use of the imagery.

After the scanning process (1996 and 2001), images were georefenced using ArcInfo 7.2.1. The georeferencing process consisted in common points on images and on the photogrammetric surveys of 1996 and 2001. After the georeferencing process all data was compared with other georeferenced ones for validation. The average residual error was $\pm$ 3 m. Figure 26.2, shows the results achieved by the georeferencing process for the 1996 survey.

Table 26.1 Acquired data

Type of image	Format	Georeferenced	Scale	Year
Aerial	Digital	No	1:15000	1965
Aerial	Digital	No	1:15000	1967
Ortho	Digital	Yes	1:10000	1979
Ortho	Digital	Yes	1:10000	1988
Aerial	Digital	No	1:15000	1989
Ortho	Digital	Yes	1:10000	1991
Aerial	Digital	No	1:15000	1992
Ortho	Digital	Yes	1:40000	1995
Aerial	Hardcopy	No	1:8000	1996
Aerial	Hardcopy	No	1:8000	2001

26.3.3 Analysis and Results

Pais-Barbosa, *et al.* (2003) presented a first approach using GIS tool analysis, on the basis of one aerial image from 1995, 1996, 2001 and 2003. Some important considerations can be withdrawn from the preliminary results achieved.

This analysis, designated here by visual analysis, has encompassed all the three stretches in which the coastal segment in analysis was divided and the other

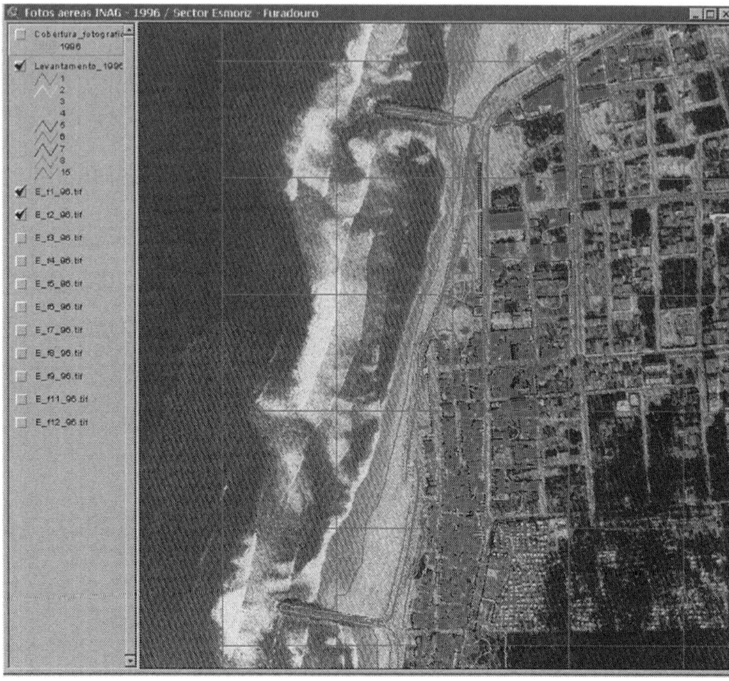

Fig. 26.2 1996 aerial survey after georeferencing (See also Plate 42 on Page 415 in Color Plate Section)

surveys. Such visual analysis consisted in the observation and digitalisation of coastline shapes in accordance with the run-up line, the cusps and mega-cusps, the dune cliffs, and the submerged bars noticed in the images (Fig. 26.3). To establish the correlation between such morphodynamic elements and the wave climate, the tide level and sediment movement patterns is one of the major goals of the present study.

Hence, patterns of longshore and rip currents, as well as morphological forms, were analysed and wave climate changes around them were identified. Whenever possible, the limit of sediment dispersion was also identified (Fig. 26.3).

In addition, a geometric analysis was done. It consisted in the measurement of the morphological forms in presence, the size of cusps and mega-cusps, the distance between the groups of cusps and mega-cusps, and of the length of rip current groups (Fig. 26.3).

The identified elements are correlated with the tide level and wave characteristics, as well as the relation between the length and the width of such elements (Fig. 26.3).

After the end of this visual analysis task, it will be possible to overlap, to compare and to make a spatial analysis of the computed data, in order to determine shoreline evolution, namely the evolution of the forms, beach and dune cliff erosion, submerged bar, as well as rip currents changes.

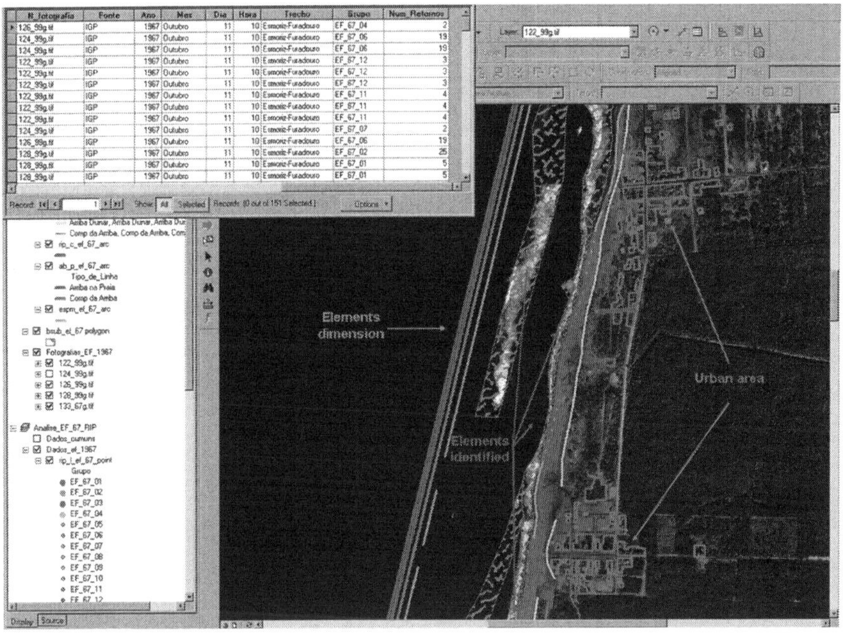

Fig. 26.3 Visual analysis of the 1965 survey (See also Plate 43 on Page 416 in Color Plate Section)

In parallel to the morphological analysis, it will also be analysed human development along the coast, focusing especially on recently built-up areas. It should be noted that all the urban areas of the three stretches in study are currently defended against wave action by groins and sea walls, but Torreira.

Another type of analysis was prepared through Matlab 7.0, using mainly the toolbox image processing. The main goal of such analysis is to compute wave crests, wave breaking areas and some coastal forms. Figure 26.4 summarises the structure of this analysis process.

To accomplish this purpose an existing Matlab algorithm was modified. The concept of this algorithm is based on Matlab filters able to analyse images. The following step is to identify and create a boundary in image where the values of colour change. The pixels with such value are converted to a matrix and transformed to another one of points with coordinates (x, y). Finally this last matrix is converted to a text file, which is then exported to be overlapped with the image in GIS database (Fig. 26.5).

From left to right, in Fig. 26.5 it can be observed: the original aerial image; the boundary points; the matrix points with x, y coordinates; and the overlapping with the original image stored in a GIS database. In the former image it is possible to verify that the wave crests and breaking area can be easily identified together with some of the bigger forms. None of the smaller forms were identified.

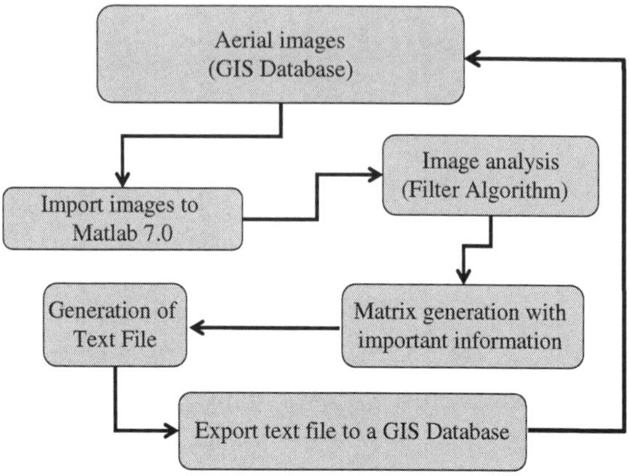

Fig. 26.4 Structure of Matlab analysis tool (See also Plate 44 on Page 416 in Color Plate Section)

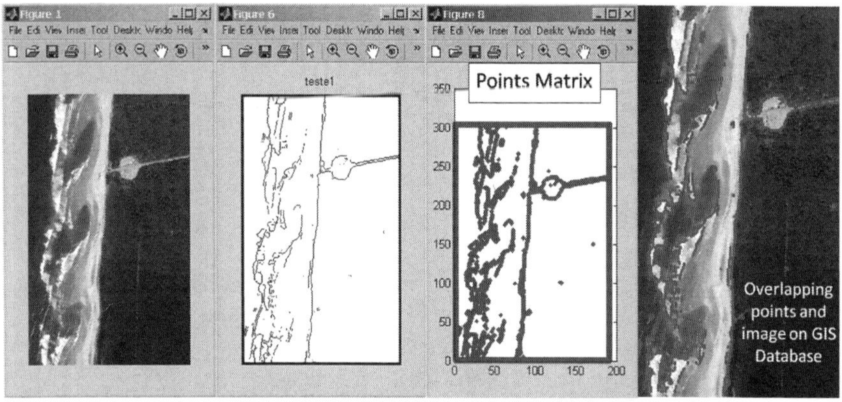

Fig. 26.5 Preliminary results of Matlab analysis and overlapping in GIS database (See also Plate 45 on Page 417 in Color Plate Section)

This tool is though still unable to process high resolution images; nevertheless, the algorithm is currently under modification in order to make it able of processing imagery analysis disregarding its resolution.

26.4 Discussion

GIT are very helpful to observe phenomena that occur in the surf zone, whilst enabling the analysis of the morphological forms based in aerial imagery.

When fully developed this GIS imagery database and analysis will be able to store comprehensive data from several years; and therefore making possible the overlap and comparison of the data resulting from the sets of image analysis, as well as the observation of coastal zone evolution, namely in matters of urban areas vulnerable to erosion hazards.

The Matlab tool, despite being unable to process images with high resolution has demonstrated that it could be very helpful on the identification of wave crests and wave breaking zones.

From the visual analysis, and following preliminary results presented in Pais-Barbosa, *et al.* (2003) rip currents and rip heads have been detected, which according to Short (1991) and Morang and Parson (2002) has never been reported in coasts with tidal ranges higher than 3 m. Cusps, mega-cusps, submerged bars, beach and dune erosion cliffs were also identified from this analysis.

Hydroforms in this Portuguese northwestern coastal segment seem very similar when compared with the classification presented by Wright and Short in 1984.

In parallel to the developments previously presented, a user friendly access tool is under development. This tool is being designed to be mainly used by non GIS specialists. The tool includes three modules and is being developed using Microsoft Visual Basic 6.0 and MapObjects 2.3; the first module is for visualization and data overlap, both vector and raster; the second one is for aerial imagery and all alphanumeric information visualization; and finally a search engine.

References

Anon. (1993). *Plano de Aproveitamento da Zona entre o Molhe Sul e a Ponte-Cais No 3*, Vol. 1, Estudos de Agitação no Porto de Leixões, IHRH Report.

Anon. (2003). Eurosion Project – *Guidelines for developing local information systems,* Vol 1, IHRH Report.

Green, D.R. and King, S.D. (2002). *Remote Sensing and Imagery for Coastal Environments*, URISA, United States.

Gourlay, M.R (1968). Beach and dune erosion tests. Delft Hydraulics Laboratory.

Morang, A. and Parson, L. (2002). Coastal Morphodynamics, in: Coastal Engineering Manual, Chapter IV-3, *US Army Corps of Engineers*, Washington.

Pais-Barbosa, J., Veloso-Gomes, F., and Taveira-Pinto, F., (2003). "Indicators of a GIS-based Aerial Images Analysis of the Portuguese Northwest Coastal Dynamics", *Proc. Fifth International Symposium on GIS and Computer Cartography for Coastal Zone Management, Integrating Information in Coastal Zone Management.*

Short, A.D. (1991). "Macro-meso Tidal Beach Morphodynamics – An Overview", *Journal of Coastal Research*, Vol 7, No 2, pp. 417–436.

Short, A.D. (1999). Wave-Dominated Beaches, in: Andrew Short, (ed), *Handbook of Beach and Shoreface Morphodynamics*, John Wiley & Sons, LTD, pp. 173–229.

Veloso Gomes, F., Taveira Pinto, F., das Neves, L., Barbosa, J.P., and Coelho, C. (2004). "Erosion risk levels in the NW Portuguese coast: The Douro mouth – Cape Mondego stretch", *Journal of Coastal Conservation*, Vol 10, pp. 43–52.

Wright, L.D., Short, A.D., (1984). Morphodynamic variability of surf zones and beaches: A synthesis. Marine Geology, Vol. 56, pp. 93–118.

Chapter 27
Paros Island (Cyclades, Aegean Sea) Coastal Zone: Natural Processes and Dynamics

N. Evelpidou, S.E. Poulos, and A. Vassilopoulos

Abstract The present contribution deals with the natural environment of Paros island coastal zone. More than 75% of the coastal zone consists of high and lower cliffs, whilst the relatively flat coastal plains are rare. The island receives a substantial amount of wave energy flux. The largest rocky and of high slopes coastal part is situated at the northeast and northwest part of the island being developed on hard (resistant to weathering processes) metamorphic rocks. Costal plains (slopes <2.5%) have been formed on Quaternary clastic formations. Furthermore, sandy and/or pebbly beach zones have been formed by the deposition of terrigenous (mainly) and marine sediments, along relatively sheltered from the wave activity parts of the coastline, i.e. the bays of Drios, Kephalos, Plastiras, Piso Livadi and Alikes. Palaeogeographic reconstruction with respect to sea-level rise within Holocene reveals that Paros island, at 10,000 yr BP, was much larger (some 1,025 km^2) combined with the islands of Antiparos, Naxos and several islets, while at 6,000 yr BP was only connected to Antiparos island covering 264.3 km^2. Finally, the predicted ca. 0.5 m of sea level rise for the year 2100, due to climatic change, will definitely threaten the socio-economy of the island as all touristic beaches will be deteriorated as well as several constructions along its coastline.

Keywords Paros Island · Coastal geomorphology · Tourist beaches · Sea-level change

27.1 Introduction

Coastal areas in Greece as well as islands have significant environmental values, which can be the basis for economic development and, consequently, of a high quality of life for the inhabitants. The 1/3 of the population of Greece lives no

N. Evelpidou (✉)
Department of Geography and Climatology, Faculty of Geology & Geoenvironment, University of Athens, Panepistimiopolis, Zografou 157 84, Athens, Greece
e-mail: evelpidou@geol.uoa.gr

D.R. Green (ed.), *Coastal and Marine Geospatial Technologies*, Coastal Systems and Continental Margins 13, DOI 10.1007/978-1-4020-9720-1_27,
© Springer Science+Business Media B.V. 2010

farther than 2 km from the coastline, whilst 85% of the population is established at a distance of <50 km from the coast. Greece has a great number of populated islands that host the majority of the touristic activities especially during the summer period (Poulos, 1998).

Common problems in coastal zone management have been recognized to be related to un-planned or miss-planned development, decline of traditional and environmental friendly socio-economic sectors, coastline retreat (beach erosion) and lack of appropriate transport networks, particularly on islands. It has also been adopted that a successful coastal zone management should be based upon a better understanding of the natural processes that are referred to both terrestrial and marine environment (E.C., 1999).

This study is focused on Paros island, as a representative "sample" of the Cyclades complex of islands and one of the most touristic island of Greece. It is also worth mentioned that the Cyclades are very small islands and therefore "coastal zone" refers not only to a narrow coastal strip but it encompass the whole island; this specific characteristic could be seen as an advantage when dealing with development issues. The aim of the present investigation is to provide all the appropriate information for the natural environment of its coastal zone that may form the base of the development of an integrated coastal zone management plan for this and other islands of the Aegean archipelago. Hence, geological and geomorphological conditions that dictate coastal geomorphology of Paros island are investigated in association with the existing climatic and coastal oceanographic characteristics. Special emphasis is placed upon the formation of the shore zone that hosts all touristic beaches, with respect to its recent (Holocene) geological evolution and the impact of future climatic changes (i.e. sea level rise).

27.2 Physical Setting

27.2.1 Geography

Paros island belongs to Cyclades complex of islands (Aegean Sea) located between the islands of Naxos and Sifnos being at a distance of ~90 nm from the port of Piraeus. The shape of the island is elipsoid and its surface covers an area of 196 km^2, being the third in size Cycladic island after Naxos (430 km^2) and Andros (374 km^2). The longest axis of Paros island is 22 km with a NE-SW direction, whilst its shorter axis, oriented NW-SE, has a length of 14 km. The length of its coastline accounts some 108 km.

27.2.2 Geology

The island of Paros belongs to Cyclades Plateau located at the back-arc basin of the Hellenic Arc-trench system; the later represents the dymanic tectonic boundary

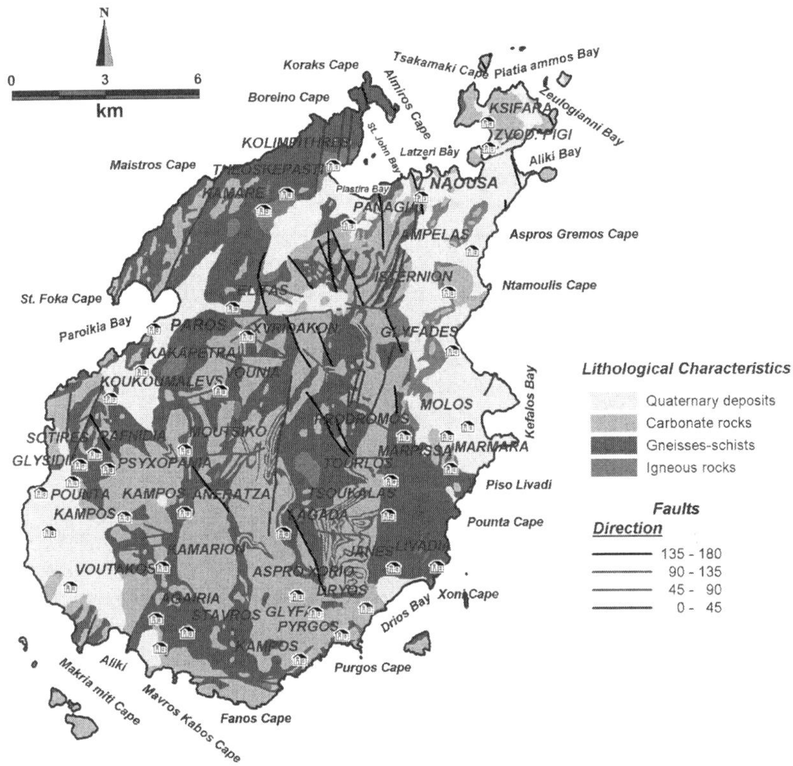

Fig. 27.1 Geological (i.e. lithology, fault lines) characteristics of Paros Island (See also Plate 46 on Page 417 in Color Plate Section)

between the southern limit of the micro-tectonic Aegean plate and the African Plate. Furthermore, the Cyclades Plateau belongs to the AttikoCycladic isotopic zone that consists mainly of metamorphic rocks; it is further divided to the northern and southern tectonic units with respect to the dominance of the blue-schist and green-schist metamorphic phase, respectively. In particular, Paros island belongs to the southern Cyclades geotectonic unit, dominated principally by metamorphic rocks: marbles, crystallic limestones and gneiss-schists plus some Neogene and Quaternary formations that cover only a small part of its coastal zone (Fig. 27.1).

The geological structure of the island resulted from the rise of the Miocenic granites, is an anticlinal dome of a N-S direction with a few internal thrusts. This tectonic evolution has resulted to the formation of isoclinic folds and thrusts with a general direction N-S being crossed in places by younger ones of a NE-SW and a E-W direction (Papanikolaou, 1977, 1980).

In terms of seismisity, Paros belongs to the central part of Cyclades plateau, which has been characterised as an "aseismic" area (Morelli et al., 1975) with only a few earthquake shocks (< 4 Richter) during the historical times (Papazachos, 1990).

Fig. 27.2 Geographical and morphological characteristics of Paros Island (See also Plate 47 on Page 418 in Color Plate Section)

27.2.3 Geomorphology

The largest part of Paros' coast is ragged and rocky with relatively steep slopes with sandy and/or pebbly beaches forming only a small part of the coast. The central part of the island is quite mountainous, with the highest peak of St. Ilias at an altitude of 771 m (Fig. 27.2). Coastal plains (slopes < 2 .5%) are those of Naousa, Marmaron, Dryos and Pounta. Water depths reach 100 m to the N and NE part of the island, but around the rest of the island the submarine relief rarely exceeds the 50 m in altitude.

Differential weathering induced by the variable lithology is one of the principal morphological factors; this enhanced by the sparse vegetation cover due to dry climatic conditions (Sabot, 1978, 1981; Evelpidou, 1997). Thus, high altitudes, observed in the central part of the island, are associated with the presence of marbles (primarily) and of schists (secondarily). In contrast, low relief areas (<100 m), usually coastal and quite flat, are related to clastic and schist formations.

The river drainage network presents an asymmetrical geographical distribution having relatively more elongated branches at the eastern and southern part of the island compared to those of the western and northern regions. Most of the river network has been developed on gneish-shists formations due to their higher erodibility and their extended dominance as they cover almost the 21.5% of the island. Furthermore, river network is controlled by the tectonic setting, having its various characteristics (e.g. branch type and length, abrupt chance branch course) in accordance to the existing tectonic "lines" i.e. faults, synclines (Evelpidou, 1997).

27.2.4 Climate

The climate is of the "Mediterranean" type with a dry and warm summer period and a mild winter season. Mean annual air temperature varies from 16.5 to 19.0°C, whilst at mountainous areas of the island temperature is lower by 2–3°C. On a monthly basis the hottest month is July (26.8°C) and the coldest is February (9.3°C). Mean annual precipitation level is 360 mm (Theocharatos & Tselepidaki, 1990) with December and January to present the highest values (~70 mm each one), whilst during the summer period precipitation is rather absent.

The wind field of Paros island as in the rest of the Cyclades is characterised by the persistence of the N-NW and secondarily by the W winds. The N-NW winds present a double fluctuation: a primary maximum during winter (December–February), with a secondary maximum during summer (July–August); the latter is also associated with the presence of the Etesians that persist for extended periods from May to September and often reaches gale force in strength (Metaxas, 1973). The monthly characteristic of the wind field in southern Cyclades is presented on Table 27.1.

Table 27.1 Monthly wind characteristics of the southern Cyclades Plateau, on the basis of measurements from the nearby Milos Island (after Rossiadou et al., 2002)

	Jan	Feb	Mar	Apr	May	Jun	Jul	Aug	Sep	Oct	Nov	Dec
WI	4.6	4.7	4.1	3.5	3.0	2.9	3.7	3.7	3.8	4.0	4.2	4.3
SW	11.8	11.0	8.1	4.6	2.5	2.8	6.7	6.3	7.1	8.3	8.8	10.0
S	3.2	2.5	1.9	0.5	0.1	0.1	0.3	0.2	0.6	1.8	1.6	2.5
C	1.3	1.2	1.6	2.0	3.1	3.4	1.5	1.7	1.9	1.9	1.8	1.6

Key: WI: mean monthly wind intensities in Beaufort; SW: mean monthly number of days with strong winds ($\geq$ 6 B); S: mean monthly number of days with storms ($\geq$ 8 B); and, C: mean monthly number of calm days.

27.2.5 Coastal Oceanography

The main factor affecting sea surface temperatures (SST) are the seasonal atmospheric conditions, which are reflected in the air temperature (AT) (Table 27.2). Monthly variations in SST and AT show that, on an annual basis, the sea temperature

Table 27.2 Monthly variation of sea surface temperature (SST) and atmospheric temperature (AT) in degrees Celcius (after Theocharatos and Tselepidaki, 1990)

	Jan	Feb	Mar	Apr	May	Jun	Jul	Aug	Sep	Oct	Nov	Dec
SST	16.4	16.1	16.0	16.9	19.3	20.3	22.0	23.1	22.4	20.7	18.9	17.4
AT	4.8	5.9	8.3	13.1	18.2	23.0	25.6	24.7	21.0	15.3	10.4	6.8

appears to lag in relation to that of the air, by one month with the annual maximum SST value occuring in August (23.1°C) and the minimum value in March (16.0°C).

Sea level variability in the Aegean Sea is mainly in response to storm surges, radiational tides and wind set-up; these may cause a temporal rise of the sea-level >0.5 m (HNHS, 2005). Astronomical tidal oscillations in the open Aegean Sea are generally semi-diurnal with amplitudes rarely exceeding the 12 cm (Tsimplis, 1994).

Wave regime in the region around the island of Paros is not available due to the lack of long-term measurements of wave characteristics (height, direction and period). Nevertheless, information for the region's wave climate originates from the Wind and Wave Atlas of the NE Mediterranean Sea (Athanasoulis and Skarsoulis, 1992). Thus, the offshore percentages of occurrence of different in amplitude wave heights in relation to the principal wind directions are given on Table 27.3. In Cyclades Plateau, in general, the offshore wave climate is characterised by the presence of waves approaching from W, NW and N directions associated with wave heights <1.5 m (>60%). The occurrence of waves higher than 5 m is small, <0.4%.

Table 27.3 Annual percentages of direction and height of wind-generated waves (according to prevailing winds) (after Athanasoulis and Skarsoulis, 1992)

	Direction								
	N	NE	E	SE	S	SW	W	NW	Calm
(%)	21.8	8.2	5.7	4.7	3.3	7.6	24.7	20.0	4.0
	Height (m)								
	0–0.5	0.5–1	1–1.5	1.5–2	2–2.5	2.5–3	3–4	4–5	>5
(%)	36.6	24.1	17.1	9.3	5.8	3.3	2.6	0.8	0.4

27.3 Methodology

Data collection concerning the coastal geomorphology was collected through fieldwork with the combined use of GPS and GIS. Especially for the creation of interpolation models for the study area morphology evolution, due to the sea level change, Vertical Mapper software and the Triangulation with Smoothing algorithm were used. Furthermore, in order to calculate the total area of the island, which is in danger because of possible sea level rise, topographical maps scale 1:5.000 of HAGS were used.

The calculations of wave's characteristics were based on the wind data taken from the "Wave and Wind Atlas" (Athanasoulis and Skarsoulis, 1992) and using the standard equations provided by CERC (1984) in the case of fetch limited offshore wave propagation. Specifically, the determination of mean significant wave height (s), period (), phase speed (C), and total energy flux () was achieved by the use of the equations: $H_s = 5.112 \times 10^{-4} W \cdot F^{0.5}$, $T = 6.238 \times 10^{-2}\,(W \cdot F)^{0.33}$, $C = gT/2\pi$ $Po = 0.125 \times \rho \times g \times Ug \times H_s^2$, where W is the wind force (m/s), F: the wave fetch (m), ρ: the water density (1025 kg/m^3), $g = 9.81$ m/s^2 and Ug: group velocity (for deep waters ≈ C/2).

27.4 Results and Discussion

27.4.1 Coastal Morphology

The geomorphological characteristics of the coastal zone vary spatially and are influenced by terrestrial and marine processes. The most important parameters related to terrestrial processes are those related to basic geological structure of the island and in particular its lithology, which in association with the climatic conditions governs the sediment flux towards the coastline. The latter has been enhanced by the abandonment of cultivated terraces, which has caused increased soil-erosion. This sediment eventually reaches the coastline and after interacting with near-shore marine processes forms the coastal zone of the island incorporating sea level fluctuations (especially during the Holocene).

Rocky coasts occupy >20% of the coastline (Fig. 27.3), often characterised by high slopes and composed of carbonates and gneiss-schists. They are observed in various places along the coast of the island but mostly at the NW, as well as, at the NE part. Coasts of medium (9.1%) and low altitude (22.2%) are observed in the remainder part of the island; the latter usually consist of Quaternary formations that are more vulnerable to erosion phenomena (Evelpidou et al., 2002). Coasts with medium slopes (<20%) represent the 28.9% (Fig. 27.3) and have been developed mostly in Quaternary deposits (Table 27.4).

The terrestrial part of the coastal zone (slopes <2.5%) represents only the 19% of the island, covering an area of 36.4 km^2. Its width is generally smaller than 476 m

Table 27.4 Coastal lithological and geomorhological characteristics of the Paros Island

Lithology	(%)	Coastal slopes	(%)	Type of coast	(%)
Quaternary	29.9	Low: <10%	31.5	Rocky coasts (mostly cliffs)	77.4
Carbonates	36.8	Medium:10 – 20%	21.5	Mixed coast (sand-rock)	4.7
Gneiss-schists	30.4	High: >20%	47.0	Beach zone	17.9
Igneous rocks	2.8			Sandy	6.8
				Sandy-Pebbly	9.7
				Pebbly	0.9

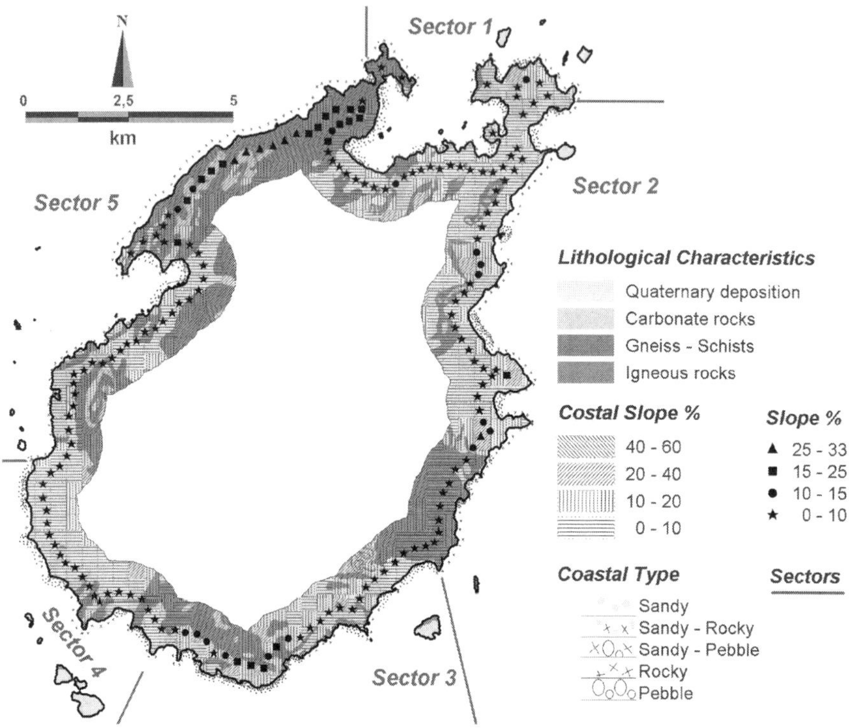

Fig. 27.3 Coastal slopes, coastal types and the sectors defined according to effective fetch of incoming wind-generated waves (See also Plate 48 on Page 419 in Color Plate Section)

with its highest values at the SW, E and NE part of the coast and consisting of Quaternary formations. Furthermore, shore zone and associated beaches in Paros Island are present in less of the 25% of its coastline length. They consist of mixed-sized material (sand, pebbles, and boulders) and have been developed in front of coastal slopes with low to medium slopes (<20%). Furthermore, purely sandy beaches, associated with low coastal slopes (<5%) occupy only a small part, less than 10% of the coastline (Fig. 27.3); the latter is related usually to the presence of the mouth of an ephemeral stream and having the characteristics of a "pocket" beach zone.

27.4.2 Coastal Oceanographic Conditions

Coastal oceanographic conditions could be divided to those related to nowadays-hydrodynamic conditions and those associated with recent (geologically) relative sea level changes. The first is basically governed by the incoming flux of wave energy (due to minimal tidal fluctuation), whilst the second is imposed by the

Table 27.5 Offshore wave power (Po), significant wave height (Hs) and water depth for shallow water propagating conditions (Ds) for the five sectors of Paros Island coastal zone

	Fetch (km)	Freq. (%)	U m (m/s)	Hs (m)	T (sec)	P (W/m)	P total (W/m)
Sector 1							
NW	52,5	26,3	7,3	0,95	4,69	1096,0	2492,3
N	37,5	22,9	8,4	0,96	4,45	925,0	(32,9%)
NE	150	4,3	6,2	1,34	6,25	471,2	
Sector 2							
NNE	150	4,3	6,2	1,34	6,25	235,6	287,3
E	10	3,7	5,5	0,30	2,42	7,8	(3,8%)
SSE	22,5	7,0	7,1	0,60	3,50	43,9	
Sector 3							
SE	75	7,0	7,1	1,10	5,23	436,9	677,5
S	30	7,8	6,9	0,68	3,82	134,6	(8,9%)
SW	37,5	6,0	6,3	0,67	3,95	106,0	
Sector 4							
S	30	7,8	6,9	0,68	3,82	67,3	285,7
SSE	75	7,0	7,1	1,10	5,23	218,4	(3,8%)
Sector 5							
W	52,5	15,5	6,2	0,78	4,38	402,2	3830,1
NW	90	26,3	7,3	1,25	5,61	2248,3	(50,6%)
N	45	22,9	8,4	1,05	4,73	1179,5	7572,9
							(100%)

upper Quaternary eustatic movements, assuming that neo-tectonic activity within Holocene is of limited importance.

The estimated flux of wave energy (wave power, P) approaching the coastal zone with the related average offshore significant wave heights (Hs) and periods (T) for the five sectors of Paros' coastline are presented on Table 27.5; these five sectors (see Fig. 27.3) have been defined according to coastline morphometry and associated wave fetch distances that as it is dictated by the geographical location of the surrounding islands.

The highest amount of wave power (>3.800 W/m) is received by Sector 5 (NW part) when the Sectors 2 and 4 receive the smallest amount of wave power (<300 W/m) as they are partially sheltered by the nearby islands of Naxos and Antiparos, respectively. Sector 1 (NE part) undergoes relatively high energy flux (>2400 W/m) due to the prevailing northerly winds. Finally, Sector 3 (E part) receives relatively low influx of wave energy (~680 W/m) as it is protected partially by the Naxos Island.

On the basis of the above, the formation of beach (shore) zones although is primarily attributed to the coastal morphology (including the presence of ephemeral streams) and lithology seems to be favored in these coastal sectors that are less exposed to wave activity, i.e. Sectors 2 and 4.

27.4.3 Palaeogeographic Evolution of the Coastal Zone

The overall morphometry of the coasts and the coastal bottom topography were shaped mainly during the upper Quaternary when sea level had gradually reached its present stage with tectonism to be of limited importance. Thus, according to Lambeck (1996) the relative sea level stage for the Cyclades plateau was lower that its present level by: –124 m (18,000 yr BP), –54 m (10,000 yr BP), –6 m (6000 yr BP) and –1.5 m (2000 yr BP). The values referred to the Late Holocene (past 6000 years) are in a good agreement with those introduced by Poulos et al (2008) in the case of the Attico-Cycladic massif.

On the basis of the above, at 10,000 yr BP Paros Island was connected with the islands of Antiparos, Naxos (Fig. 27.4a) and with many other islets covering an area of some 1,025 km^2 and having a coastline length of 206.5 km. Subsequently, at 6,000 yr BP, when sea level was approximately 6 m lower than nowadays (Fig. 27.4b) Paros was combined with Antiparos island and some other islets, covering an area of 264.3 km^2 and having a coastline length of 200.4 km.

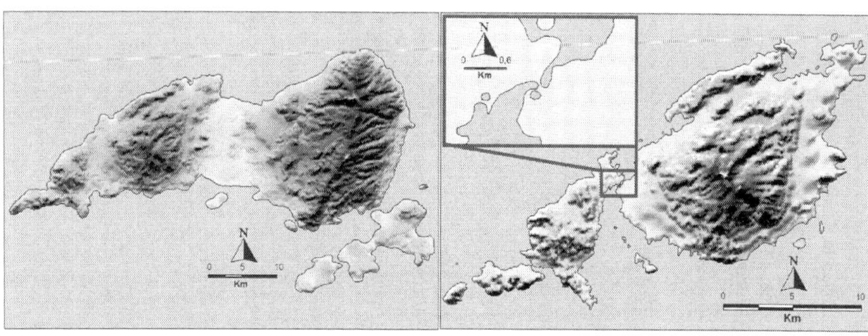

Fig. 27.4 The morphology of Paros Island at (**a**) 10,000 yr BP, when sea level was ca. 54 m lower than its present stage (*left*) and (**b**) 6,000 yr BP, when sea level was ca. 6 m lower than its present stage (*right*) (See also Plate 49 on Page 419 in Color Plate Section)

27.4.4 Coastal Zone Vulnerability to Future Sea-Level Changes

Coastal low-lands are obviously susceptible to any sea level change. On a global scale sea level over the last 100 years has been rising with a rhythm of 1–2 mm/yr and it is expected to continue rising in the next century due to thermal expansion of the oceans and the ice melt caused by global warming (Church et al., 2001; IPCC, 2007). Similarly, altimetric measurements after the mid-90s suggested a rapid rising of sea level in the eastern Mediterranean of the order of 1.8 mm/yr associated with sea surface temperature rise (Cazenave et al., 2001). According to the most recent IPCC (2007) report the sea level is expected to rise by the year 2100 from as much as 18 cm up to 68 cm (moderate estimate: 38 cm).

According therefore to the concept of a future rise of sea level and knowing that the 36.4 km^2 of Paros' coastal land have a slope of <2.5%, a rise of 0.5 m will flood approximately the 1/5 of this low-lying coastal area (some 7.3 km^2). Furthermore, this coastline retreat will affect primarily the east and south coast of Paros Island where most of the touristic beach zones will be reduced (if not disappear) in size, whilst other establishments along the coastline such as ports and marinas will be threatened also.

27.5 Conclusions

The largest part of the Paros' coast is rocky and of high slopes, especially at the northeast and northwest part of the island. The morphometry of the coastal zone is governed primarily by the overall geological evolution of the island incorporating the upper Quaternary eustatism and recent wave regime.

Palaeogeographic reconstruction with respect to sea-level rise within Holocene reveals that Paros island, at 10,000 yr BP, was much larger (some 1,025 km^2) combined with the islands of Antiparos, Naxos and several islets, while at 6,000 yr BP was only connected to Antiparos island covering 264.3 km^2.

Coastal morphology, is characterised by the presence of steep coastal cliffs being developed on hard (resistant to weathering processes) metamorphic rocks and coastal plains (slopes < 2.5%) formed on Quaternary clastic formations. Furthermore, sandy and/or pebbly beach zones have been formed by the deposition of terrigenous (mainly) and marine sediments, along relatively sheltered from the wave activity parts of the coastline, i.e. the bays of Drios, Kephalos, Plastiras, Piso Livadi and Alikes.

The predicted ~0,5 m of sea level rise for the year 2100, due to climatic change, will definitely threaten the socio-economy of the island as all touristic beaches will be deteriorated as well as several constructions along its coastline.

References

Athanasoulis G.A. and Skarsoulis E.K. (1992), Wind and Wave Atlas of north-east Mediterranean Sea, *Laboratory of Nautical and Marine Hydrodynamics*, NTUA, Athens.

Cazenave A., Cabanes C., Dominh K. and Mangiarotti S. (2001), Recent sea level changes in the Mediterranean Sea revealed by TOPEX/POSEIDON satellite altimetry, *Geophysical Research Letters*, 28(8), pp.1607–1610.

CERC (Coastal Engineering Research Centre), (1984), Coastal Engineering Manual, *US Army Corps of Engineers*, Washington DC 21314.

Church J.A., Gregory J.M., Huybrechts P., Kuhn M., Lambeck K. and Nhuan M.T. (2001), In: Qin D. and Woodworth P.L., Chapter 11 of the *Intergovernmental Panel on Climate Change Third Assessment Report*, Cambridge University Press.

E.C. (European Communities) (1999), Towards a European Integrated Coastal Zone Management (ICZM) Strategy: General Principles and policy Options. Luxemberg, Office for Official publications of the European Communities pp. 31.

Evelpidou, N. (1997), Geological and Geomorphological observations in Paros island (Cyclades) using Photointerpration and GIS techniques, *Master Thesis,* University of Athens, Geology Department, pp. 181.

Evelpidou, N., Poulos, S. and Vassilopoulos, A. (2002), Geomorphological and Hydrological setting of the touristic beaches of the Paros island, *Proceedings of the 1st Conference 'Oceanographical aspects of a sustainable Mediterranean', Athens, September 27–29,* in CD.

HNHS (Hellenic Navy Hydrographic Service), (2005), Sea level in gauged stations (Tides of Hellenic Ports, *Hellenic Navy Hydrographic Service*, Athens, 54 pp.

IPCC (Intergovernmental Panel on Climate Change) (2007), The physical science basis, Contribution of Working Group I to the fourth assessment report of the Intergovernmental Panel on Climate Change: *Cambridge University Press*, Cambridge, United Kingdom and New York, N.Y., USA.

Lambeck, K. (1996), Sea-level change and shoreline evolution in Aegean Greece since Upper Palaeolithic time, *Antiquity*, 70, pp. 588–611.

Metaxas, D.A., (1973), Air-Sea interaction in the Greek Seas and resulted Etesian Wind characteristics, University of Ioannina, Technical Report No 5, 32 pp.

Morelli, C., Pisani, M., Gantar, C., (1975), Geophysical studies in the Aegean Sea and in the Eastern Mediterranean, *Bolletino Geofisical Teoritica & Applicata,* 18(6), 127–167.

Papanikolaou, D. (1977), On the structural geology and tectonics of Paros island (Aegean Sea), *AGPH,* 28, pp. 450–464.

Papanikolaou, D. (1980), Contribution to the geology of Aegean Sea: The island of Paros, *AGPH,* 30/1, pp. 65–96.

Papazachos, B.C., (1990), Seismisity of the Aegean and surrounding area, *Tectonophysics*, 178, 287–308.

Poulos S.E. (1998), Coastal Zone: natural environment, socio-economic activities and human impact, *Geotechnical Scientific Issues,* 9(3), pp. 104–109.

Poulos S.E, Ghionis G. and Maroukian H., (2008), Sea-level rise trends in the Attico-Cycladic region (Aegean Sea) during the last 5000 years, *Journal of Geomorphology* (in press; doi 10.10.16/jgeomorph 2007.05.022).

Rossiadou K., Makroyiannis T., Balafoutis C. and Floka E. (2002), Geographical distribution of the winds over Aegean. *Proceeedings of the 6th Pan-Hellenic geographical Conference*, I, pp. 323–330 (in Greek).

Sabot, V. (1978), La geomorphologie et la geologie du Quaternaire de l'ile de Naxos, Cyclades-Greece, *Thesis*, Brussel.

Sabot, V. (1981), Variations dun niveau marin pendant le Quaternaire dans la region des Cyclades, refletees par la morphologie du fond de la mer entre les iles Naxos et Paros, Commum C.O.S., Bruxelles.

Theocharatos G.A. and Tselepidaki I.G. (1990), The temperature of the sea-surface in the region of the Aegean and its relation to air temperature, *Theoretical and applied climatology*, 42, pp. 117–119.

Tsimplis M. (1994), Tidal oscillations in the Aegean and Ionian seas, *Estuarine Coastal and Shelf science*, 39, pp. 201–208.

Chapter 28
Palaeogeographic Evolution of the Cyclades Islands (Greece) During the Holocene

K. Gaki-Papanastassiou, N. Evelpidou, H. Maroukian, and A. Vassilopoulos

Abstract The Cycladic islands are located in the central Aegean Sea (Greece) forming a partly submerged plateau separated into two parts: the eastern shallower one (Andros, Tinos, Mykonos, Naxos, Paros, Syros, Ios, Sikinos, Folegandros) which formed one big island (6.978 km^2) at the end of the last glacial period; the western islands (Kea, Kythnos, Serifos, Sifnos, Milos) which remained separated during the same period. The eastern islands constitute an erosional plateau which is the end product of a Neogene palaeosurface that was partially submerged due to thinning of the crust during the Quaternary. The presence of numerous Neolithic sites both on land and submerged indicates the existence of an advanced civilization in the area for thousands of years. The location of the lost Atlantis could be found in this area probably between Naxos, Paros and Antiparos.

28.1 Introduction

The Cycladic islands are located in the central Aegean Sea (Greece), forming the so called Cycladic plateau (Fig. 28.1), having a relatively low seismicity area.

There are about 20 large islands, the largest one being Naxos (430 km^2) followed by Andros (380 km^2) and Tinos (196,5 km^2), all of them located in the eastern part of the plateau. There are four islands aligned along the western end (Kea, Kythnos, Serifos, Sifnos) which are much smaller. Between these two groups of islands are Syros, Paros, Ios, and Mykonos. Around the southern edge of the plateau are located the three islands of Milos, Santorini and Amorgos which are peripheral extensions of the plateau.

K. Gaki-Papanastassiou (✉)
Department of Geography and Climatology, Faculty of Geology and Geoenvironment, University of Athens, Panepistimiopolis, Zografou 157 84, Athens, Greece
e-mail: gaki@geol.uoa.gr

D.R. Green (ed.), *Coastal and Marine Geospatial Technologies*, Coastal Systems and Continental Margins 13, DOI 10.1007/978-1-4020-9720-1_28,

© Springer Science+Business Media B.V. 2010

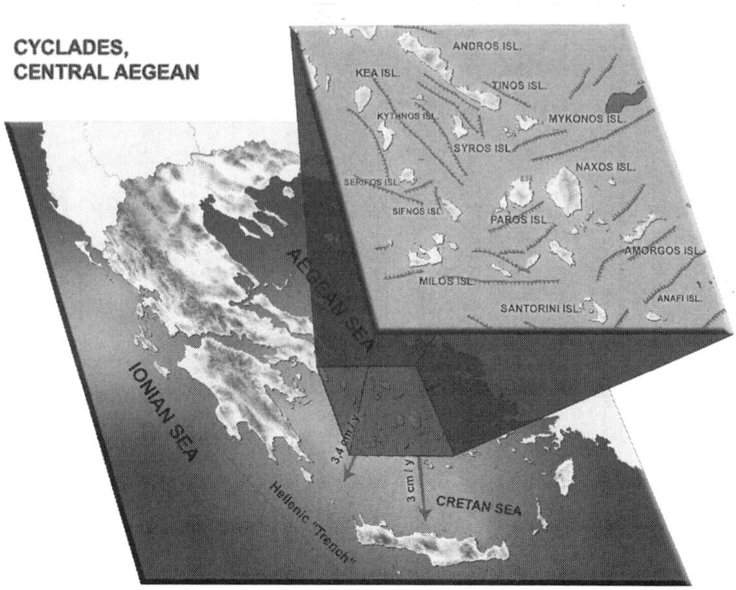

Fig. 28.1 The study area situated in the Central Aegean, showing the rates and direction of motion of Cyclades during the Quaternary and the major submarine fault zones (See also Plate 50 on Page 420 in Color Plate Section)

The largest islands of Naxos, Andros, Tinos, have the highest elevations reaching 1.001, 995 and 657 m respectively. Bathymetrically, although the relief of the plateau is gentle, there is a general slope towards the west from about −100 m down to nearly −200 m. The aim of this study is to present the evolution of the Cyclades area during the Holocene. Particular attention is given in the presentation of the morphological differentiation of the plateau and to its development especially since the last glacial period.

28.2 Methodology

This study derived data from existing maps, bibliography, aerial photos, satellite images and field work in the area of Cyclades Islands. The used maps were: a) geological b) topographic (scale 1:50.000) and c) the bathymetric of the south Aegean (1:50.000 and 1:150.000). The maps were georeferenced and then on screen digitization took place, in order to create information layers concerning the drainage network, the geology and the topography (contour lines, altitudes and depth points, isobaths) of the area.

The information layers that contained the contour lines, the isobaths, the altitudinal and depth points were analyzed with the use of Vertical Mapper software in order to create a digital elevation model (DEM). The algorithms that were used for the creation of the DEM were: the algorithm "Triangulation with Smoothing" for the Altitude Elevation Model, where the source data were dense, and the algorithm "Natural Neighbour" for the Depth Elevation Model where the source data were sparse. The two DEMs were united into one through Vertical Mapper software. The cell size of the DEM's grid was set to 200 m × 200 m.

Using the DEM's basic characteristic (each grid cell corresponds to a specific altitude) we created the isobath line of −120 m and thus estimated the shape of the coastline 10.000 years ago. The next step was to create two new DEMs, one for the slope per cent values and one for the aspect values of the area.

The final DEMs (relief, relief over 10.000 years, slope, aspect) were used within GIS Mapinfo for statistical and geographical analysis in the area. Statistically, we estimated the mean depth and height values of specific areas of the case study. For the geographical analysis we also used the information layers that concerned the drainage network so as to estimate the geographical distribution of possible palaeo-valleys in the area. The results of this study appear in maps, also prepared with the use of the DEMs and information layers, through GIS Mapinfo software.

28.3 Geotectonic Setting

The geology of the Cycladic area consists mainly of metamorphic rocks such as mica schists, marbles, gneisses, amphibolites, glaucophane schists and plutonic rocks (Papanikolaou et al., 1981, Papanikolaou, 1987). The main metamorphic events are of Tertiary age (Eocene, Lower Miocene) as documented by radiometric studies (Marakis, 1972, Altherr et al., 1977, Andrienssen, et al., 1979). The structures in the metamorphic rocks are dominated by isoclinal folding, thrusting and refolding during Eocene – Oligocene.

The recent structural history of the Cycladic complex starts with the alpine orogeny during the Eocene when a period of compressional tectonism was dominant. A group of rocks was metamorphosed under high pressures and low temperatures and became known as blueschists. The compressional period changed to tensional during Oligocene or Miocene times in the form of shallow normal faults. During the Miocene the Central Aegean was a very shallow domain with extensive emerged regions and small elongated basins, as a consequence of prolonged intense compression (Angelier, 1976, 1979, Mercier et al., 1976, 1979) which resulted in the formation of steep graben faults that rapidly uplifted the blueschists. In this period lower pressures metamorphosed a new group of rocks which formed the greenschists. There followed the intrusion of plutonic rocks (granites) through the old faults which reached the surface of the Cycladic mass after the faulting has ceased. About 5 million years ago, the south Aegean volcanic arc was formed and extended from the Corinth area (Sousaki) to the island of Nissiros (Dodecanese islands). In the area of Cyclades volcanism occurred mainly in the islands of Milos and Thera. The crust in the Cycladic area is rather thinner than normal. The Cycladic plateau which is generally less than 200 meters deep lies mostly underwater because of the thinner crust. The thinning is a result of the widespread Neogene and Holocene spreading of the crust of this area as well as the westward extrusion of the Anatolia block. This area is presumably under an extensional tectonic regime behind the modern volcanic arc at the centre of the Aegean plate and possesses a relatively thin continental crust of about 28–30 km. Recent studies have found that the crustal thickness is 25 km underneath the Cyclades and less (22–23 km) southwest of the islands. This is owed to two factors, the gravitational collapse of the Aegean crust due to the southward retreat of subduction front during the Cenozoic and westwards extrusion of the Anatolian block in the Aegean during the Neogene (Tirel et al., 2004). The relative motion of the Cyclades during the Holocene is towards the south and south-west with a rate of about 3 cm per year (Peterek and Schwarze, 2004) (Fig. 28.1).

The general trend of the major submarine fault tectonism is from east to west, exhibiting a curved shape and coinciding with the volcanic and back arc Cycladic area. In recent years several major offshore faults have been identified suggesting the presence of tectonic depressions with horsts (Lykousis et al., 1995, Doutsos and Kokkalas, 2001, Tirel et al., 2004) (Fig. 28.1).

28.4 Terrestrial and Submarine Geomorphology of the Cyclades

The Cyclades are affected by folding in different periods (Eocene – Miocene) which resulted in broad asymmetric anticlines and synclines and together with ensuing fracturing formed parallel horsts like Andros, Tinos, Mykonos, and Kea, Kythnos, Serifos grabens, the latter being underwater today.

Riedl (2004) maintains that the Cycladic plateau originated only in a few cases by fault- block dynamics. In most cases he recognized piedmont benchlands originating by phases of tetonic stability. In Greece, there exist peneplains with different ages and at different altitudes where the older peneplains are situated higher than the younger ones. Palaeopedologic investigations on the peneplains showed deeply weathered mantle with high contents of kaolinite and hematite. In general, the geomorphological assemblage of Greek peneplains is derived from a relict complex formed by strong chemical weathering of the bedrocks and sheetflood erosion. The relict Neogene complex, destroyed recently by linear erosion (fluvial), indicates a tropical or subtropical summer humid paleoclimate at the time of its development.

The morphology of the Cycladic plateau is inherited from Miocene times when climatic conditions where different (much warmer and more humid) and the extent of this area was more widespread as the Aegean realm at that time included much more extensive land than today (Riedl, 2004). There followed the Holocene fault tectonism which broke the Aegean area into smaller parts.

The geomorphology of the islands presents some characteristic features. It is believed that they form an undulating plateau which consists of high elevations in the east and dips towards the west where the smaller and lower islands are located. Most of this plateau lies underwater down to a depth of about 100 meters but is separated by the western islands (Kythnos – Serifos horst and Kea) by a deeper elongated N-S running tectonic depression. It would thus be better to refer to the Cycladic plateau only to the islands which are located east of this depression. The island of Kea seems to be an extension of Attika – Eubea plateau.

The Cycladic plateau is surrounded by steeper slopes at the edge of the 200 m isobath which coincides well with the continental slope. However, there are some steep slopes formed by fault tectonism like those south of Amorgos island. Additionally, steep submarine slopes are observed around the volcanic islands such as Thira.

The stability of the Cycladic plateau during the Holocene is proven by the complete lack of morphological coastal features indicative of uplift such as marine terraces or benches, beachrocks, marine notches, raised aeolinites and Plio-Pliostocene deposits. We should emphasise the complete absence of raised Holocene coastal deposits. On the contrary, we have submerged beachrocks down to a depth of about 4 m at Dilos (Fouache and Dalongeville, 2003), Syros, Andros, Tinos and Naxos (Fig. 28.2a). Another significant evidence of submergence is coastal archaeological remains like those at Kea (Karthea – drowned quay), Tinos (Kionia – submerged walls), Kythnos and Naxos (Fig. 28.2b). Flemming and Webb (1986) mentions three submerged sites in Paros, one at a depth of –5 m having an age of 7.000 years and two at depths of – 3 having an age of 2.400 years. In the city of Naxos, next to the port (Grotta site) a submerged Mycenaean coastal road has been discovered next to an eroding Mycenaean wall at sea level (Fig. 28.2c). All the previously mentioned submerged features are only at a scale of a few meters and are almost entirely due to eustatic sea level changes.

The morphotectonic – palaeogeographic evolution of the Cyclades area, is not affected only by the short term tectonism of the area but it is more complicated. Located at the back arc area of the active Aegean subduction it is influenced by the existence of the volcanic arc, the westward extrusion of the Anatolia block in the Aegean, as well as the long term evolution of the Aegean arc which has resulted in the thinning of the Cycladic massif, being an Eogene erosional palaosurface with residual uplands in the form of inselbergs. During the Holocene, this plateau dipped

Fig. 28.2 From *left* to *right*: (**a**) Eroding beachrock in Syros. (**b**) Submerged Roman fishtanks in Naxos. (**c**) Eroding Myceanean wall at sea level in the city of Naxos (See also Plate 51 on Page 420 in Color Plate Section)

to the west, submerged a few tens of meters leaving the higher elevations in the form of islands.

The peripheral islands of Kea, Kythnos, Serifos and Sifnos with Amorgos, are most probably formed by Late Pleistocene-Holocene fault tectonism.

28.5 Geoarchaeology of the Cyclades During the Holocene; Does the Underwater Morphology Provide Clues for the Lost Antlantis?

During the Holocene, on the front of the Aegean Arc local and regional tectonic movements are dominant, while in the back arc basin where the Cyclades are located, there is a continuous submersion, related to the hydro-isostatic component amplified by a slow subsidence on a large scale.

Most of the Holocene period fault tectonism seems to be the important shaping factor of the central Aegean area. However, in the Late Holocene eustatism starts playing a more dominant role in the shaping of the Cyclades. It is worth mentioning that tectonic movements take hundreds of thousands of years to change the landscape of an area, but recurring vertical eustatic changes last only tens of thousands of years and usually are morphologically more spectacular.

The morphological picture at the end of the last glacial period, 18–20.000 years ago, was completely different from the present situation (Fig. 28.3). Sea level drop affected only part of the so called Cycladic plateau which is believed to extend down to a depth of 200 m and includes all the previously mentioned islands except Milos, Santorini (Thera) and Amorgos which seem to be peripheral to Cyclades.

In the case of the first two islands they belong to the volcanic arc which was active from Pliocene times until today. Amorgos, being outside the Cycladic massif has evolved differently and is affected by fault tectonism mainly in the Late Pleistocene – Holocene.

Today the underwater part of the previously large island (5.282 km^2) has a mean depth of –75,72 m while the islands have an area of 1.696 km^2 with a mean height of 206,37 m with the highest elevation at Naxos (1.001 m). On the other hand, the lowest depression (–120 m) in the underwater plateau lies south of Mykonos.

It is very possible that the famous ancient Atlantis was one of the flourishing city states on the large Cycladic island that was drowned following the rapid sea level rise between 18.000 and 7.000 years ago. The story of Atlantis is based on the ancient scripts of Plato. A lot has been written about whether this legend is true or not and many probable sites have been proposed for the location of Ancient Atlantis. The disappearance of Atlantis may not be owed to tectonic reasons (sudden submergence) but to eustatic ones (marine transgression). If we postulate that Atlantis was a city-state flourishing around 10.000–9.000 years B.P., that it was located in the central Aegean sea, and that the "old" Cycladic island was diminishing quickly due to the rapid sea level rise, it becomes obvious that many coastal Neolithic settlements were drowned by the sea. At that time, the remaining land was between the islands of Naxos, Paros, Antiparos and Heraklia in the south. It is worth mentioning

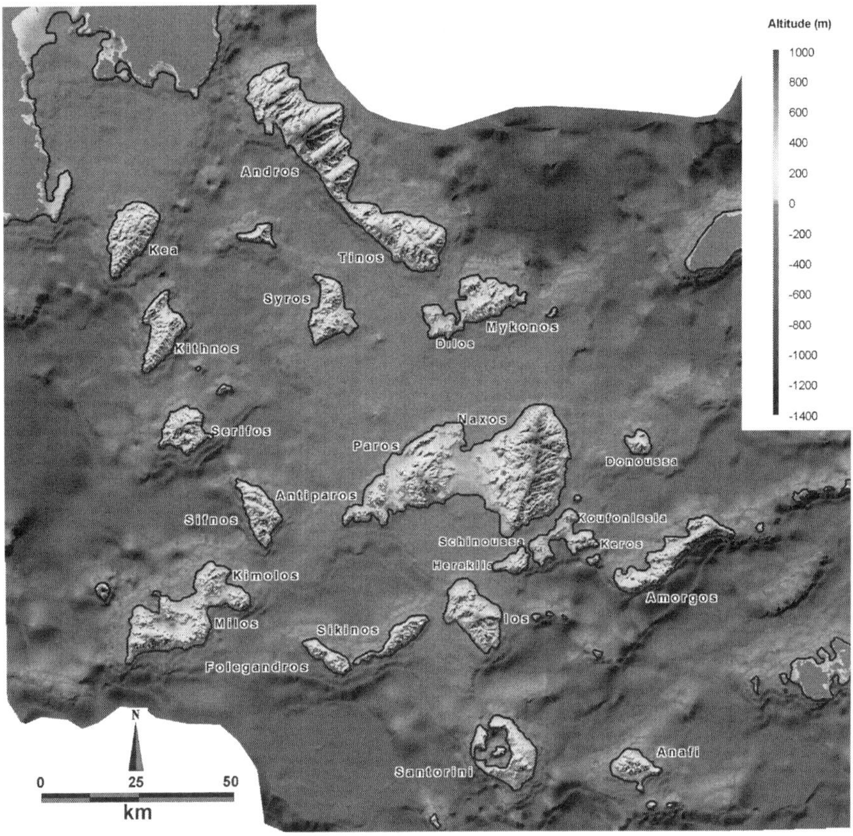

Fig. 28.3 The palaeocoastline of Cyclades 10.000 years ago indicated by the 40 m isobath

that the oldest Neolithic settlement found in the Cyclades is located on the isle of Saliagos (7300–6700 BP) between Paros and Antiparos which comprised one land.

At that time a large number of the eastern islands, (Andros, Tinos, Mykonos, Siros, Paros, Antiparos, Naxos, Ios, Folegandros, Sikinos and some smaller islands), had formed one large island when sea level had dropped 120 m, having an area of 6.978 km^2.

The presence of numerous statuettes on several Cycladic islands dating back to 5200 BP indicates the existence of an advanced civilization known as ProtoCycladic which could have been derived from the older civilization of Atlantis.

References

Altherr R, Keller J, Harre, Hohndorf A, Kreuzer H, Lenz H, Raschka H, Wendt I and Wagner G (1977) Miocene metamorphism and related plutonism within the Attic-Cycladic crystalline complex. VI Coll. Geol. Aegean Region, vol. I, pp. 345.

Andrienssen P, Boelrijk N, Hebeda E, Priem H, Vermurmen E and Verhure R (1979) Dating the events of metamorphism and granitic magmatism in the Alpine orogen of Naxos (Cylades, Greece). Contib. Mineral. Petrol., vol. 69, pp. 215–225.

Angelier J (1976) Sur l'alternance mio-plio-quaternaire de mouvements extensifs et compressifs en Egee orientale: L'ile de Samos (Grece). C.R. Acad. Sci., Paris, Ser. D., vol. 283, pp. 463–466.

Angelier J (1979) Neotectonique de l'arc Egeen. These Doct., Univ. Paris VI.

Doutsos Th and Kokkalas S (2001) Stress and deformation patterns in the Aegean region. J. Structural Geology, vol. 23, pp. 455–472.

Flemming NC and Webb CO (1986) Tectonic and Eustatic Coastal changes during the last 10.000 years derived from archaeological data, Z. Geomorphol. N.F., vol. 62, pp. 1–29.

Fouache E and Dalongeville R (2003) Neotectonic impact on relative sea-level fluctuations over the past 6.000 years: examples from Croatia, Greece and southern Turkey, Human records of recent geological evolution in the Mediterranean. CIESM Workshop Monographs, No 24, pp. 43–47.

Lykousis V, Anagnostou C, Pavlakis P, Rousakis G and Alexandri M (1995) Quaternary sedimentary history and neotectonic evolution of the eastern part of Central Aegean Sea, Greece. Marine Geology, vol. 128, pp. 59–71.

Marakis G (1972) Datation des roches des zones internes de la Grece. C.R. Soc. Phys.Hist. nat. Geneve, vol. 7, no. 2–3, pp. 52–58.

Mercier JL, Carey E, Philip H and Sorel D (1976) La neotectonique Plio-Quaternaire de l'arc Egeen externe et de la mer Egee et ses relations avec la seimicite. Bull. Soc., Geol. Fr., vol. XVIII, no. 7, pp. 355–372.

Mercicr JL, Delibassis N, Gauthier A, Jarrige JJ, Lemeille F, Philip H, Sebrier M and Sorel D (1979) La neotectonique de l'arc Egeen. Rev. Geol. Dyn. Geogr. Phys. vol. 21, pp. 67–92.

Papanikolaou D (1987) Tectonic evolution of the Cycladic blueschist belt (Aegean Sea, Greece). In: H.C. Helgeson (Edt), Chemical Transport in Metasomatic Processes, Kluwer, Dordrecht, pp. 429–450.

Papanikolaou DJ, Sabot V and Papadopoulos T (1981) Morphotectonics and seismicity in the Cyclades, Aegean Sea. Z. Geomorph. N.F., vol. 40, pp. 165–174.

Peterek A and Schwarze J (2004) Architecture and Late Pliocene to recent evolution of outer-arc basins of the Hellenic subduction zone (south-central Crete, Greece). J. Geodynamics, vol. 38, pp. 19–55.

Riedl H (2004) New contributions to the geomorphology of the Aegean archipelago in particular consideration of dating peneplains and pediments and their paleoclimatic conditions. Ann. Geol. Pays Hellen., Vol. XXXX, pp.11–50.

Tirel C, Gueydan F, Tiberi C and Brun JP (2004) Aegean crustal thickness inferred from gravity inversion. Geodynamical implications, Earth and Planetary Science Letters, vol. 228, pp. 267–280.

Chapter 29
Structural Control of Geomorphological Evolution of Meganissi Island (Ionian Sea) Coastal Zone and Natural Hazard Risk Detection Based on Fuzzy Sets

Th. Gournelos, N. Evelpidou, A. Vassilopoulos, and S. Poulos

Abstract This chapter describes the study of the structural control of the Meganissi island coastal zone and its geomorphological evolution; based on these, it presents a coastal hazard risk map. The coastal geomorphology is affected by the lithological configuration, the tectonism, the wider seismotectonic status and the sea activity. For this study, a database has been created in GIS. using all data deriving from the above characteristics as well as from detailed fieldwork, aerial photos, satellite images and pre-existing maps. Data were analysed and processed in order to understand the geomorphological evolution of this island and its close connection with the structural evolution.

Finally, we proceeded to develop a coastal hazard risk map using the slope gradient, the wave energy and the structural characteristics of this island as input variables. This has been achieved via a fuzzy inferences model using a loosen coupling between the Mat lab software package and the MapInfo GIS.

Keywords Structural · Geomorphology · Coastal Zone · Hazard · Risk · Detection

29.1 Introduction

Meganissi island is situated eastern of Lefkas island at the Ionian Sea (Fig. 29.1) and its coastal zone is influenced by the tectonic activity of the wider area. Meganissi is situated among the co-ordinates (20° 43′ 49", 38° 40′ 51"), (20° 48′ 49", 38° 40′ 51"), (20° 48′ 49", 38° 35′ 18") and (20° 43′ 49", 38° 35′ 18") and it has an area of 20.3 km^2, with a coastal length of 52.89 km. The western part of the island is mountainous and the highest peak is Raches (297 m), while the east part is of mild relief and rarely extends the altitude of 100 m. The drainage system of the island is composed by 168 branches, the total length of which reaches 45.86 km.

Th. Gournelos (✉)
Department of Climatology-Physical Geography, Faculty of Geology and Geoenvironment, University of Athens, Panepistimiopolis, Zografou 157 84, Athens, Greece
e-mail: gournelos@geol.uoa.gr

D.R. Green (ed.), *Coastal and Marine Geospatial Technologies*, Coastal Systems and Continental Margins 13, DOI 10.1007/978-1-4020-9720-1_29,
© Springer Science+Business Media B.V. 2010

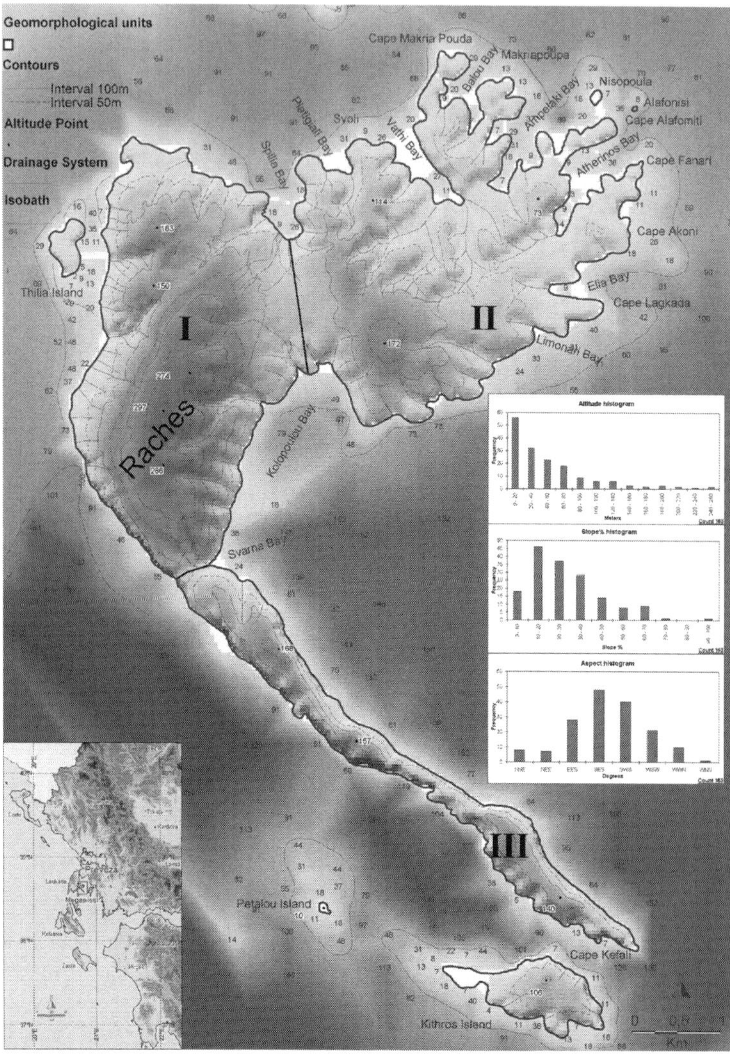

Fig. 29.1 The topography, the drainage system and the geomorphological units of Meganissi island (See also Plate 52 on Page 421 in Color Plate Section)

The rocks, forming the structure of the island, are mainly limestones of Malmium to Upper Eocene with small appearances of cherts and marls and marly limestones of Lower Burdigalian to Tortonian (Aubouin and Dercourt, 1962, Bornovas, 1960, IGME, 1994).

In this chapter the geomorphological evolution of Meganissi island has been studied and several maps, including the coastal hazard risk map, have been developed. Coastal hazards study involves a series of parameters that are concisely focused on the structural, the morphological (morphological slopes, drainage system characteristics, etc), the lithological, the climatological and the wave characteristics.

Emphasis is placed on the structural control of the whole geomorphological evolution at this island. In order to estimate the hazard risk factor of the coastal zone, the geomorphological, the lithological and the tectonical characteristics of the studied area were initially considered. Secondly, the origin of each part of the coastal zone was studied. Finally, rules based on fuzzy logic were developed and applied in GIS environment in order to extract the final hazard risk map.

29.2 Methodology

The study of the geomorphological evolution in the coastal zone of the island entails various tasks (Fig. 29.2) up to generating the results and the derivative maps:

- *First stage*: During the first stage, data was collected from bibliographical resources and extensive fieldwork. To obtain accurate topographical data we used topographic maps and GPS. Primary data was also derived from analogue maps, aerial photos and satellite images. Aerial photos of 1:33.000 scale and satellite images (Landsat) have been photointerpretated. Topographical maps of Hellenic Geographical Military Service (H.G.M.S.) of 1:50.000, and 1:5.000 scales and the geological map of Institute of Geological and

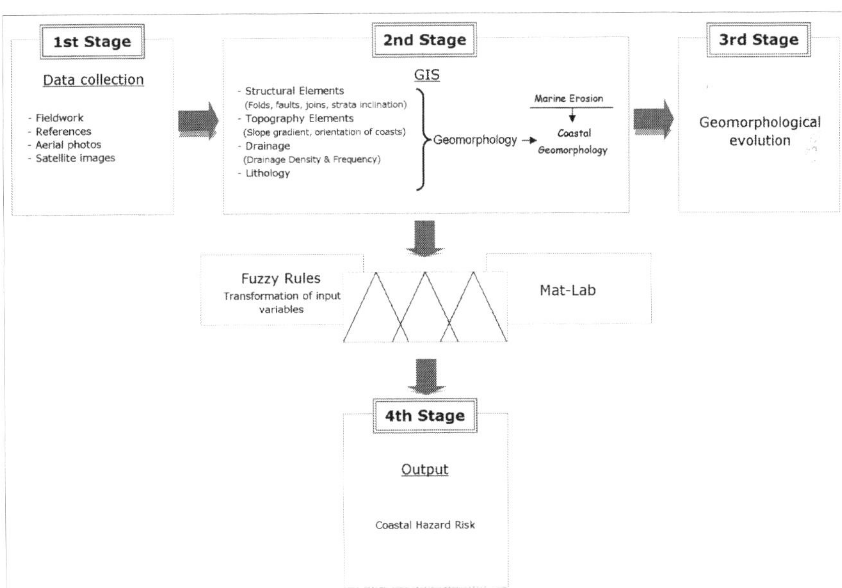

Fig. 29.2 The main stages of the methodology of this chapter

Mineralogy Research (I.G.M.E) have been digitized in a GIS environment (MapInfo Professional, v.8). Information layers attributing geological, altitudinal, geomorphological, structural, hydrological and other topographical data have been created and stored in the spatial database. All the data were imported in a grid of 500*500 m.
- *Second stage*: At the second stage, a database in Geographic Information System was developed and it is continuously updated. Processing the primary information layers resulted in a series of derivative maps, such as the lithological map, the drainage density and the drainage frequency map, the vegetation map, the slope gradient map, the structural elements distribution map and the wave energy map.
- *Third stage*: Both observations from fieldwork and the resulted thematic maps have been used to study the geomorphological evolution in this island. More precisely, the influence of the structural and the erosional processes have been studied in order to interpret the recent landscape of this island.
- *Fourth stage*: In this stage a hazard risk map was developed. For this purpose we have chosen the most important coastal variables such as slope gradient, wave energy and the existence of a favorable structure to landsliding. Within the above variables, two of them (the slope gradient and the wave energy) have been treated as fuzzy variables (Dubois and Prade 1980, Klir and Yuan 1995, Yager et al., 1987, Zadeh 1965, 1987, Zimmermann 1991) because of the obvious relatively spatial uncertainty of their values and boundaries. As regards the third variable, the existence of a favorable structure to mass movement, it is referred to structural elements (strata or faults) which favor landslides and it is treated as Boolean taking two values: yes or no. This procedure is implemented in a MATLAB environment.

A fuzzy inference mechanism has been developed in order to transform the three input variables to the output one, the coastal hazard risk. For this reason we have established a number of logical rules based on the majority of observations concerning mass movements (Table 29.1). A similar fuzzy inference mechanism has been developed to construct erosion risk map (Gournelos et al., 2004).

29.3 Geomorphology and Coastal Hazard Risk of Meganissi Island

The Meganissi island is characterized by three geomorphological units (Fig. 29.1): (I) The central part of this island, where the hilly area of Raches is dominant. (II) A relatively low relief is present in the northern part. (III) Finally, the third one is the very elongated southern part.

The drainage system is mainly developed in the first two geomorphological units, with small streams of relatively high gradient. The dominance of the limestones is responsible for the poor development of the drainage system. We have proceeded to

Table 29.1 Logical fuzzy rules to transform the coastal input variables to the output one (hazard risk)

If	Slope	Is	High	and	Wave energy	is	Medium	and	Favorable structure	is	Yes	Then	the	CHR	Very high
If	Slope	Is	Medium	and	Wave energy	is	Medium	and	Favorable structure	is	Yes	Then	the	CHR	High
If	Slope	Is	High	and	Wave energy	is	Very Low	and	Favorable structure	is	Yes	Then	the	CHR	High
If	Slope	Is	Medium	and	Wave energy	is	Low	and	Favorable structure	is		Then	the	CHR	Medium
If	Slope	Is	Low									Then	the	CHR	Low

the analysis of the orientation of the drainage system. It is obvious (Fig. 29.1) that two principal directions dominate: N60° and N100°.

Next the topography of the island was analysed. We have studied the altitudinal distribution, the aspect and the slope gradient (Fig. 29.1). Low and medium slopes dominate in the island. The aspect is, as expected, mainly orientated S-SE (24,45%), E-SE (17,18%) and S-SW (24,54%).

The next step was to study the orientations of the coast (Fig. 29.3) and to map all the coastal caves. The systematic study of the coastal caves, located in the southern part of the island (Fig. 29.3), has shown the domination of the structural control to the development of these erosional forms. Most of the caves are parallel to the discontinuities of the rocks (faults, joints, stratification).

The orientation of both the coastline and the bays is differentiated in the three parts of the island. In the north-eastern part, where the bays represent a percentage of 67,8% of the total bays of the island, a homogeneous orientation distribution is observed (Fig. 29.3). Orientations within the range of 90–120° are more often in this part. In the north-western part of the island there are few bays (only 17,8%) and the orientation of the coasts lies between N0-60°. In the southern part of the island the majority of coasts are orientated from N10° to N150°.

The geomorphology of Meganissi island is mainly controlled by its structural evolution (Fig. 29.3). Thus, the last recognized deformation phases are responsible for:

1. The general synclinic structure of the island on NW-SE directions.
2. One anticlinic structure which is posterior with the same direction.
3. A discontinuous deformation phase which produced the fault distribution of the Meganissi island.

We proceeded in analyzing all the structural elements involved in this area such as stratification, joint direction and inclination, fault direction and inclination, fold axes and axial planes. Figure 29.3 shows the fault directions. The main dominant directions (NW-SE and NE-SW) are very clear.

This superposition of the deformation phases is the main factor which sculptured the relief of Meganissi; the last faulted period is, especially, responsible for the final topography of the island. It must be noted that the structural control (normal faults) of the southern part of the island defines this elongated area (Fig. 29.3). In general, it is concluded that most of the coastal forms are controlled by the neotectonic faulting.

The last step involved taking into account all the above levels of information to produce a coastal hazard risk map (Fig. 29.3). This map has been produced by combining the structural elements, the slope gradient distribution map and the wave energy characteristics. The most evident factor was the tectonic elements which primarily influenced coastal mass movements (rock falls).

As we have already mentioned in the methodology, we have produced the coastal hazard risk map using a fuzzy logic model to transform the three input variables (slope gradient, wave energy and favorable structure) to the output one, the risk map (Fig. 29.2, Table 29.1). The aim of this model is to map zones of different coastal

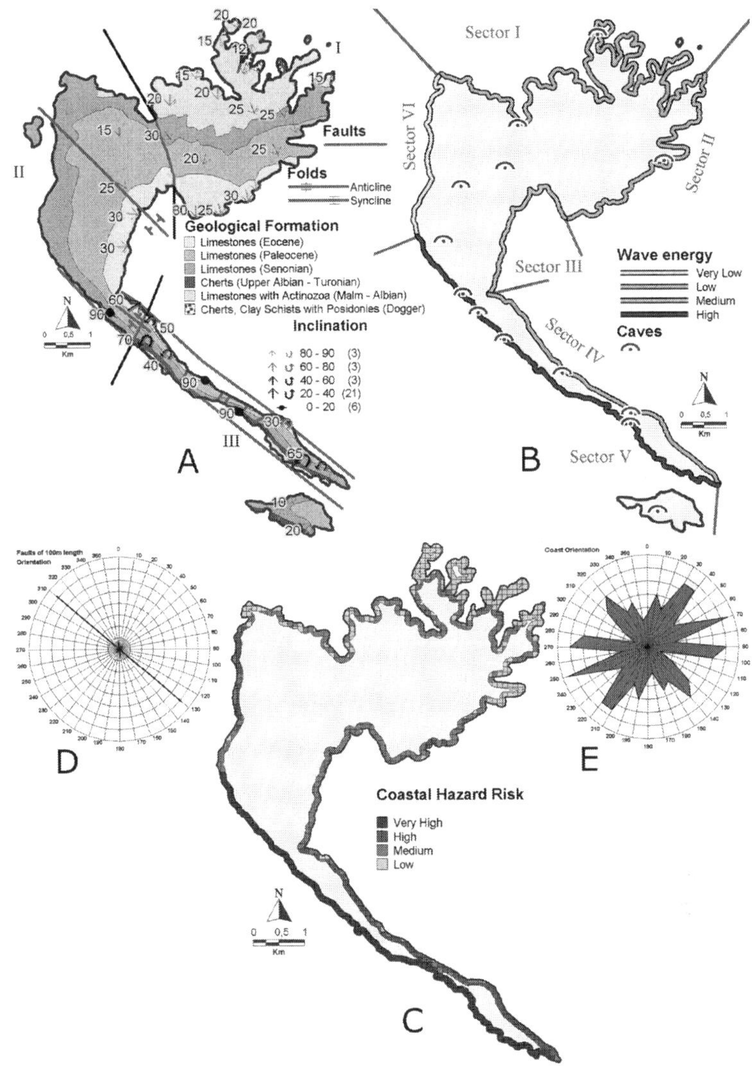

Fig. 29.3 (**A**) The geological map of the Meganissi island (I.G.M.E., 1963, 1994), modified by the authors. (**B**) The wave energy map, (**C**) The final coastal hazard risk map of Meganissi island, (**D**) A rodograme of faults orientation, (**E**) A rodograme of coast orientation (See also Plate 53 on Page 422 in Color Plate Section)

hazard risk. Hazard risks in rocky coasts are usually mass movements, landslides, cliff failures and rock falls, depending on slope gradient, wave characteristics which provoke rocks undercutting and discontinous structures (stratification, faults, joints) with deep inclinations seaward.

The slope gradient variable has been extracted from the topographic map of this island (Fig. 29.1). As it is expected, very high slope gradients are in the southern faulted part of this island.

Table 29.2 Effective fetches for the dominant direction of wind induced waves for the six sections (I–VI) of the coastal zone of the Meganissi Island

Wind	Speed (m/s)	Frequency (%)						
B: Beaufort		N	NE	E	SE	S	SW	W
Weak (1–2B)	1.7	1.50	3.17	2.43	5.97	1.16	1.02	5.15
Moderate (3–5B)	6.7	0.55	1.87	1.20	3.84	2.20	1.55	5.26
Strong (6–7B)	13.9	0.08	0.21	0.30	0.45	0.33	0.17	0.27
Very strong (=B)	20	0	0.02	0.04	0	0	0.01	0.00

For the calculation of wave characteristics along the coast of Meganissi Island, the coastline has been divided into six sectors according to its exposure to the different wind directions (Fig. 29.3). The wind data, originating from the meteorological station located at the city of Lefkada, has been abstracted from Ginis (1974) and presented on Table 29.2.

The predicted significant wave height and period of wave propagation in offshore conditions are based on CERC (1984) equations:

$$H_s = 5.112 \times 10^{-4} W \cdot F^{0.5}$$

$$T = 6.238 \times 10^{-2} (W \cdot F)^{0.33}$$

where W is the wind stress ($W = 0.71\ \mathrm{U}^{1.23}$) (U the wind speed) and F the effective wave fetch (m) (listed on Table 29.2). Subsequently, the approaching offshore wave power (Po) per metre of wave crest has been calculated using the equation:

$$Po = 0.125 \times \rho \times g \times C_o \times H_s^2$$

where $C_0 = 1.56 \times T$, ρ: the water density (1025 kg/m^3) and $g = 9.81$ m/s^2.

The estimated flux of wave energy approaching the coastal zone, related to wave heights are presented on Table 29.3 for each of the 6 Sectors of the coastal zone, as defined according to effective fetch of wave conditions and predominant wind conditions. The overall wave regime is rather weak; the Island receives on an annual basis 1500 W/m in total, when the incoming wave power at Kyparissiakos Gulf reaches 16,000 W/m per year (Poulos et al., 2002). This difference of one order of magnitude is due to the limited wave fetches, as the Island of Meganissi is located in between the Island of Lefkada and the West Coast of the Greek mainland. Furthermore, Table 29.3 shows that only along the sector 5, we have relatively significant wave heights and values of the incoming wave power. Thus, along the coast of sector 5 the marine processes are expected to contribute to the evolution of the coastal zone; the latter is expected to be associated mostly with cliff erosion, due to the steep coastal slopes and the absence of substantial riverine / terrestrial sediment inputs. Based on the above observations, we divided the wave energy distribution on three fuzzy classes: medium, low, very low.

Table 29.3 Annual incoming wave power (P), mean and maximum significant wave heights approaching the various sectors of the Meganissi coastal zone

	P (W/m)	Hs(m)	
		Mean	Max.
SECTOR 1	21.7	0.22	0.45
SECTOR 2	17.6	0.25	0.45
SECTOR 3	138.5	0.35	0.65
SECTOR 4	27.7	0.22	0.45
SECTOR 5	1047.2	0.47	1.15
SECTOR 6	1.5	0.08	0.35
	1254.2		

SECTOR	Predominant wind direction	Fetch (m)
I	N	10470
	NE	13570
II	E	12305
III	E	13640
	SE	35250
IV	NE	14045
	E	7375
V	SE	13695
	S	47555
	SW	195420
VI	W	1340

Regarding the third variable, the existence of favorable structures to coastal hazards, has been obtained by combing field data and the structural map (Fig. 29.3). It is obvious that such structural elements favorable to coastal mass movement are present in the southern elongated part of Meganissi island.

Then, we produced the final hazard risk map (Fig. 29.3) by applying the fuzzy inference mechanism based on the rules of Table 29.1. In this map we may observe that the southern elongated part of the island is the most risky zone. Especially risky is its southern facing coast where dominate high slope gradient, high wave energy and favorable structures to mass movements.

Next comes the northern facing coast with high slope gradient existence of favorable structures to landslides but low wave energy. Finally, zones of medium or low risk dominate in the rest of the island (Fig. 29.3).

29.4 Conclusions

It is concluded that the tectonic processes are the main factors contributing to the formation of the present landscape of this island. Today, the marine erosion processes modify day by day the morphology of the coastal zone creating various geomorphological characteristics, the most important of which are the coastal caves at the western part of the island.

Only a small part of the coastal zone originates from terrigenous deposition, while terrigenous erosion coasts cover a relatively large part of the coastline, forming multivariable bays.

The coastal hazard map is divided into four risk zones: (1) A very high risk zone located in the south-external coast where high to very high slope gradients are observed. The existence of marine erosion (caves) and the discontinuities of rock (fault, joints, stratification) accelerated coastal erosion and have been producing mass movements (rock falls). (2) A high risk zone located in the southern – internal coast, where the wave energy is relatively lower than in the first zone. (3) The third zone of medium coastal risk is situated in the eastern part of Meganissi. (4) The fourth zone is of low risk and is located in the northern part.

Furthermore, the erosion of the cliffs, due to its resistant lithology, is expected to be attributed primarily to mechanical erosion, whilst at the remaining island's coast, marine erosion is expected to be associated with chemical rather mechanical abrasion due to limited wave activity.

References

Aubouin J, Dercourt J (1962) Zone Preapoulienn, Zone Ionienne et Zone de Gavrovo en Peloponnese occidentale. Bull. Soc. Geol. France, 4, pp 785–794.

Bornovas J (1960) Obsarvations nouvelles sur la geologie des zones preapulienne et ionienne (Grece occidentale). Bull.Soc. Geol. France, Ser.7, 2, pp 410–413.

CERC (Coastal Engineering Research Centre) (1984) Shore Protection Manual. US Army Corps of Engineers. Washington DC 21314.

Dubois D, Prade H (1980) Fuzzy Sets and Systems Theory and Applications, Academic Press, New York.

Ginis S (1974) Wind conditions of the Ionian Sea. Unpublished PhD thesis, University of Athens, pp. 144.

Gournelos Th, Evelpidou N, Vassilopoulos A, (2004) Developing an Erosion risk map using soft computing methods (case study at Sifnos island). Natural Hazards. Vol.31, No 1, January 2004. Kluwer Academic Publishers, pp 39–61.

Klir GJ and Yuan B (1995) Fuzzy Sets and Fuzzy Logic theory and applications. Prentice-Hall, New Jersey.

Poulos SE, Voulgaris G, Kapsimalis V, Collins M and Evans G (2002) Sediment fluxes and the evolution of a riverine-supplied tectonically-active coastal system: Kyparissiakos Gulf, Ionian Sea (eastern Mediterranean). In: Jones S.J. and Frostick L.E. (eds) Sediment Flux to Basins: Causes, Controls and Consequences. Geological Society, London, Special Publications, 191, pp 247–266.

Yager RR, Ovchinnikov S, Tong RM and Nguyen HT (1987) Fuzzy Sets and Applications, selected papers by L.A. Zadeh, Wiley, New York.

Zadeh LA (1965) Fuzzy sets. Information and Control, Vol 8, pp 338–353.

Zadeh LA (1987) The concept of linguistic variable and its application to approximate reasoning. In: Yager R.R., Ovchinnikov S., and Tong R.M., Nguyen H.T., (eds) Fuzzy Sets and Applications. Wiley, New York, pp 293–329.

Zimmermann HJ (1991) Fuzzy Set Theory and Its Application. 2nd ed., MA:Kluwer Academic.

Chapter 30
A Simple and Efficient Methodology to Assess Long Term Shoreline Evolution – Case Study

Filipa S.B.F. Oliveira

Abstract The present chapter describes an efficient methodology to predict the impact of a marina on the adjacent indented beaches planform: initially, the impact of the marina on the hydrodynamic agents that induce sand transport in the adjacent indented beaches; finally, the beaches planform response to the modification of the hydrodynamic conditions. The innovative aspect of this analytical methodology concerns the process of estimating the dominant wave energy flux in front of the beaches, which is based on a statistical analysis of the incident wave regime. The methodology was successfully validated through the comparison of the shoreline configurations obtained analytically with the ones extracted from aerial photographs, before and after the construction of the marina.

Keywords Modelling · Beach · Shoreline · Wave data

30.1 Introduction

The objective of the present study is to establish and validate a methodology to analyse and evaluate the impact of a marina in the nearshore hydrodynamics and adjacent beaches planform. The present methodology can be reliably applied in coastal environments with similar physics to predict the long-term impact of coastal structures on the beaches at their vicinity.

30.1.1 Study Area

The study area is located between two singularities of the west coast of Portugal: the Raso headland (and simultaneously, inflection point of the coastline direction) and the Tejo estuary mouth (Fig. 30.1). The littoral stretch under analysis contains

F.S.B.F. Oliveira (✉)
Laboratório Nacional de Engenharia Civil, Av. do Brasil 101, 1600-077 Lisboa, Portugal
e-mail: foliveira@lnec.pt

D.R. Green (ed.), *Coastal and Marine Geospatial Technologies*, Coastal Systems and Continental Margins 13, DOI 10.1007/978-1-4020-9720-1_30,
© Springer Science+Business Media B.V. 2010

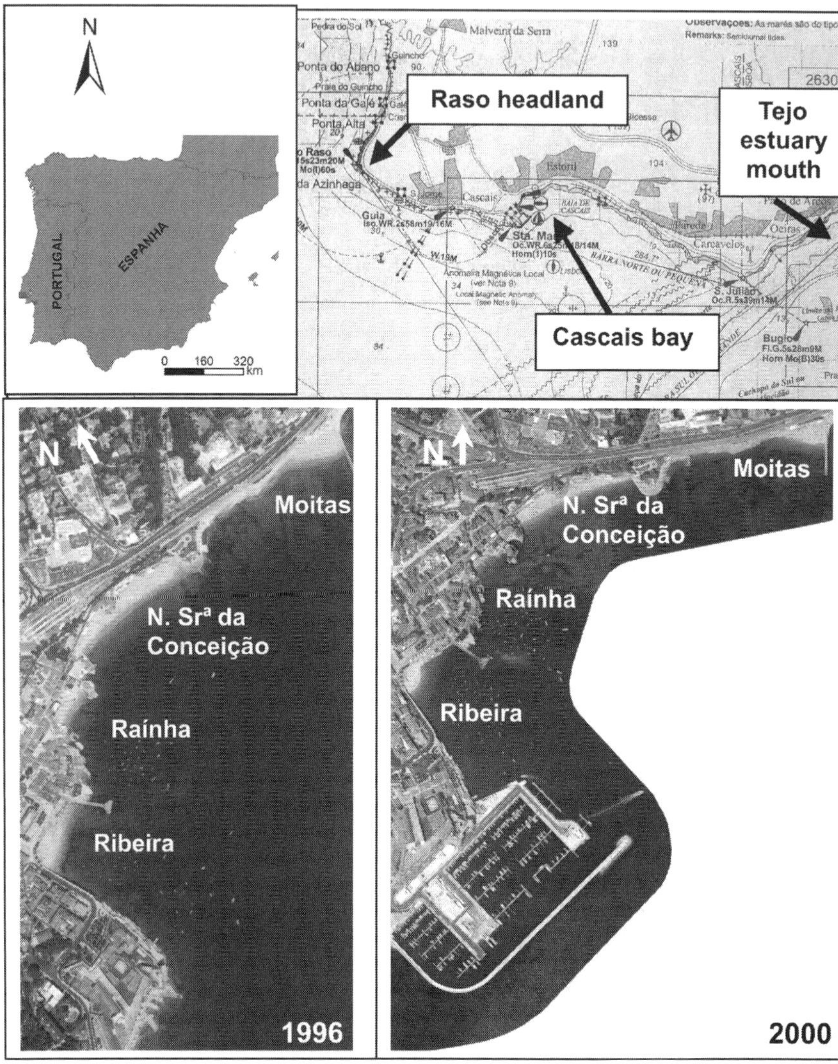

Fig. 30.1 Location of the study area and aerial photographs of Cascais bay beaches before and after the construction of the marina (See also Plate 54 on Page 423 in Color Plate Section)

Cascais bay and is presently densely occupied. It has been changed, from its natural characteristics, through the construction of marginal infrastructures and cliffs stabilization. It has a rather irregular coastline configuration and is alternately composed of cliffs and small indented beaches.

Hydrographic surveys prior to the construction of the marina (concluded in 1998) show that the seabed bottom contours were parallel to the shoreline configuration. Bellow approximately 10 m depth relatively to the hydrographic Chart Datum (CD), which origin level is approximately 2.00 m bellow the mean sea level (MSL), the sea bottom slope was between 0.6 and 1% (Consulmar 1989). In what concerns the seabed material, surveys showed that above 10 m depth bellow CD, discontinuous areas of sandy bottom, with sediment mean grain size, D_{50}, between 0.24 and 0.29 mm, seat over hard rock material (Oliveira 1967).

The study area is under a meso-tidal regime (tidal range about 2.7 m) due to a semi-diurnal astronomic tidal cycle. The tidal currents in Cascais bay are of low intensity, smaller than 10 $cm.s^{-1}$ (Consulmar 1989), and do not induce significant sediment transport when compared with the wave induced currents. The nearshore wave action, as the main agent responsible for the sediment dynamics, before and after the construction of the marina, is presented in more detail in the next section.

Finally, the littoral sediment transport must be negligible because, besides the inexistence of local sediment sources in this coastal area, like creeks and cliff erosion, it was not observed trapped sediment due to littoral drift after the construction of the marina.

30.2 Methodology and Results

The methodology was applied in two phases. The first step of the first phase was the statistical analysis of the offshore wave regime, which consisted in the calculation of the distribution, per classes of occurrence, of the wave parameters height, period and direction of incidence, and the establishment of an offshore simplified representative wave regime; the second step was the simulation, based on numerical modelling of wave transformation, of the propagation of this regime components to the nearshore region, to acknowledge the wave climate in front of the beaches; finally, this second step was followed by the calculation of the intensity and direction of the annual average wave action, weighted based on occurrence, along an isoline of bathymetry in front of the beaches, at the entrance of the surf zone. The second phase consisted on the validation of these results through the comparison of the predicted shoreline geometry with the real shoreline configuration (extracted from aerial photographs). The methodology was applied for the two situations, before the construction of the marina (BM) and after the construction of the marina (AM), and allows predicting the impact of the marina in the nearshore hydrodynamics and in the evolution of the adjacent beaches planform.

30.2.1 Offshore Wave Regime

The offshore regime, recorded from January 1984 to December 1995 at a wave rider station located 92 m depth bellow CD, 170 km north of Cascais, was statistically analysed in order to obtain the histograms of occurrence, per classes of 10°, 0.5 m and 2 s, for the wave characterisation parameters direction, height and period, respectively (Fig. 30.2a–c). Based on the analysis of the above results, three waves,

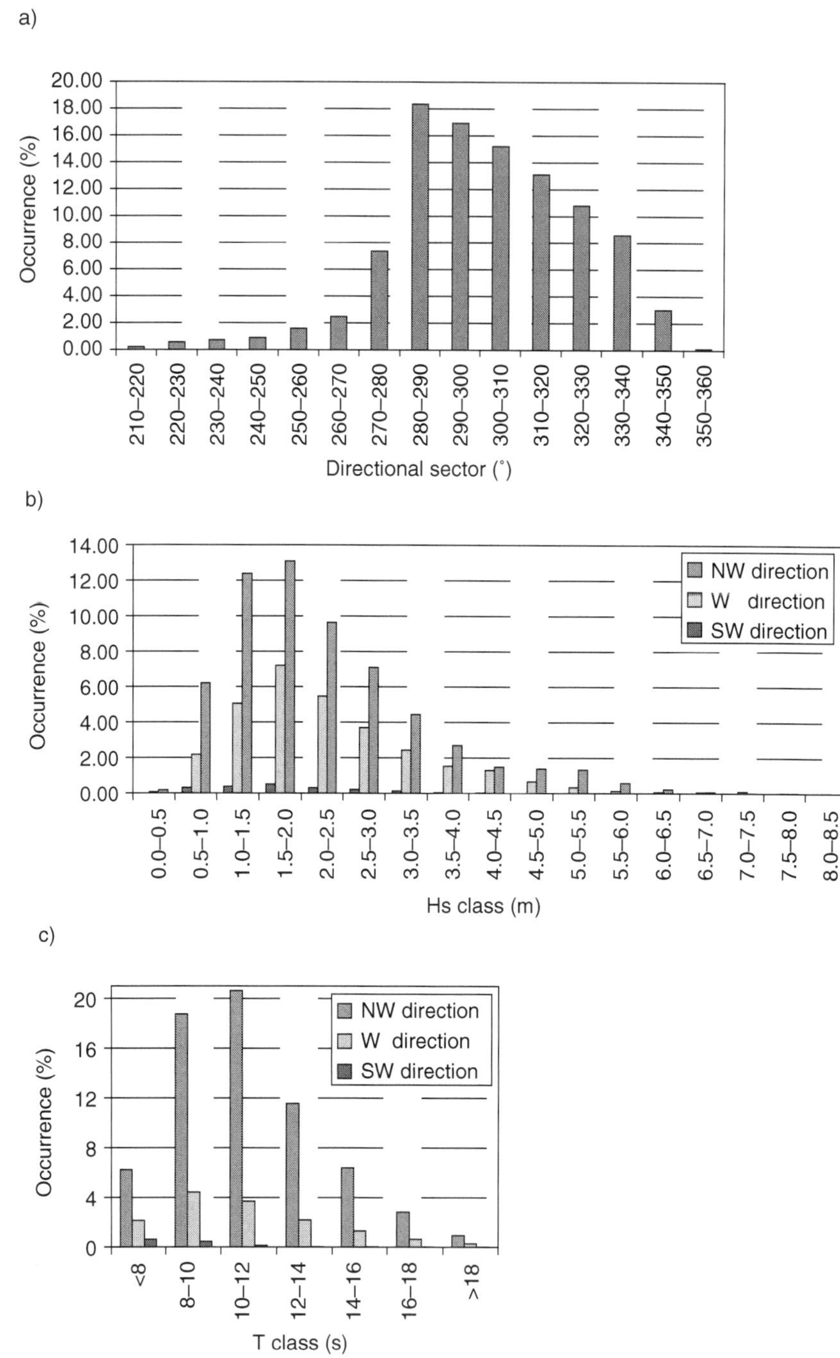

Fig. 30.2 Histograms of occurrence for the offshore wave regime parameters: **a**) direction; **b**) significant height; and **c**) period (See also Plate 55 on Page 424 in Color Plate Section)

Table 30.1 Components of the representative offshore wave regime

Direction (°)	Hs (m)	T (s)
SW (N225°)	1.5	8
W (N270°)	2.0	10
NW (N315°)	1.5	10

with characteristics described in Table 30.1, were considered as components representative of the offshore wave regime.

30.2.2 Nearshore Wave Climate Changes

30.2.2.1 Before the Construction of the Marina

The results of wave transformation from offshore to the shore, calculated at MSL with a wave propagation mathematical model, show that the wave with SW (southwest) offshore direction is the one that reaches the shoreline less refracted and is therefore associated to the highest wave amplification factor, $f_a = H/H_0$ (where H is the local wave height and H_0 is the offshore wave height), in front of the beaches. As the offshore direction of incidence turns clockwise the wave amplification factor decreases in the study area, reaching the lowest values, lower than 0.20, for the offshore wave incoming from NW (northwest).

The wave amplification factor and the wave direction, alongshore, in the isoline of 4.0 m depth bellow CD (6.0 m bellow MSL), at the entrance of the surf zone, in front of the four beaches, for the three representative waves, can be seen in Fig. 30.3a and b, respectively. It is shown that the highest gradient of energy occurs for the SW incoming wave and that the energy gradient decreases as the offshore wave turns clockwise. As expected, the SW incoming wave is the one, from the three representative waves, that reaches the surf zone with the direction most normal to the shoreline.

30.2.2.2 After the Construction of the Marina

Similarly to the situation before the construction of the marina, the numerical results of wave propagation show that, in the study area, the wave amplification factor decreases, as the offshore wave turns clockwise. However, in the present situation, the marina generates a shadow zone, which extension includes the total length of the shoreline of the first adjacent beach, Ribeira beach, in front of which the wave amplification factor does not vary significantly with the incident wave.

Once again, the wave amplification factor and the wave direction, along the isoline of 4.0 m depth CD, at the entrance of the surf zone, in front of the four beaches, are presented (Fig. 30.3c and d, respectively). The results show the protection of Ribeira beach against wave action, caused by the presence of the marina.

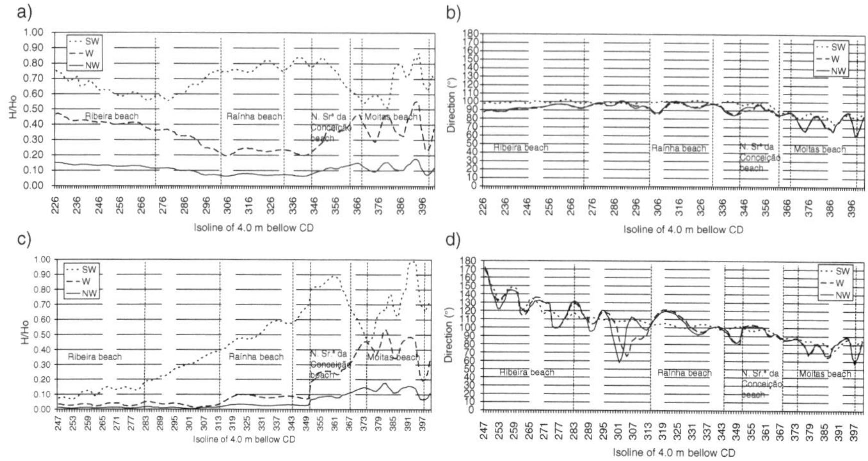

Fig. 30.3 Wave amplification factor and wave direction: **a**) and **b**) before the construction of the marina; and **c**) and **d**) after the construction of the marina

The effect of protection extends gradually alongshore, from WNW to ESE, as the offshore waves turn clockwise. The range of the wave direction values in front of the beaches is three times wider in the present situation than before the construction of the marina, since now the waves reach Ribeira beach affected by the diffraction process caused by the marina breakwater.

30.2.3 Beach Planform Evolution

The comparison of the orientation of the shoreline main alignment, for each beach, in the two situations BM and AM, allows to conclude about the beaches planform evolution due to the effect of the marina. Despite the existence of seasonal variations of the beaches planform and profile along the year, due to the mutual adjustment between the hydrodynamic conditions and the morpho-sedimentological characteristics of the beaches, two aerial photographs were compared, because they were both taken during the maritime summer, one in June of 1996 and the other in July of 2000 (Fig. 30.1). Although the beaches planform was extracted with confidence from the aerial photographs, the beaches width did not allow deducing reliable conclusions, because the aerial photographs do not correspond to the same sea level: in the first, the sea level is approximately 1.84 m above CD, whereas in the second, it is approximately 3.00 m above CD.

As it is shown in Table 30.2 the parameter orientation of the beach main alignment, reveals an approximate rotation of 17°, 10°, 6° and –1° of the shoreline for the beaches Ribeira, Raínha, N. Srª da Conceição and Moitas, respectively. The aerial photographs show that the first beach was clearly the most affected by the presence of the marina, the two following adjacent beaches (Raínha and N. Srª da Conceição)

Table 30.2 Orientation of Cascais bay beaches main alignment (extracted from aerial photographs)

Beach	Before the marina	After the marina
Ribeira	N40°	N23°
Raínha	N60°	N50°
N. Srª da Conceição	N75°	N69°
Moitas	N98°	N99°

suffered nearly half the impact of the first one and Moitas beach was not affected by the construction of the marina.

The fact that the aerial photographs correspond to different sea levels and the beach profile is not uniform, either in time or along each beach, affects its main alignment. However, for the situation AM, three aerial photographs, one of 2000 and two of 2002, were used to analyze the parameter orientation of the beach main alignment.

30.3 Discussion

The average wave amplification factor and the wave direction were calculated along the isoline of 4.0 m depth CD, in front of each beach, for the representative components of the wave regime and for the two situations, BM and AM.

Based on these results, the dominant wave power, $\bar{P}$, defined as

$$\bar{P} = \bar{E}C_g, \tag{30.1}$$

was calculated, for each situation, as being the weighted average of the energy flux per unit of wave crest width of the representative components of the wave regime, where the weighting factor was the occurrence, according to the methodology described next.

The total average wave energy per unit surface area, or energy density, $\bar{E}$, was estimated as

$$\bar{E} = \frac{E}{L} = \frac{\rho g H^2}{8}, \tag{30.2}$$

where E is the total energy of a wave system (sum of its kinetic and potential energy), or energy density, L is the wavelength, ρ is the mass density of salt water, g is the gravitational acceleration and H is the wave height.

The group velocity, C_g, defined as

$$C_g = \frac{1}{2}\left[1 + \frac{4\pi d/L}{\sinh\left(4\pi d/L\right)}\right]C, \tag{30.3}$$

where d is the water depth and C is the phase velocity, was approximated, for shallow water, as

$$C_g = C \approx \sqrt{gd}. \tag{30.4}$$

Thus, the wave power alongshore, in front of each beach, was estimated as

$$\bar{P} = \frac{\rho g\,(H_0 f_a)^2}{8}\sqrt{gd} = \frac{\rho g^{3/2}\,(H_0 f_a)^2\,d^{1/2}}{8}, \tag{30.5}$$

where f_a is the local wave amplification factor, as defined before.

Associating the frequency of occurrence of each component of the wave regime to the respective wave power, $\overline{P_{SW}}$, $\overline{P_W}$ and $\overline{P_{NW}}$, the resultant in the local reference system (x, y), where the y axis is aligned with the geographic North, is:

$$\begin{aligned}\overline{P_x} &= \overline{P_{SW}}\cos\theta_{SW}0.0192 + \overline{P_W}\cos\theta_W 0.3033 + \overline{P_{NW}}\cos\theta_{NW}0.6775\\ &= \left(\frac{\rho g^{3/2} d^{1/2}}{8}\right)[\,(H_0 f_a)^2_{SW}\cos\theta_{SW}0.0192\\ &\quad + (H_0 f_a)^2_W\cos\theta_W 0.3033 + (H_0 f_a)^2_{NW}\cos\theta_{NW}0.6775]\end{aligned} \tag{30.6}$$

and

$$\begin{aligned}\overline{P_y} &= \overline{P_{SW}}\sin\theta_{SW}0.0192 + \overline{P_W}\sin\theta_W 0.3033 + \overline{P_{NW}}\sin\theta_{NW}0.6775\\ &= \left(\frac{\rho g^{3/2} d^{1/2}}{8}\right)[\,(H_0 f_a)^2_{SW}\sin\theta_{SW}0.0192 + (H_0 f_a)^2_W\sin\theta_W 0.3033\\ &\quad + (H_0 f_a)^2_{NW}\sin\theta_{NW}0.6775]\end{aligned} \tag{30.7}$$

where θ_{SW}, θ_W and θ_{NW} are the local average direction of the representative waves and $(H_0 f_a)_{SW}$, $(H_0 f_a)_W$ and $(H_0 f_a)_{NW}$ are the average wave height of the representative waves.

The comparison of the components, in the local reference system (x, y), of the variable factor of the resultant wave power, defined as

$$\overline{P_x}\bigg/\left(\frac{\rho g^{3/2} d^{1/2}}{8}\right) \tag{30.8}$$

and

$$\overline{P_y}\bigg/\left(\frac{\rho g^{3/2} d^{1/2}}{8}\right), \tag{30.9}$$

for each situation, BM and AM, and for each beach, allows to evaluate the variation, caused by the construction of the marina (Fig. 30.4), on the intensity and direction of the resultant of the energy flux.

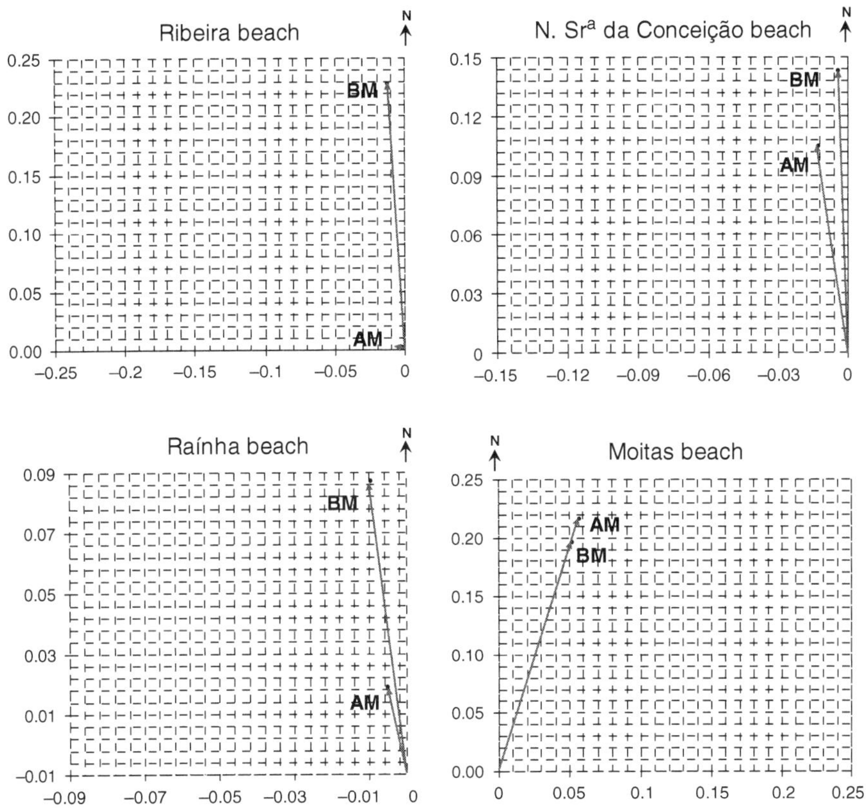

Fig. 30.4 Variable factor of the resultant wave power, for the situations BM and AM, in front of the beaches: Ribeira, Raínha, N. Srª da Conceição and Moitas (See also Plate 56 on Page 425 in Color Plate Section)

From BM to AM, at each beach, it is observed a decrease of intensity and rotation anticlockwise of the resultant of the energy flux. The magnitude of this variation increases from N. Srª da Conceição beach to Ribeira beach. For the most eastern beach, Moitas beach, the direction of the resultant of the wave power remains unchanged although it can be observed a slight increase of its magnitude. These results show that the marina protects the three adjacent beaches against wave action with a sheltering effect that increases from ESE to WNW. The resultant of the incident wave action in front of the beaches turned, anticlockwise, an angle that also increases from ESE to WNW. These modifications of the nearshore wave climate were the cause of the rotation of the shoreline that occurred in the first three beaches adjacent to the marina.

Despite the difficulty found for Ribeira beach, due to its nearly circular planform at high sea level, as it was the case of the last aerial photograph, the beaches planform observed in the aerial photographs allowed drawing a main shoreline alignment. The comparison of this parameter in both photographs revealed a rotation of approximately 17°, 10° and 6° of the shoreline for Ribeira, Raínha and N. Sr.ª da Conceição beaches, respectively. For Moitas beach this parameter remained nearly unchanged. Based on the assumption that the resultant wave power is normal to the shoreline main alignment, when comparing the above values with the analytically estimated rotation of 39°, 9° and 5° respectively (Fig. 30.4), it can be concluded that the impact of the marina on the beaches agrees quite well for the beaches Raínha and N. Sr.ª da Conceição and was correctly predicted as null for Moitas beach. The only exception is the first adjacent beach, Ribeira beach, where occurs a local phenomenon, not taken into account in the methodology applied, that affects the beach morphology: it is the wave reflection from the non-dissipative vertical structure that limits the west side of the beach, that does not allow the sediment to settle, but pushes it parallel to the shoreline, affecting the beach planform. Thus, with this particular exception, whose possible cause was pointed out, the methodology applied was proved to be suitable and efficient to predict the long-term shoreline configuration.

Despite being in the presence of beaches with different lengths, a parameter that plays a determinant role in the readjustment of the beach planform to the modifications of the direction and magnitude of the dominant wave action, the results do not allow to conclude on its effect on the beaches planform modification or rate of readjustment.

30.4 Conclusions

A methodology to assess long-term shoreline evolution was implemented and verified in a coastal stretch of indented beaches of small longshore extension, on the central part of the west coast of Portugal, where the longshore drift is negligible. The incident wave climate was represented by the resultant weighted average wave action in order to account for the annual variability of the incident wave regime. The construction of a marina introduced significant changes on the hydrodynamics of the nearshore adjacent beaches, which were quantified through this methodology. For each beach, the rotation of the shoreline was estimated based on the rotation of the resultant of the wave action caused by the construction of the marina. The predicted shoreline was verified against the observed shoreline extracted from aerial photographs and the results show a good agreement, validating the methodology.

Acknowledgments The author thanks to her sister Ana for the revision of the English and to their colleagues from the NPE division for assisting with the wave propagation numerical model runs.

References

Consulmar (1989) Marine of Cascais. Environmental impact study. Final Report. Vol.1 – Text. Consulmar, Lisboa, Portugal (in Portuguese).

Oliveira R (1967) Contribution to the study of Tagus estuary – Sedimentology. Report n° 296, LNEC, Lisboa, Portugal (in Portuguese).

Chapter 31
Study of Human Induced Recent Geomorphological and Land Use Changes of the Acheloos Delta Area in Western Greece Using GIS

D. Green, Th. Gournelos, A. Vassilopoulos, and N. Evelpidou

Abstract This chapter studies the counteractive dynamic marine and river environments responsible for a river mouth's and a delta's modulation. The Acheloos River (West Greece) was chosen as the case study due to the recent changes taken place in its Delta area; the geotechnical constructions and the agricultural planning have altered the basin's conditions. The aim of this study is to analyze the present dominant environment and the impact of human activity on this balance with the application of modern technological tools; emphasis is placed on the geomorphological along with the land use changes taking place in the three district parts of the Acheloos' delta within a period of four decades (1960–2000): delta plain, delta front, pro delta. Bibliographic references, geographical and geological maps, aerial and satellite imagery of different seasons and dates, were utilized in addition to extensive fieldwork measurements and mapping. A geographic database was developed and it is continuously updated.

31.1 Introduction

The study area is focusing on Acheloos' river delta in west Greece close to the Ionian coastal zone and west of Pindos mountain (Fig. 31.1). The geographical growth of Acheloos' basin is worth mentioning; Acheloos springs from Lakmos mountain (central Pindos), flows at the limits of prefectures Arta and Karditsa, in western Thessaly and at the limits with prefecture Evritania is linked with the rivers Tavropo and Agrafioti, separating Evritania from Etoloakarnania, and flows into

D. Green (✉)
Department of Geography and Environment, University of Aberdeen, Elphinstone Road, AB24 3UF, Aberdeen, UK
e-mail: d.r.green@abdn.ac.uk

This chapter is a revised version of an article that was published earlier in the Journal of Coastal Conservation 11:209–213 (2007).

D.R. Green (ed.), *Coastal and Marine Geospatial Technologies*, Coastal Systems and Continental Margins 13, DOI 10.1007/978-1-4020-9720-1_31,
© Springer Science+Business Media B.V. 2010

Fig. 31.1 The location of the study area, the development of drainage basin and the Dams of Acheloos basin (See also Plate 57 on Page 426 in Color Plate Section)

Ionian Sea shaping an extensive deltaic field. In its wider basin of flow, four natural lakes exist (Trichonida, Lisimachia Amvrakia, Ozeros), and several dams have been created (the main are: Kremaston, Kastrakiou, Stratos I and Stratos II).

Acheloos is the main river of Western Central Greece, the first in water contribution and the second in length, found in the Greek territory, with a total length of

220 Km. It constitutes the main source of water both for the flat and the semi mountainous regions of Etoloakarnania. The shape of its basin is oblong with a maximum axis of 147 Km length and 63 Km width. The basin covers a total area of 6.226 Km^2 and it is defined by the following geographic coordinates: north 21.08°, 38.29° and south 21.92°, 39.73°. The maximum altitude of the area reaches 2.420 m.

Acheloos river's action, as well as Evinos river's (Piper and Panagos, 1981), resulted in the development of lagoons such as Mesologi and Etoliko. Its extrusions compose a unique ecosystem (protected by the Ramsar treaty) of significant ecological value. Besides, the flora and the fauna of the region present particularity.

The climate in the Acheloos basin belongs to the mild Mediterranean (average temperature: 18,6°C) type and tends to the Continental type (average temperature: 21,4°C) in territorial north and north-eastern regions (Liakouris, 1971). In particular, the warmer month of the year is July (average temperature: 27,4°C), while the coolest is January (average temperature: 8,3°C). The annual height of rain is approximately 1.100–1.200 mm.

The drainage basin of Acheloos may be divided into three individual subbasins: (a) the upper part, which is defined by the administrative boundaries of Trilofo community, (b) the middle part, which is delimited in the south by Stratos village and includes Tavropos, Kremasta, Kastraki and Stratos artificial lakes, and (c) the lower part, which includes the Trichonida, Lisimahia, Amvrakia and Ozeros natural lakes, the allouvial plain and the Mesologi, Etoliko and Klisova lagoons (Liakouris, 1971).

The Acheloos drainage basin covers three isopic geological zones: the Ionian one in the western region, the Gavrovo one in the central part and, finally, the Pindos zone in the eastern and north-eastern part (Aubouin, 1959, Doutsos et al., 1987).

The area of Acheloos delta is defined by the following geographic coordinates: north 21.080, 38.290 and south 21.320,38.420 and it covers a total area of 175,2 Km^2. Acheloos delta belongs to the lobolate type, as it has the shape of a lobo with a solid head and a columella front. Its evolution depends on the predominance of the two environments: the continental and the marine one.

31.2 Methodology

The study of the geomorphological and land use changes entailed a series of different stages: bibliographical references, field-work observations and processing of analogue maps, aerial photographs and satellite images in a Geographical Information System. All primary data, deriving from different sources, were imported in the G.I.S. (MapInfo Professional 8.0) by direct digitizing on the basis of two different aged (1960 and 1986) aerial-photos (Hellenic Military Geographical Service – H.M.G.S., scale: 1:33000), satellite images (Landsat, 2000), topographic (H..M.G.S. scale 1:50.000) and geological maps (I.G.M.E., scale: 1:50000).

In order to increase the accuracy of photointerpretation, aerial photos were geo-referenced and finally transformed to ortho-aerial photos. These functions, along with the stereoscopic observation of the aerial photos, were supported by photogrammetric software (Image Analyst), enabling the user to perform on screen stereoscopic observation and instant on screen digitization to the GIS database (MapInfo Professional 8.0). The photo-interpretation of different dated aerial photos and satellite image was, thus, of essence for the comparison of the geomorphological and land use changes in the two different periods.

An environmental spatial database was developed and it was continuously updated. Information layers holding geological, geomorphological, tectonical, lithological, hydrological, land use and anthropogenic data were created and processed. Data were analysed quantitatively and qualitatively resulting in a variety of thematic maps depicting geomorphological and anthropogenic alterations.

31.3 Geomorphological and Land Use Changes in the Delta Area

Acheloos delta is a very dynamic system influenced primarily by fluvial and marine processes, as well as human interaction. Some of the delta's features have been enormously changed since historic times (Piper and Panagos, 1980, Vott et al., 2002, Sabot et al., 2002). n Acheloos delta area, three main geomorphological units are distinguished: the delta plain (I), which is the dominant geomorphological feature, the delta front (II) and the Prodelta (III) (Fig. 31.1).

The major anthopogenic intervention in Acheloos river is the construction of six hydroelectric dams (Fig. 31.1). The biggest one is at Kremaston, which is situated three Km south-western of the junction of the rivers Tavropos, Agrafiotis and Acheloos. It is the first dam developed in Acheloos drainage basin (1966) and the most significant hydroelectric construction in Greece. The Kastraki is the second dam developed in Acheloos drainage basin (1969) and it is placed thirty five Km southern of the Kremaston one. Stratos I and Stratos II dams, developed in 1989, are situated eight km southern of Kastraki dam and sixty two km of Acheloos mouth.

The aim of this chapter is to focus on geomorphological and land use changes after the constructing of the dams. It is obvious that such a major human intervention provoked multiple effects on the whole geomorphological and the environmental system. Both the geomorphological and the land use changes are examined on the basis of the above defined three units.

Firstly, the subaerial delta area has been studied and the corresponding geomorphological and land use units of 1960 and 2004 have been mapped (Fig. 31.2A and B). In general, the abandonment of the old drainage system (ox-bow lakes, old delta mouth) is prominent. Most of the old drainage system, which used to be organized with cultivations, became inactive when a complex system of canals took its place, mostly for irrigation purposes. As far as land use is concerned, most of the deltaic field appears to be subjected to human intervention, focusing on the stabilization of the area and the growth of organized cultivations along with the road network. Furthermore, an expansion of the agricultural land and a decrease of the wet ground are noticeable.

The above mentioned environmental changes become very clear by observing the aerial photos of 1960 and 1986. In the aerial photos and satellite image

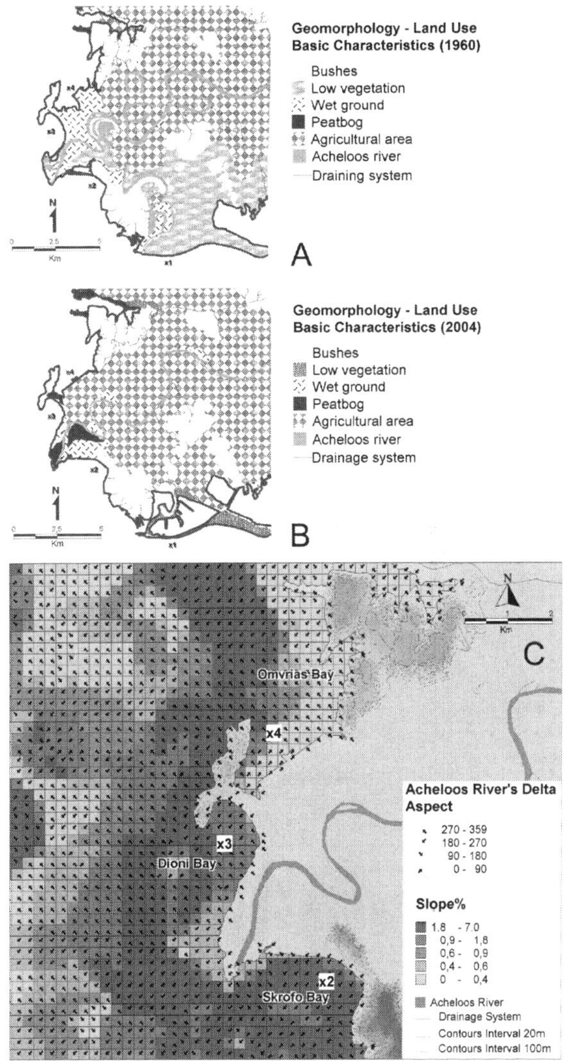

Fig. 31.2 The delta plain. Geomorphological and land use changes from 1960 (Fig. 2A) to 2004 (Fig. 2B), based on aerial photos and satellite images interpretation, and field mapping. Fig. 2C: Digital model of the submarine environment (See also Plate 58 on Page 427 in Color Plate Section)

(Fig. 31.3), where Aloupotripa and Lagades are located, the small pools, derived from the meandrous cutting off, are active in 1960, while in 1986 they had already dried up and in 2000 their marks are hardly identified. The land is, nowadays, dominated by the agricultural activities.

- All the areas with closed meander neck have disappeared in 2004.
- Severe decrease of wet land has been occurred since 1960.
- Bushes have slightly expanded in 2004.
- Agricultural land covers all of the fluvial area in 2004.

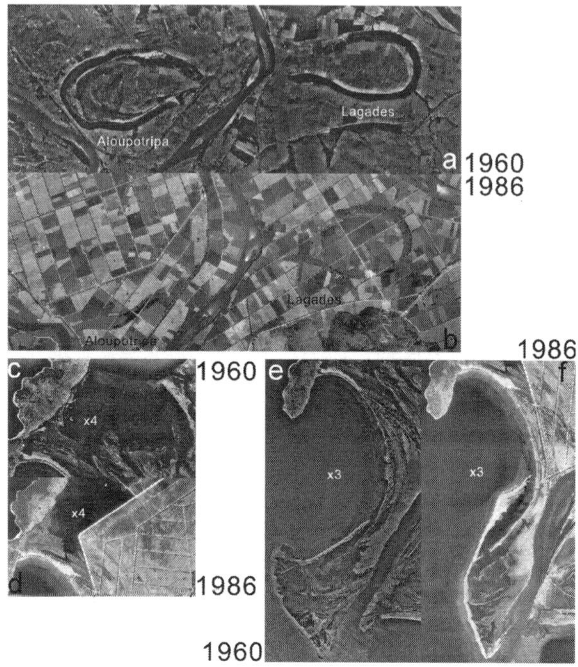

Fig. 31.3 Zooming on geomorphological and land use changes as have been observed by aerial photos and satellite images (1960–2000), in delta plain (See also Plate 59 on Page 428 in Color Plate Section)

The coastline geomorphological changes are also very clear. The shape of the coastline has been altered during those years due to wave and the current action. The delta coastline is very sensitive to any geomorphological and environmental change. The main changes of this area may be observed in Figs. 31.2 and 31.3; more precisely we reported:

- The delta front has also been extensively modified in the past decades. This is apparent by comparing the Fig. 31.3b and aerial photos (1960 and 1986). Some of these changes are localized in the southern part (point x1, x2 in Fig. 31.3b) and in the northern part (point x3, x4).
- The shape of the deposition tongue and the coastline has changed. The deposition tongue shown in the western part of the area has been eroded at its west part and appears more elongated (Fig. 31.3b).
- Peat-bogs have appeared in 2004 in the place where low vegetation and dry lakes used to exist four decades ago.
- Dunes fields are, nowadays, replaced by low vegetation or even wet grounds.

The pro-delta area reflects the changes of the old delta front and its adjustment to the new geomorphological situation.

- In order to locate and estimate the coastal zone changes in the pro-delta area, a digital model of the submarine environment was developed (Fig. 31.2C). It is remarkable how the area near the river's mouth has been altered; a bog used to be in the place of today's petbog.
- The old front delta is nowadays submarine. This may be visualised by observations in the submarine morphological slopes.

Some of the delta front changes have been observed by comparing aerial photos and satellite images from 1960 to 2000. Thus, the study of Acheloos delta evolution showed that the balance of this dynamic system changed due to dams' constructions. The fluvial supply has been progressively decreased and nowadays the marine processes predominate over the delta front.

31.4 Conclusions

This chapter should be considered as the beginning of a project handling the design and the implementation of a spatial database that stores the geomorphological and environmental data of Acheloos' delta. Input of various data, deriving from diverse sources (fieldwork, bibliography, maps, photointerpretation, etc.), is viable. In a platform such as GIS, both the processing of the data and the combinatory analysis are feasible; the geographical database enables the user to retain a life-long database by updating it and expanding it both in terms of space and time. Especially studying a long-term phenomenon, such as the evolution of a drainage basin, utterly connected both to human landscape and environmental settings within a temporal sequence, GIS are a pre-requisite.

In Acheloos' delta it is obvious that human induced factors have modified the physical environment to a great extend. The dams' construction has resulted in a progressive reduction of the fluvial sediments. Diverse geomorphologic and land use changes concerning the coastal plains and the coastal area are clearly detected. The degradation of this coastal area is accelerated by the salt water's intrusion, the over-exploration of the groundwater and the expansion of the touristic structures.

Thus, there is a great demand for sustainable development policies; the scope of this chapter is to stimulate the necessity of approaching environmental coastal problems by using modern technological tools and developing a long-term monitoring mechanism for local and regional policy makers aiming at concrete and feasible solutions.

Acknowledgments This work is funded by the Greek Ministry of Development and the British Council in the frames of the bilateral Greece – UK project entitled "Study of the environmental alterations at the Acheloos river delta (W. Greece) that are caused by anthropogenic interference with the use of Remote Sensing and GIS methods", which is gratefully acknowledged by the authors.

References

Aubouin, J., 1959, Contribution a l' etude geologique de la grece septentrionale: les confines de l' Epire et de la Thessalie, PhD thesis, Paris. Ann. Geol. Pays Hellen., Vol. 10, p. 526.

Doutsos, T. Kontopoulos, N., Frydas, D., 1987, Neotectonic evolution of northwestern-continental Greece, Geologische Rundschau 76: 433–450.

I.G.M.E., Geological Map of Astakos, 1986, scale 1:50.000, Institute of Geology & Mineral Exploration.

I.G.M.E., Geological Map of Echinades, 1989, scale 1:50.000, Institute of Geology & Mineral Exploration.

I.G.M.E., Geological Map of Evinochorio, 1991, scale 1:50.000, Institute of Geology & Mineral Exploration.

I.G.M.E., Geological Map of Greece, 1983, scale 1:500.000, Institute of Geology & Mineral Exploration.

I.G.M.E., Geological Map of Messologi, 1996, scale 1:50.000, Institute of Geology & Mineral Exploration.

Image Analyst, Z/I Imaging Corporation, Huntsvill, 1999, USA

Liakouris, D., 1971, Geomorphological and geological research in the area of Acheloos River (upper part)', PhD thesis, School of Sciences, University of Athens, Athens.

MapInfo Professional, 2005, MapInfo Corporation, Troy, New York.

Piper, D.J.W. and Panagos, A.G., 1981. Growth patterns of the Acheloos and Evinos deltas, western Greece, Sedimentary Geology, 28: 111–132

Sabot, V., Evelpidou, N., Vassilopoulos, A., 2002, Study of environmental and geomorphological consequences at Acheloos delta (West Greece) due to anthropogenic interferences, using GIS, Congress of Remote Sensing for Environmental Monitoring, GIS Applications, and Geology II, Crete, 22–27 September, Vol. 4886, p.p. 381–389.

Vott, A, Handl, M., Bruckner, H., 2002, Rekonstruktion holozaner Umweltbendingugen in Akarnanien (Nordwestgriechenland) mittels Diskriminanzanalyse von geochemischen Daten. Geological et Palaeontological 36: 123–147, Marburg.

Part V
The Coastal Environment

Chapter 32
Multi-Functional Assessment of Coastal Landscapes with Climate Change

Iain Brown

Abstract A strategic-level evaluation tool for comparing coastal management policies against different future scenarios of climate change is described. The effect of sea-level rise on tidal and flood limits is mapped using a high-resolution digital elevation model (DEM) and an adapted hydrological routine. Implications for the coastal landscape are then determined using empirical thresholds encapsulated in transition rules for land use and land cover change. The resulting spatial data are analysed using a multi-criteria assessment procedure to facilitate an evaluation of policy options against the provision of multiple coastal services. Analysis suggests that small-scale managed realignments of the coast, although providing a political compromise, do not score highly on performance criteria compared to either large-scale realignment or maintaining the existing position. This approach also allows consideration of a broader range of environmental benefits and ecosystem services than a conventional analysis based upon property and land values.

Keywords Climate change · Landscape change · Digital elevation model (DEM) · Transition rules · Multicriteria assessment (MCA)

32.1 Introduction

Coastal managers are becoming increasingly aware that climate change implies that current policies for the coastal zone may be unsustainable. In particular, maintaining a fixed coastline and static patterns of land use, despite the increased hydrodynamic loading on defence structures associated with sea level rise, presents a particular challenge. Alternative strategies that aim to encourage a more proactive

I. Brown (✉)
Macaulay Institute, Craigiebuckler, Aberdeen, AB15 8QH, UK
e-mail: i.brown@macaulay.ac.uk

D.R. Green (ed.), *Coastal and Marine Geospatial Technologies*, Coastal Systems and Continental Margins 13, DOI 10.1007/978-1-4020-9720-1_32,
© Springer Science+Business Media B.V. 2010

and dynamic response to these pressures, notably by allowing some migration of the coastline inland ("managed realignment"), have often met with opposition (e.g. Ledoux et al., 2005a). A change in management policy therefore clearly requires acceptance by local communities in order to be successfully implemented, necessitating a full and ongoing dialogue with all stakeholders to assess and review implications. GIS-based tools can be a key facilitator in this engagement process because in addition to providing a mechanism for integration and manipulation of scientific data, they can also encourage a dialogue with stakeholders via interactive visualisation linked to the appraisal of policy options and the relevant spatial plans.

This chapter outlines how a large-scale assessment of the impacts of climate change on the coastal landscape can be used to facilitate a rapid and strategic evaluation of the long-term implications of different management policies. It can therefore act as a screening tool, by eliminating unfeasible options from a later more detailed assessment. The scale of assessment is consistent with the littoral cell concept now adopted by Shoreline Management Plans (SMPs) for the UK; cells are defined as systems in which sediment movements are mainly considered internal to that unit (Leafe et al., 1998). It also means that policy reviews consider the regional context for management as well as specific local issues. The assessment explores the influence of tidal flooding frequency on landscape change through both land-use decisions (related to socio-economic factors) and the modification to land cover (ecological factors). At present, management decisions in most areas are made primarily on economic criteria relating to coastal defence, thereby protecting land and property from flooding and erosion, but a more sustainable approach requires inclusion of a wider range of environmental and social factors. By adopting a multi-criteria assessment approach, it is possible to derive indicators of these additional factors by reference to the changes in land use and land cover that are integrated within the GIS database for each policy option.

32.2 Study Area

The area of study is the North Norfolk coast of eastern England (Fig. 32.1), which is characterised by a "barrier" system of dunes and shingle ridges fronting a low-lying hinterland consisting of both inter-tidal areas and land reclaimed from the sea for agriculture (Andrews et al., 2000). The risk of flooding has meant that extensive sections of the coast have embankment structures to protect land and property, and these are maintained at considerable expense. Large areas of land have high ecological value as recognised by their designation as protected sites within the EU Natura2000 network, including both the inter-tidal wetlands (as Special Areas of Conservation: SAC) and parts of the reclaimed areas that have now become freshwater wetlands (as Special Protection Areas: SPA).

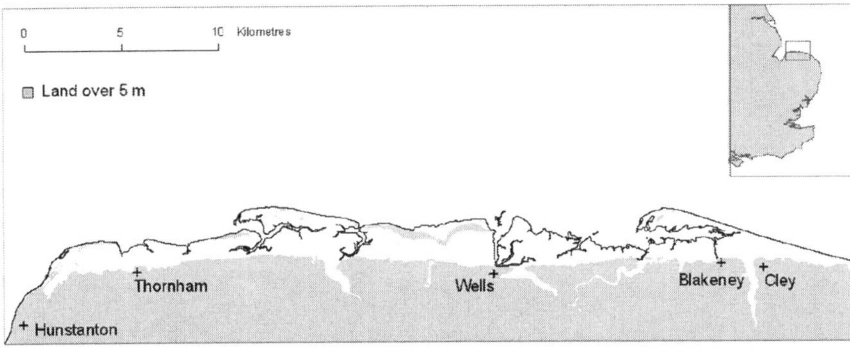

Fig. 32.1 The North Norfolk coastline

Four alternative policies were explored:

Policy Option 1 – maintain the present defences (i.e. "hold the line");
Policy Option 2 – small-scale managed realignment, with a limited movement of defences inland at appropriate locations whilst still protecting most land and all settlements;
Policy Option 3 – large-scale managed realignment allowing a more dynamic coast with defence structures only protecting major settlements;
Policy Option 4 – removal of all man-made defences from the coastline.

Policy Options 1, 2 and 3 were defined based upon documents already available as part of policy review procedures, particularly the local SMP and the more recent Coastal Habitat Management Plan (CHaMP) (Pethick and Cottle, 2003). Defences were assumed to be built or maintained to keep pace with the rate of sea level rise and therefore provide similar standards of service to those available today.

32.3 Key Datasets

32.3.1 Elevation

Three elevation data sources were evaluated for use in this assessment:

(i) Ordnance Survey (OS) contour-derived products;
(ii) Light Detection and Ranging (LiDAR) survey data;
(iii) Interferometric Synthetic Aperture Radar (IfSAR) survey data.

The coarse resolution (vertical accuracy ~3 m) of the conventional OS data compared to LiDAR (vertical accuracy ~0.3 m) and IfSAR (vertical accuracy ~0.5 m) meant that it proved too imprecise to model sea level rise adequately. In addition, the LiDAR data, despite its high quality, was not available seamlessly for the whole study area. Therefore the IfSAR data, which had been developed through the NextMap initiative specifically for flood risk assessment (Intermap, 2005), was chosen to provide the baseline digital elevation model (DEM) for the study area. The IfSAR data had been post-processed to remove features such as buildings and trees, therefore indicating the true ground surface, but retained current defence structures such as embankments, which proved a valuable feature for defining and validating Policy Option 1.

32.3.2 Land Use / Land Cover (LULC)

A baseline LULC 25 m raster grid was derived from the classified LandSat satellite imagery of the Land Cover Map of Great Britain 2000 (LCM2000; Fuller et al., 2002). Validation of the LCM2000 against field data, aerial photographs and other imagery for the study area noted some significant areas of misclassification at detailed levels, therefore the data was generalised into a set of broad landscape classes: saltmarsh, mudflat/sandflat, shingle/dune ridges, managed grassland, rough grass (including freshwater marsh and semi-managed grassland), arable agriculture, woodland, built land, sea, and lake.

32.3.3 Buildings

Buildings data were derived from the Ordnance Survey ADDRESSPOINT dataset. This provided both spatial data and associated attributes describing the type of building.

32.3.4 Tidal and Flood Levels

Present-day reference levels for specific tidal and flood events were obtained as elevation values from SMP documents (Table 32.1). To provide future projections related to changes in sea levels these levels were then incremented by values provided by climate change scenarios (Brown, 2006; Hulme et al., 2002). For the area of study and the year 2050, these scenarios indicate a global rise in sea level of 8 cm (low scenario) and 42 cm (high scenario).

32.4 Mapping Land Inundation from Sea-Level Rise

To determine the spatial distribution of flooding under different combinations of climate change scenario and policy option, a procedure adapted from the hydrological tools available within specialist GIS packages was utilised (Brown, 2006); in this case study, ARCGIS GRID and AML scripts were utilised. The objective was to

Table 32.1 Height data (above standard datum) for key tidal and coastal flood limits in the case study area

Tidal or flood limit	Height
Mean low water (MLW)	0.0 m
Mean high water neaps (MHWN)	1.7 m
Mean high water springs (MHWS)	3.5 m
Annual flood limit (1/1)	4.0 m
10 year extreme flood (1/10)	4.4 m
50 year extreme flood (1/50)	4.7 m
100 year extreme flood (1/100)	5.0 m
200 year extreme flood (1/200)	5.3 m

define the present and future tidal floodplain at different magnitude/frequency levels at a step more advanced than a simple "naive" implementation of the contour method which would simply define all areas below a threshold value. The contour method is particularly inappropriate for the study area because, being a "barrier coast", it features a series of dune and shingle ridges which along with the engineered defence structures act to prevent the lower-lying hinterland from flooding. Therefore the contour method would considerably overestimate the flood risk. At a more advanced level hydrodynamic methods can provide a detailed simulation of water movement and levels during particular events, but these methods are considerable more data-intensive and time-consuming to apply over a large area or for multiple scenarios.

Hence, the procedure adopted was a compromise between accuracy and analysis time, consistent with the rapid strategic scale of the assessment (Brown, 2006). It involves running a routine which detects internal basins within a DEM, these being zones ("sinks") which are surrounded on all sides by higher ground; "sinks" are then "filled" with the "sill" value of the lowest of the higher cells (the FILL routine in ARCGIS GRID provides a method for this). Once this procedure is completed, then defining floodplain land below a reference contour value is much more realistic for "barrier" coasts. However, robust application of this approach tends to be reliant on a good-quality high-resolution DEM because other DEMs are likely to contain "sink" features that are more related to data artefacts from the coarse sampling resolution rather than being genuine internal basins. Tests of the NextMap IfSAR data against LiDAR data and against field knowledge showed that, whilst there were some minor deficiencies, it contained no major inaccuracies.

Tidal and flood inundation limits were mapped using this procedure for the future sea-level rise increments on a sequence of DEMs defined from Policy Options 1, 2, 3 and 4. The DEM for Policy Option 1 was similar to the baseline DEM except with defence heights increased. For the other Policy Options, abandonment of the current defence line was represented by defence features in the baseline DEM being "breached" by editing of the elevation data at key defined locations. Policy Options 2 and 3 also required superimposition of new defences on the baseline DEM after conversion from the original vector lines in policy planning documents.

At this strategic scale of assessment, the assumption was made that the primary driver on changes in the coastal landscape was the frequency of inundation by the sea. Therefore, a series of threshold values were described and utilised to explore expected changes in LULC. These were encapsulated within a system of transition rules and, for implementation within a GIS, defined with a lookup table (Brown, 2006). An example of land use change is the abandonment of agricultural land when flooding exceeds a certain threshold (e.g. 1 in 10 years), whereas land cover change is typified by the conversion of saltmarsh to bare mudflat when sea-level rise causes land of a certain elevation within the tidal frame to become inundated too frequently for marsh vegetation to subsist (usually below MHWN level).

The major advantages of the rule-based method accrue from its simplicity. This means that it is inherently flexible, and different threshold values can be rapidly explored through sensitivity analysis if necessary. In addition, the modelling assumptions are clearly transparent to users and can be subject to further testing with stakeholders and scientists, improving interaction and dialogue. It is for example possible to develop the rule base further to include other data which may influence landscape change where available (e.g. slope, salinity, sediment availability), although such data is often not universally available at the regional scale.

The methodology can be validated by comparing the baseline LULC map with a simulated version using Policy Option 1 and a 0 cm sea level rise, as the latter combination should not produce any landscape change. This validation test for the study area showed some areas of divergence which in nearly all cases could be attributed to a misclassification of the baseline LULC map, particularly in the transition zone between terrestrial and intertidal areas. The simulated baseline can therefore act as a contextual filter to improve the accuracy of the actual baseline dataset (Figs. 32.2 and 32.3).

32.5 Multi-Criteria Assessment of Policy Options

Evaluation of policy options is normally based upon an economic assessment developed through cost-benefit analysis (CBA), with the most favoured options having a high ratio of benefits to costs. However, a conventional CBA generally fails to account for the more intangible benefits that may accrue because they are difficult to quantify in purely economic terms, notably the socio-cultural and environmental advantages of schemes. By contrast, developing a multi-criteria assessment (MCA) using a standardised performance scoring system per criteria (e.g. from 0–100) can combine several disparate categories together into one overall assessment (Janssen et al., 2005). Furthermore, it can also potentially involve a broader range of stakeholders in the decision-making process.

By using the information on LULC change and location of settlements, coastal defences etc. in the GIS database, a MCA approach has developed that allows the role of a wide range of different criteria to be compared in one strategic evaluation. However, one of the potential pitfalls of any CBA or MCA procedure stems

Fig. 32.2 Current land use and land cover for the Blakeney-Cley section of the study area (See also Plate 60 on Page 428 in Color Plate Section)

Fig. 32.3 Simulated 2050 land cover/land use for Policy Option 3 and the high sea level rise scenario: same area as Fig. 2 (See also Plate 61 on Page 429 in Color Plate Section)

from errors due to "double-counting", whereby separate criteria are not in reality independent of each other and scores in one criteria can be replicated in another. In this prototype assessment, criteria have therefore been grouped into five broad categories. For each category, cumulative scores are estimated spatially from the areal GIS data then normalised relative to a standard performance score from 0 (worst performing policy) to 100 (best performing policy) using linear interpolation. The future time horizon for the policy evaluation is assumed to be the year 2050. Each of the categories is briefly outlined below, then the issue of producing a summary across all categories is discussed, including the degree of weighting between categories, before we provide indicative results.

(i) *Category 1: Protection of properties at risk from flooding.* This has been derived from an intersection of flood limits and building data. An indicative risk index was devised with a scoring system as follows (the highest score representing the largest risk):

- buildings in highest risk zone (flood frequency < 1 in 10 years) = 5 points;
- buildings in medium risk zone (< 1 in 100 years) = 3 points;
- buildings in lowest risk zone (<1 in 200 years) = 1 point

(ii) *Category 2: Protection of land assets at risk from flooding.* This is the economic value of the land that would need to be acquired to allow any managed or unmanaged realignment; in the study area, this mainly refers to agricultural land that would be lost from production. Notional land values (£ /ha) were applied to each LULC class in the GIS database using guidance from recent economic studies (for further discussion see Dickie and Pilcher, 2001; Ledoux et al., 2005b).

(iii) *Category 3: Costs for Coastal Defences.* These were estimated based upon the length of defences needed for a particular Policy Option, using topological and attribute data from the GIS. The scoring system is inversely related to the cumulative costs and includes costs incurred for both new build (assumed to occur in 2010) together with annual maintenance costs for existing defences, depending on policy and degree of sea-level rise (see Turner et al., 2007).

(iv) *Category 4: Environmental benefits.* This is the most difficult category to quantify because the perceived value of these benefits typically varies subjectively and significantly between individuals. Nevertheless the potential benefits cannot be disregarded and include scheme advantages for landscape character (e.g. aesthetics), recreation, biodiversity, cultural heritage, and water quality; however, these values are interdependent and therefore considered here as a single index value. Recommended areal values for wetland areas and agricultural land were derived from the central estimates of the recent valuation meta-analysis (Department for Communities & Local Government, 2006), which confirms that improved agricultural land has less benefits of this type compared to unimproved land and inter-tidal areas.

(v) *Category 5: Carbon Sequestration.* Recent ratification of the Kyoto Protocol on greenhouse gas emissions means that increased attention is now being directed at carbon sequestration as a means to reduce overall emissions, and wetlands can have a significant role in this process. Although this benefit is not considered in conventional assessments, this category is increasingly likely to become an important consideration in coastal areas in the future, particularly through the extent of intertidal areas (Shepherd et al., 2007). Scores were therefore related to the areas of inter-tidal zones together with varying sedimentation rates relative to sea level rise scenarios (low: 1 mm/yr; high:

4 mm/yr) and assuming bulk density values of 1.6 for mudflat and 2.6 for saltmarsh (Parkes, 2004).

(vi) *Weighting of Criteria.* A key feature of MCA is the weighting of criteria to produce the final summary value per option, once the individual criteria scores have been determined. This is often the topic of greatest divergence in opinion but also potentially the most instructive in understanding and reconciling different stakeholder opinions. Hence the approach adopted in the present study was not to try and develop some form of "consensus" but rather to present a range of different weighting systems in order to encourage diverse stakeholder groups to appreciate the range and validity of other opinions and positions. Figure 32.4 therefore provides indicative results using different weighting schemes and using the method adopted by Janssen et al. (2005) to communicate findings from a GIS-based MCA procedure.

(vii) *Summary of Findings.* Although the majority of stakeholders would agree that the most important criteria for coastal management policy is defending property (and by inference, lives and livelihoods) from the risks of flooding and erosion, the assessment suggests that the justification for continuing with Policy Option 1 as the current management policy is only sustained if the value of agricultural land is equally strongly weighted. Policy Option 1 would therefore maintain current patterns of land use. If other criteria in addition to property and land values are included on comparative weightings, either individually or cumulatively, then the score for Policy Option 1 declines and Policy Option 3, which has high environmental benefits and low defence costs (because defences are concentrated around major settlements) becomes the highest scorer. Policy Option 4 is unlikely ever to be feasible, because of the high loss of property in major settlements.

Perhaps surprisingly, Policy Option 2 is not the strongest scorer under any weighting system. This is because it requires extensive sections of new defence to protect agricultural land and freshwater wetlands whilst not gaining any major advantage over Policy Option 1 in terms of protected assets or Policy 3 in terms of environmental benefits. It is interesting to note, however, that small-scale managed realignment schemes such as that represented by Policy Option 2 have been advocated as by some public bodies as the most viable solution, particularly because they purport to provide a political compromise that implies a more sustainable future whilst meeting local concerns over flooding and erosion. The evidence from this initial assessment suggests that this response seems to be despite (or perhaps, in ignorance of) poor overall performance scores for this option. Exploration of the MCA scores for the future projections with the two different sea level rise scenarios does not significantly alter these conclusions, except to highlight the reduced scores that Policy Option 1 achieves with a higher rate of sea level due to the extra defence costs. As a consequence, if the rate of sea-level rise accelerates from its current rate (equivalent to the low scenario) towards higher rates (as represented by the high scenario) then Policy Option 3 consistently provides the best overall performance across the range of criteria identified here.

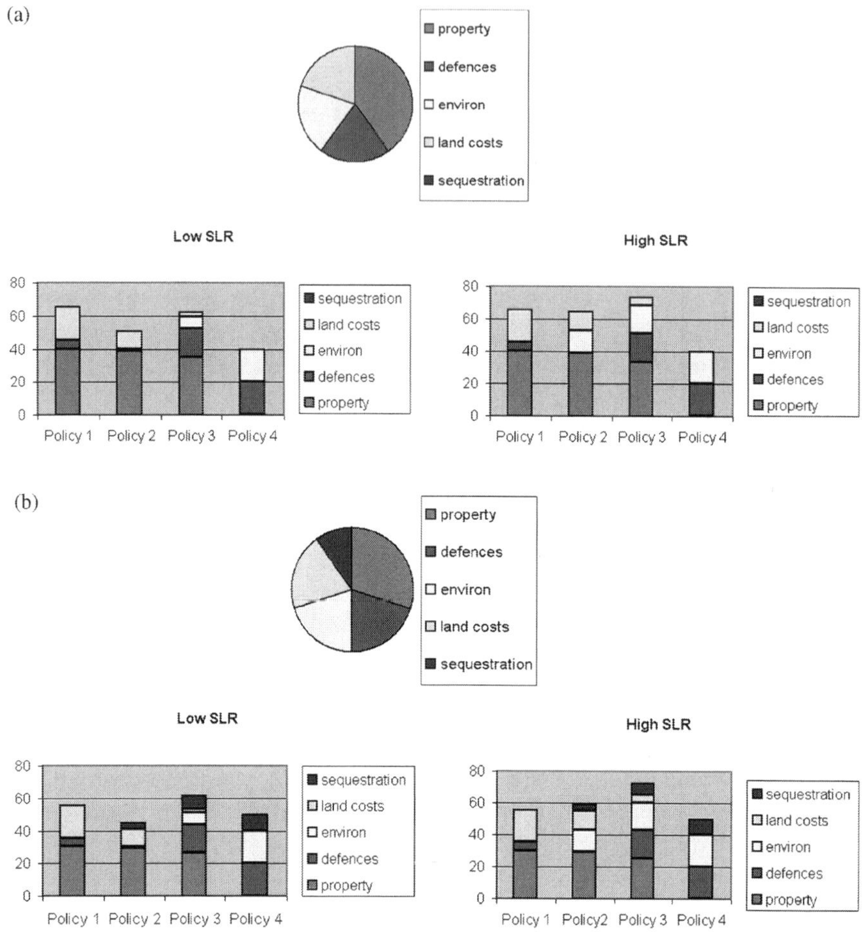

Fig. 32.4 Sample results from the MCA with different criteria weighting systems (**a**) and (**b**). The weighting systems are illustrated in the pie-chart and the resultant scores for each policy shown for two different sea level rise (SLR) scenarios (See also Plate 62 on Page 429 in Color Plate Section)

32.6 Conclusions

The present study has explored the role of flooding in influencing landscape change on a regional scale through a modified form of hydrological analysis linked to a series of key datasets and transition rules in GIS. High-resolution DEM have a particularly important role to play in assessing the risks to coastal areas from sea level rise and climate change. Scenario analysis using this GIS database has been used to develop an integrated assessment of future management policy options using a spatial MCA approach. Further work is now expanding the empirical rule-based concept to explore the influence of changing geomorphological processes for sustainable coastal management. This becomes increasingly important as the scale of assessment moves from the regional to the local scale, due to the influence of sediment supply on patterns of erosion or accretion, and therefore on the natural response of coastal landforms to sea-level rise. The MCA approach is particularly useful for considering the broader environmental benefits that different land use systems or habitats provide, together with their potential buffering role against the more extreme effects of climate change. This can also facilitate increased awareness amongst stakeholders of the direct and indirect benefits from maintaining "ecosystem services" for human well-being (Millennium Ecosystem Assessment, 2005).

Acknowledgements This work was funded by the Tyndall Centre for Climate Change Research. Diane Burgess and Simon Jude provided very useful advice. Data was kindly provided by the National Earth Observation Data Centre and Intermap Ltd. (NEXTMap), the Centre for Ecology & Hydrology (LCM2000), Ordnance Survey (ADDRESSPOINT) and Environment Agency (coastal defences).

References

Andrews, J.E., Boomer, I., Bailiff, I., Balson, P., Bristow, C., Chroston, P.N., Funnell, B.M., Harwood, G.M., Jones, R.W., Maher, B.A., and Shimmield, G. (2000). Sedimentary evolution of the north Norfolk barrier coastline in the context of Holocene sea-level change. In: Shennan, I, Andrews, J.E. (Eds.), *Holocene LandOcean Interaction and Environmental Change Around the North Sea*. Geological Society Special Publication 166, London, 219–251.

Brown, I. (2006). Modelling future landscape change on coastal floodplains using a coastal GIS. *Environmental Modelling and Software*, 21, 1479–1490.

Department for Communities & Local Government (2006). Valuing the external benefits of undeveloped land. http://www.communities.gov.uk/documents/planningandbuilding/pdf/158136.pdf

Dickie, I., and Pilcher, R. (2001). Coastal Flood Defence: Recent Economic Evidence. RSPB Discussion Paper.

Fuller, R.M., Smith, G.M., Sanderson, J.M., Hill, R.A., and Thomson, A.G. (2002). The UK Land Cover Map 2000: construction of a parcel-based vector map from satellite images. *The Cartographic Journal*, 39, 15–25.

Hulme, M., Jenkins, G.J., Lu, X., Turnpenny, J.R., Mitchell, T.D., Jones, R.G., Lowe, J., Murphy, J.M., Hassell, D., Boorman, P., McDonald, R, and Hill, S. (2002). Climate change scenarios for the United Kingdom: The UKCIP02 Scientific Report. Tyndall Centre for Climate Change Research, UEA, Norwich.

Intermap, (2005). Product Handbook and Quick Start Guide.! http://www.intermap.com/images/handbook/producthandbook.pdfo.

Janssen, R., Verhoeven, J.T.A., Omzigt, A.Q.A., Maltby, E., Goosen, H., and Verhoeven, M.L. (2005). Decision support for integrated wetland management. *Environmental Modelling and Software*, 20, 215–229.

Leafe, R., Pethick, J., and Townend, I. (1998). Realising the benefits of shoreline management. *Geographical Journal*, 164, 282–290.

Ledoux, L., Cornell, S., O'Riordan, T., Harvey, R., and Banyard, L. (2005a). Towards sustainable flood and coastal management: identifying drivers of, and obstacles to, managed realignment. *Land Use Policy*, 22, 129–144.

Ledoux, L., Beaumont, N., Cave, R., Turner, R.K., (2005b). Scenarios for integrated river catchment and coastal zone management. Regional Environmental Change, 5, 82–96.

Millennium Ecosystem Assessment (2005). *Ecosystems and Human Well-Being: Biodiversity Synthesis*. World Resources Institute, Washington DC.

Parkes, D. (2004). Storage and cycling of organic carbon and nutrients in Holocene coastal sediments. Ph.D. Thesis University of East Anglia.

Pethick, J.S., and Cottle, R. (2003). *North Norfolk Coast Habitat Management Plan (CHaMP)*. English Nature, Peterborough.

Shepherd, D. Burgess D, Jickells, T., Andrews, J., Cave, R., Turner, R.K., Aldridge, J., Parker, R., and Young, E. (2007). Modelling the effects and economics of managed realignment on the cycling and storage of nutrients, carbon and sediments in the Blackwater estuary UK. *Estuarine, Coastal and Shelf Science*, 73, 355–367.

Turner, R.K., Burgess, D., Hadley, D., Coombes, E., and Jackson, N. (2007). A cost-benefit appraisal of coastal managed realignment policy. *Global Environmental Change*, 17: 397–407.

Chapter 33
A GIS for Managing Past Knowledge in Coastal Defence Planning

Edi Valpreda, Michele Rotunno, Umberto Simeoni, Simona Gragnaniello, and Maurizio Farina

Abstract The chapter describes the work for a coastal defence monitoring program that has been promoted by Regional Authority in Northern Italy (Emilia Romagna). With the aim to organize knowledge on coastal defenses (i.e., presence, state and management) and make analyses comparable with coastal evolution of both emerged and submerged beach, a geodatabase has been implemented; this last has further been improved with the collaboration of technicians that are involved on coastal defense management. The final aim of the work was to supply administration with a powerful and innovative tool for assessing the efficacy of coastal defenses in time. The work also proposes a way to solicit attention on data effectiveness (by way of metadata) and on positional accuracy for coastline evolution assessment.

Keywords: Coastal defences · Shoreline evolution · Coastal planning · GIS · Metadata

33.1 Introduction

Coastal management and planning in Italy is nowadays extremely fragmented and depends on the different geographic districts. The first law on beaches guard dates back to 1907. Municipalities that require coastal defences have a National Government contribution for the construction of the defensive works (3/4 of the total amount). Privates who get benefits from the defences can share expenses for the defences building; the management costs are in charge to the Municipality. That is still the regulation in force at national level; a devolution system started since 1977. Some Regional Councils began to apply rules on the coastal defences built

E. Valpreda (✉)
National Agency for New Technologies Energy and the Environment (ENEA), Environment Department, 40128 Bologna, Italy
e-mail: valpreda@enea.it

D.R. Green (ed.), *Coastal and Marine Geospatial Technologies*, Coastal Systems and Continental Margins 13, DOI 10.1007/978-1-4020-9720-1_33,

© Springer Science+Business Media B.V. 2010

even if, at this time, the choice of building a coastal defence was still only a National Government's responsibility. Only since 1989 (D.L. 183/89) regions have achieved full delegation about this topic and start to make their own regulations. That induces different ways to allocate coastal monitoring and coastal defences planning competences: these charges have been ascribed to Municipalities or other regional Authorities (i.e. Basin Authorities, Regional Agencies for Soil Defence). Nowadays, a fragmented management approach happens, depending from each region choice. Emilia Romagna Regional Council was the first one that has deliberated a specific regulation with regional coastal defence plans: the first plan was accomplished in 1981 and it was the first one in Italy. The Civil Defence in this Region (as part of Regional establishment) plays a central role in the coastal defences plan and management together with the Regional Agency for Environment (ARPA) and the Regional Geological Office (Idroser, 1996).

These offices need to have (and share) knowledge on the past and present coastal defences. Emilia Romagna starts to build hard coastal defences since the latter part of 1960s (Dal Cin & Simeoni, 1984). A so long tradition in coastal management improves this requirement for using past experiences to better decide the future coastal risk mitigation policies. Many times paper archives have been lost, and this because of the lack of an organic and systematic archive for managing data and let them easily accessible in coastal defences projects: these are prepared on paper, submitted to authorities and finally recorded into archives that are generally not well organised. Data are often scattered and their availability depends, mainly, from personal reminiscences. All that is a serious historical awareness waste. Besides, in this way a very important tool to assess and evaluate the past management choices is lacking. That is very important because not all the coastal defences attain, during the time, the expected outcomes. Starting from third century B.C. dikes were built in the Egyptian Mediterranean shores (Charlier et al., 2005). Since then, in the entire world men tried to protect coasts steadying the shoreline or trying to protect beaches against sea-storms constructing groins, breakwaters, seawalls. Moreover, the possible disadvantages of hard defences became aware. Since the latter part of the twentieth century, geologists and engineers for coastal protection have deployed alternative approaches and systems to hard structures. Several new procedures have been developed, both to protect the coasts and rebuild beaches ravaged by erosion with special attention to the ICZM (Integrated Coastal Zone Management) concept whereof new technologies (like the artificial nourishments) are a meaningful ingredient (Preti, 2002).

The coastal risk mitigation policies changes are mainly based on balance between beaches benefit outcomes and costs perceived by local policy makers. The possibility to do a comparison between coasts evolution and defences built up, modification or demolition, becomes a fundamental way to evaluate this balance. With this perception, in the context of a wider project founded by Emilia Romagna Region to realize a beach nourishment across a coastal segment (about 40 km long) southwards the present Po Delta apparatus, it started the idea for creating a numerical archive for manage, homogeneously, and compare, inside a GIS application, past and future knowledge on coastal evolution and coastal defences. Geodatabase was planned and

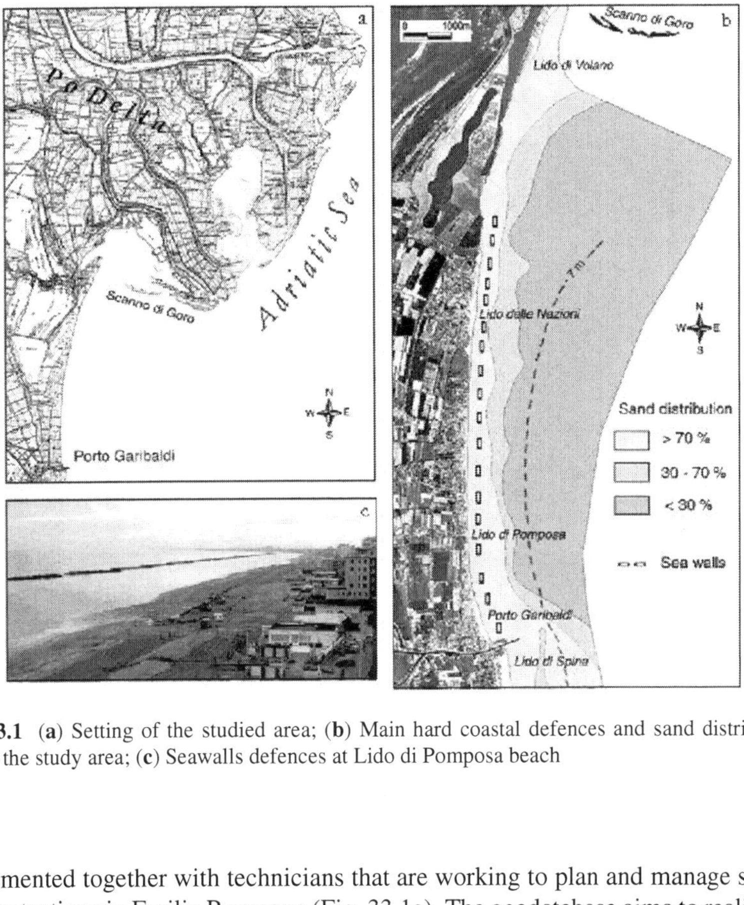

Fig. 33.1 (**a**) Setting of the studied area; (**b**) Main hard coastal defences and sand distribution within the study area; (**c**) Seawalls defences at Lido di Pomposa beach

implemented together with technicians that are working to plan and manage shoreline protections in Emilia Romagna (Fig. 33.1a). The geodatabase aims to realize an innovative outfit, simple to use and update. Data on beaches defences presence, state and management have been related to morphological and sedimentological data on emerged and submerged beach, produced inside the same project, and to coastline changes from 1950s to today. Great attention was given to data quality assessment and declaration to favourite data exchange among different technical offices working on the coastal problems.

Particular consideration on shoreline movements derived by image processing and bathymetric data has also been given in terms of accuracy and usability.

33.2 A GIS for Coastal Defence Management and Planning

The capability to propose, realize and verify the protective structures functionality to mitigate beaches vulnerability and risk strongly depends from availability and fruitfulness of former knowledge for the site.

This idea may seems to be prosaic but it assumes that former data exist, are easy accessible and their genealogy be explicit (as metadata). Besides, data have to be spatial and conceptually comparable. The spatial homogeneity of data goes beyond GIS requirements: survey data produced in different time can be differently affected by errors from interpolation and analysis models. This can greatly affects the possibility of a comparison between them with any methods. Moreover the perception of past knowledge safeguarding is a new concept linked to *e*-government society improvement and to rational and economical awareness to share geodata among public institutions, instead of duplicating them. This approach was forced by European initiatives (as INSPIRE Project). Consequently, it begins to involve also national and regional ambits. Having these European experiences and suggestions in mind, for the needs of local regional services, a vector GIS application was planned and implemented for the coastal zone between Lido di Volano and Porto Garibaldi towns. Thus, all available knowledge on the coastal defences and coastal evolution from 1950s has been considered. Moreover, results are standardized with hardware and software technologies yet available inside the regional offices.

The GIS application is mainly devoted to supply a more efficaciousness storage of data on defence operas creation and maintenance. The main objective was to support new defences planning and designing.

The geodatabase is a quite complex integrated data system that easily allows data to be edited as for example complex queries that can involve coastal interventions presence, attributes, conservation, maintenance etc, with beach emerged and submerged features, state and trend. Data available refer to coastal protections and coastal evolution analysed both from a geomorphologic and sedimentological point of view. The geodatabase was compounded by former available knowledge gathered inside Emilia Romagna Region from paper archives and technicians verbal communications. Other data comes out from investigations, conducted during the project itself. Finally a large part of data on shoreline evolution derives from image processing on maps (from 1954 to 1999), while more recent data (2000–2004) comes from DGPS (Differential Global positional System) surveys (Simeoni et al., 2004).

Defences features are been traced on more recent high detailed available images (1:10.000 scale, from 1999 year) while more recent defences are traced by DGPS relief done in 2004. Defences data are represented trough polygons that sketch each defence bounder. Shorelines are drawn as polyline. Other features depict morphological and sedimentological features on emerged and submerged beach, measure and sampling data point as derived isobaths.

Geodatabase conceptual design manages two congruent and homogeneous sectors distinctly: defences and coastal evolution. That allows editing, managing and comparing, at the same time, if required, whole geodatabase contents as only one thematic sector. Data model is based on "one to one" or "one to many" relationships: each single feature is related to at least one record in the main geodatabase. Defences maintenance trace storage and management are organised also trough relational tables that keep "many to many" relationship type and allow to link one or many recurring safeguarding interventions to each defence item. These tables are provided to let a database simple to update (also in a normal Windows environment) based on defence ID codes without loss of dataset congruence.

Defences are classified for type (11 different category are, at present, editable), for project and keep information on used materials, defence makers, building year, nowadays defence existence, defence length, whole project length, distance from coastline (and years to which refers), present distance between defences, etc.

Nourishments are classified by year of realization, length of involved littoral and deposited sands quantity, textural sands features both of deposited and native sands.

Information set that is linked to each coastal defence is useful to organize many more details than nowadays are available in the regional on paper archives. Many times any information is lacking and the "history" of some group of coastal defences has been achieved only trough technician's memory or trough successive images comparison. Having this great difference in available data a particular attention was given to data origin that is illustrated in the tables linked to defences.

Shoreline evolution derives from different sources (Regional Technical Maps, ortho-images, aerial photos, on field DGPS surveys. Shoreline trend derives from ten successive statements that have been used to identify the amount of changes represented as oldest coast line transverse transects placed at regular distance among these (Fig. 33.2a). Coastal trend assessment is very important to evaluate defences needs and defences impact on beaches evolution. Paying attention to the accuracy in determining the changes among successive coastal lines is a fundamental condition to derive correct conclusions.

Because of changes rate ranges are very wide (from 1 to over 20 m), it's required to give much attention to significance of these values. Shorelines were traced on different digital and on paper maps and have different accuracy depending from sources and from handling done to make cartographic bases consistent (scanning, georeferencing, changes of coordinate system or datum, etc). Many measured changes values can not have relevance when we consider only values overcome inaccuracy rate. The inaccuracy assessment was done comparing many known ground points, along coast, between certified cartography (made available by Emilia Romagna Region) and maps used in Volano Project.

A great attention in managing extant defences is given by local coastal public bureaux. So it is very important to have the possibility to compare defences with coastal morphologies and sedimentology distribution and composition, and coastal evolution in time.

Database meets these elements: coastal defences presence, details, maintenances and changes, etc., can easily be compared with coastal morphologies and sedimentology distribution and composition. All data and sedimentology coming from surveys done inside Volano Project (on merged and submerged beaches and up to 8 m in the sea) were organized inside the same geodatabase (coastal evolution environment) that is spatially congruent with coastal defence environment.

Isobaths lines resulting from bathymetry, samples location and morphological profile traces are available in GIS model. Relational tables were created to make all sedimentological analyses results available (spatially linked to each sample location) (Fig. 33.2c).

Fig. 33.2 Detail of coastline evolution: (**a**) changes between two successive coastline traces: 1972 (*black points*) and 1978 (*points and line*) are measured on crosswise transects that show distance values ; (**b**) beaches retreating (from 1978 to 1999) in concomitance with groins built before 1936 (the five northern groins were demolished during 2000). In *black*, a near shore seawall, built during 1999; (**c**) nearshore isobaths (*light grey lines*), samples location (*crosses*) and sedimentological profiles (*dashed lines*) done during the main Volano Project and included in the geodatabase

33.3 Coastal Management

Coastal tract inside which Volano project works (Fig. 33.1b) is in the Southern part of Po river Delta, about 20 km length, between Goro lagoon and Porto Garibaldi harbour.

Wide and continuous sandy beaches are present with a general retreating tendency. While till to nineteenth° century beaches are prograding or stable (Bondesan et al., 1978), from twentieth century beach enlargement trend begins to decelerate and, from 1950, starts a whole regressive (sometimes even parossistic) coastal dynamic. Even if there are diffuse coastal defences to mainly protect littoral from North Eastern winds, regressive continuous trend still goes on.

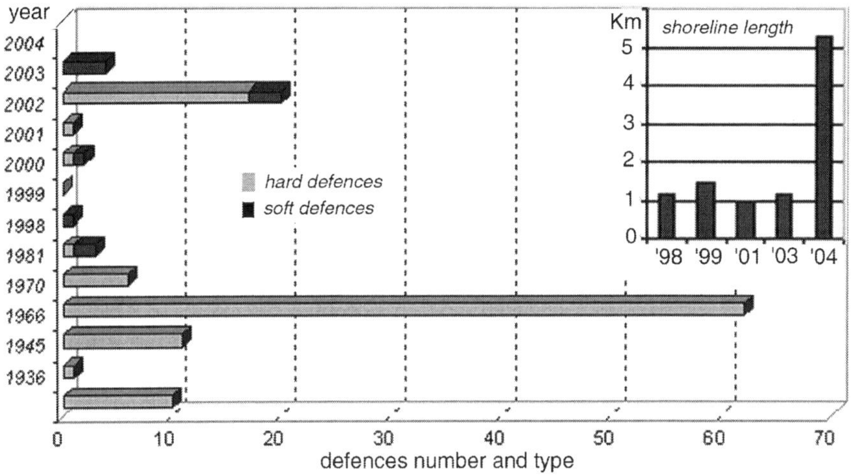

Fig. 33.3 Temporal evolution of hard and soft coastal defences building in study area (from 1936 to 2004) and (*upper right side sketch*) nourished shoreline length (kilometres) from 1998 to 2004

Coastal dunes growth mainly 100–200 years ago, are locally still preserved even if large amount of demolition occurs both by sea waves (due to beach erosion) and human actions.

Many bathing establishments have grown on the beaches from 1950s. Having a persistent beaches erosive crisis an emergency state becomes. As consequence, before 1936, Emilia Romagna Region starts to build coastal defences (Fig. 33.3). In 1968 more than 80 hard defences were yet present. Till 2002, hard defences realisations have gone on and it has reached 117 defensive works, only 95 of which still persist: 19 emerged groins (from 27), 14 submerged groins (one of which is now emerged after recharge), 74 seawalls (2 of which are still submerged). Nowadays inside coastal studied tract, groins are present for more than 9 km; seawalls for about 2 km. Re-nourishments began in 1998, sometimes together with hard defensive works, but become, in the last years, the prevailing mitigation option applied in Emilia Romagna littorals (Corbou et al., 2004). Between 1998 and 2003, 11 non traditional coastal risk mitigation interventions have been realized (8 beach nourishments projects and 3 projects coastal dunes building). During last 6 years (1998–2004) more than 5 km (of 20 studied) was nourished moving more than 200,000 m^3 of sands.

Despite many coast safeguard interventions, started around 70 years ago, in the studied littoral beaches are still retreating. Analysing coastal changes in the whole considered time (1972–2003), 7800 m of shoreline are advancing while 5800 m are retreating. During the latest 31 years, the average advancing is 41 m while the retreating one is 46 m (respectively corresponding to 1.3 and 1.5 m/years). Detailing the changes of coastlines in times, erosion does not show a linear trend: a focus happens in last of 1970s while during 1990s retreating average values are constantly lower but greater than corresponding advancing values (Fig. 33.4a).

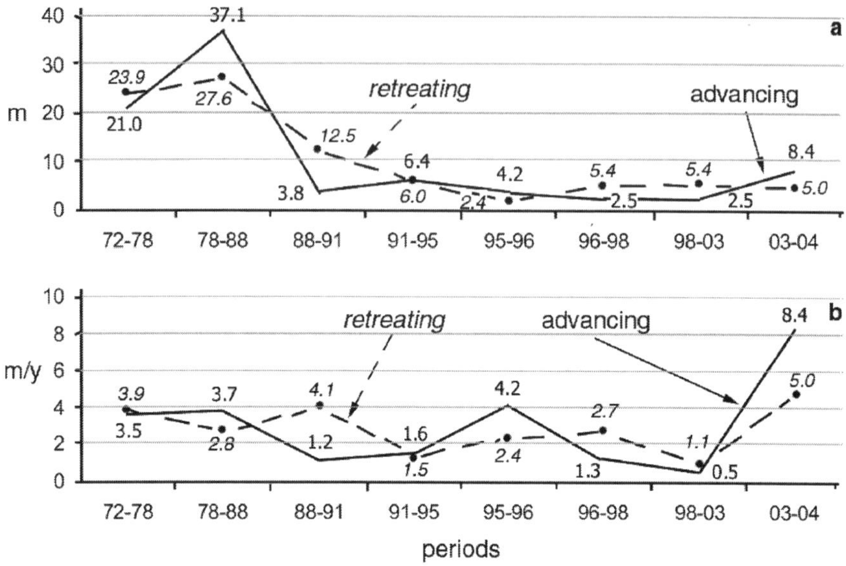

Fig. 33.4 Coastal evolution from 1972 to 2004: (**a**) averages (in metres) of total advancing and total retreating; (**b**) rates (metres/year) of coast line advancing and retreating

It is very interesting to consider advancing and retreating rates calculated in each time span (Fig. 33.4b). Beyond this, database allows analysing this phenomenon in the space. It results that erosive focuses moved in time towards littoral from southern to northern area and that some coastal erosive recurrences happens independently from defences presence.

Main factors influencing the moving back of the beaches are: sedimentary balance decreasing and modifications induced in hydro-sedimentary flow and currents by hard defences. In fact, hard defences, also where well designed, can lose their efficaciousness as consequence of changes in emerged and submerged beach morphologies or changes from project initial geometries.

The possibility to known, analyse and compare sedimentology and bathymetry is a fundamental condition to support (hard and soft) defences efficaciousness evaluation programs.

The sedimentological analysis of 256 samples indicate that, at the present, beaches are principally composed by sands with medium diameter of 1–1.5 phi. According to the Shepard's classification (1954), 16% are clayed silt, 2% sandy silt, 14% silty sand and the rest corresponds to sand. Submerged littorals deeper than 4–6 m are composed by silt and clay sediments with mud percentages (>4 ϕ) that varies from 100% to 1% (with medium value around 69%). Samples collected between 4 and 1 m water-depth show sands dominance (more than 95%). In the studied coastal area, spatial distribution of sands greatly varies (see also Fig. 33.1): data elaboration highlights a wider sands presence Southern to Porto Garibaldi harbour, where sand are found at water-depth ranging from –3 to –7 m. While in the northern tract sand deposit is narrower (limited at depth between –2 and –3 m). This effect is probably a consequence of harbour seawalls interference. Porto Garibaldi seawalls interrupt the long-shore currents and stop sediments supply coming from Reno river mouth and from the erosion of the beaches located in Southern coastal tract (Simeoni et al., 2003). Sand presence reduces rapidly seaward (getting percentage of 30% at –6 m). Shallow waters main changes happen in this area between 0 and –6 m (at beach distance from 2600 to around 900 m) and where slopes show 0.5% gradient. In the Northern surveyed area, nearest to submerged Po River delta apparatus, slopes gradient is smaller.

33.4 Conclusions

Coastal erosion mitigation projects are certainly induced benefit for coastal environments solving main risk situations avoiding beaches and mainly inner lands losses. But it is also important to determine exactly the time when defences well operate as their possible interferences with coastal morphology and sedimentology.

Until today coastal risk management is not really based on coastal processes or past choices effects awareness.

Great relevance of these coastal areas together with more and more fast evolution of these frails environments (affected by erosion, subsidence, flooding and pollution) induces the real possibility to apply ICZM approach to meet different activities that, all together, impact on coastal lands. Knowledge is a substantial factor to achieve these goals and reach solution hypotheses that don't stake a claim on the future. As consequence of this concept a new tendency in coastal guard government in Emilia Romagna has started in the last recent years: inside defences efficaciousness evaluation program, since year 2000, hard defences demolishment has also accomplished. Privates financed these actions working together with regional technical offices. Defences for coastal environments depend on local specificities and these defences are "at term", their duration depending both from maintenance done in time and coastal evolution in consequence to their presence.

All that made clear the importance to dispose of organised and homogeneous data on morpho evolutionary aspects not only to monitor defences together with coastal evolution but also to support defences design. This kind of geodatabase can easily provide all needed information on the littoral where a risk mitigation program is required. Data are generally not easily accessible or certified: this is a new concrete possibility offered to the risk mitigation designers and planners activities. Finally to collect and organize integrated information is not only useful to assess or plan actions: many times information is not written and can be forgotten. Archives on paper can be scattered and can be easily lost. Moreover metadata allow distinguish between data and data interpretation. Disposing of datasets that also explain the processes used to derive elaborations (as done with the presented geodatabase) allows the comparison between successive surveys, having the same geographic projection,

Datum, nominal scale, declaring methods used to georeferencing data, tracing isolines, etc. It is important to remember that the lack of correct and certified details is debated in all main recent coastal projects (i.e., EUROSION) as one of the main reason for the failure in truthful coastal management.

Acknowledgments Our thanks to Ing. A. Peretti who provided advice and helped with discussion and suggestions our work. Drs Bonora N., Corbau C., Schiavi C., Tessari U. and Zamariolo A. for field data gathering and elaboration. Colleagues from The Regional Geological Office in Emilia Romagna for data supplied. Sarti Company for logistics supplied.

References

Bondesan, M., Calderoni, G., and Dal Cin, R. (1978). Il litorale delle province di Ferrara e Ravenna (Alto Adriatico); evoluzione morfologica e distribuzione dei sedimenti, *Bol. Soc. Geol. It.*, 97, 247–287.

Charlier, R.H, Chaineux, M.C.P, and Morcos, S. (2005). Panorama of the history of coastal protection, *J. Coastal Res.*, 21(1), 79–111.

Dal Cin, R., and Simeoni, U. (1984). Variazioni volumetriche del delta padano nell'ultimo secolo, distribuzione dei sedimenti nei rami deltizi e caratteri granulometrici dei materiali cavati nel basso Po, in: *Atti del 2° Convegno di Idraulica Padana, Tecnografica*, Parma, pp. 255–279.

Corbou, C., Tessari, U., Simeoni, U., Bonora, N., Gragnaniello, S., Zamariolo, A., Farina, M., Peretti, A., and Sarti, L. (2004). First results of a beach nourishment in north Adriatic Sea, in: A. Micallef and A. Vassallo (Eds), *First International "Conference on the Management of Coastal Recreational Resources"* 20th–23rd Oct. 2004, Malta, pp. 31–39.

Idroser (1996). Progetto di Piano per la Difesa dal Mare e la Riqualificazione Ambientale del Litorale della Regione Emilia-Romagna; Relazione Generale. Regione Emilia-Romagna, Bologna. pp. 365.

Preti, M. (2002). – Ripascimento di spiagge con sabbie sottomarine in Emilia-Romagna. *Studi Costieri*, Firenze, 5, 107–134.

Simeoni U., Gragnaniello S., Bonora N., Farina F., Peretti A., Sarti G., Schiavi C., and Valpreda E. (2003). Effects of anthropization on the Ferrara littoral and intervention of re-equilibrium. In: Ozhan E. (Ed), *Proceedings of the Sixth International Conference on the Mediterranean Coastal Environment,* 7–11 October 2003, Ravenna, Italy, 1–3, 1605–1616.

Simeoni, U., Bonora, N., Corbou, C., Gragnaniello, S., Schiavi, C., Tessari, U., and Zamariolo, A., (2004). Il litorale tra Porto Garibaldi ed il Lido di Volano. *Rapporto Tecnico*, Regione Emilia-Romagna, Ottobre 2004, pp. 296.

Part VI
Coastal Hazards and Vulnerability

Chapter 34
Environmental Indicators GIS of the Catalan Coast

Jorge Brenner, José A. Jiménez', and Rafael Sardá

Abstract This chapter describes the environmental indicators GIS of the Catalan coast, Spain. The spatial data model is based on vector and raster layers with three main modules: the biodiversity, the general biophysical and the socioeconomic. Presently, the database has a number of pressure and impact indicators that have been used to model the components and structure of the system and are suggested to build ecological resilience. Special interest has been given to the analysis of functional groups of species that are relevant to the dynamics of the coastal system, and preliminary results are presented. This system constitutes a user-oriented analytical and monitoring tool for coastal zone managers and researchers. Although, the system is under development it is expected that resulting spatial indicators of environmental condition can be used to promote more sustainable coastal strategies and actions in the Catalan coast.

Keywords Geographic information systems (GIS) · Integrated coastal zone management (ICZM) · Catalonia, Spain · Environmental indicators

34.1 Introduction

Until recently, there has been relatively little attention to measure the interactions between the socioeconomic system and the environmental variability (Bowen and Riley, 2003), with consequent limited understanding of the linkages of between coastal systems dynamics and the social benefits associated with them for Integrated Coastal Zone Management (ICZM). From a holistic point of view the interaction

J. Brenner (✉)
Harte Research Institute for Gulf of Mexico Studies, Texas A&M University – Corpus Christi, 6300 Ocean Dr., Corpus Christi, TX 78412-5869, USA
e-mail: jorge.brenner@tamucc.edu

This chapter was published earlier in the Journal of Coastal Conservation 11:185–191 (2007).

D.R. Green (ed.), *Coastal and Marine Geospatial Technologies*, Coastal Systems and Continental Margins 13, DOI 10.1007/978-1-4020-9720-1_34,

© Springer Science+Business Media B.V. 2010

between the environmental and economic ecosystem processes determine the *environmental valuable attributes*, which can be used to measure the conservation state and other environmental functions (King, 1995). Environmental value is a concept that relies on the relationship between the functional state and quality of an element or process and the human perception, and although desirable it is not possible to determine their contribution to human society in an objective way (Cendrero et al., 2003).

Coastal zone environmental research and state assessment often requires the integration of physical, chemical, geological, biological, social, economical, governance and management dimensions of the environmental system. The use of the new information and communication technologies is powerful to integrate data from these fields in a way that new perspectives can be achieved. Geographic Information Systems (GIS) allows exploring the complexity of the coastal system in a comprehensive manner and promote synergies between different types of data thatINT-break; provide more information than that obtained by considering data from each field separately (Fornés et al., 2003). The role of GIS in ICZM has been highly appraised, e.g. Ellis (1972), Ader (1982), Fairfield (1987), Davis and Davis (1988), Townend (1990), Welch et al. (1992), Riddel (1992), Ricketts (1992), Jones (1995), Bartlett (1990, 1994, 2000), Deakin and Miment (1994), Stanbury and Starr (1999), Vallega (2003), as ICZM requires handling a large amount of spatial and tabular data. Watts (2003) proposed a set of different GIS dimensions and best practices needed to be addressed for ICZM planning; alternatively, Longhorn (2003) synthesized the current pan European initiatives on ICZM and GIS that should constitute a regional starting point for any GIS work in the field.

The advantages of applying GIS to the coast and marine environment have been well articulated by Bartlett (2000), being: (1) the ability to handle large databases and to integrate and synthesize data from a wide range of sources, (2) the encouragement for the development and use of standards for coastal data definition, collection an storage, (3) the use of shared database, especially if access is provided via a data network, facilitating the updating of records and the provision of a common set of data, and (4) the ability to model, test and compare alternative management scenarios, before a proposed strategy is imposed on the real world. However, ICZM planning process need of frameworks capable to measure and understand the coastal subsystem interactions and variability.

Li (2000) revised the available coastal and marine data models concluding that we cannot integrate efficiently multidimensional and temporal data, therefore future efforts should focus on the temporal nature of coastal–marine data and the integrity of spatial data from interdisciplinary sources. Recently, Breman et al. (2002) proposed the *ArcGIS Marine Data Model* that has been implemented for ESRI's version 8 of its GIS software. Although it constitutes a novel data model for the marine environment it is still not an integrated coastal zone system. Coastal system dynamics calls for new models and functions in order to solve real world management problems, and therefore the development of coastal GIS should align with ICZM principles. Four characteristics have been proposed by Zeng et al. (2001) as desirable of ICZM functionalities in order to succeed in a GIS application through balancing the level of application it self and the required capabilities: (1) based on a long term integrated approach; (2) zone focused or based on environmental management units; (3) multidisciplinary and multicriteria analysis based; and (4) equipped with uncertainties handling tools (fuzzy sets, probability and simulation analysis). Although some coastal-marine GIS have been implemented as decision-support systems and general data analysis tools, most applications have been developed as simple spatial data management and mapping tools (Jones, 1995; Zeng et al., 2001).

The objective of this study was to develop a spatial framework of environmental state indicators of ICZM. The spatial framework needs to be implemented in a GIS environment and be capable of supporting main indicator functions and application for the Catalan coast environmental state assessment and valuation.

34.2 The Coastal Zone Environment Indicator System

In order to conceptualize an indicator system (IS), a reasonable first step is to determine which components adequately specify a comprehensive monitoring system. (Sardá, 2003) has referred to five environmental themes as present core needs of the Catalan coast environmental condition: (1) coastal-marine natural protected areas; (2) coastal water quality; (3) economic and development areas zoning; (4) biodiversity loss reduction; and (5) commercial species regulation and monitoring.

Then an environmental indicator-based assessment and valuation framework has been designed to track the performance and progress of the Catalan coast ICZM Plan (GenCat, 2004). Similar approaches have been implemented for other coastal programs from local to national scales (Burbridge, 1997; Belfiore, 2003; Henocque, 2003; Olsen, 2003; Pickaver et al., 2004). Developing indicators capable of capturing the biophysical and socioeconomic structure and processes remains one of the most difficult challenges in ICZM, due to the enormous potential of coastal use conflicts between sectors and the natural functioning who share the same space. The developed environmental health and state indicators for ecosystem management of the Catalan coast were based on an ecosystem approach, since this constitutes an analytical and synthetic framework of multidisciplinary science-based decision-making (Kay, 1994; Crober, 1999).

Environmental indicators of coastal zone have typically been developed within the *Driver – Pressure – State – Impact - Response* (DPSIR) framework at national level. The DPSIR framework was originally developed by the Organization for Economic Co-operation and Development (OECD, 1993), later adapted to the coastal zone by Turner et al. (1998). This model has been adopted by the European Environment Agency and thus used as the analysis framework for the environmental state indicators of the Catalan coast. This causal model organizes ICZM into a more systematic evaluation cycle that couples environmental change and progress towards more sustainable forms of coastal development. The efficiency of coastal indicators of ICZM in Catalonia and elsewhere requires demonstrating progress and results on a comparable manner across spatial scales and management levels. The difficulties of this approach are more apparent whenever the spatial and temporal heterogeneity of the coast increases and complicates by the difficulty of combining indicators of the natural and socioeconomic subsystems in the assessment process.

A vector data model has been implemented as the first phase of the Catalan coast GIS in *ESRI's Arcview GIS v3.x*. The information system of the Catalan coastal zone in Fig. 34.1 includes three main modules: coastal-marine biodiversity, general natural (biophysical) and socioeconomic. The information system has the capacity of database maintenance, analysis, visualization and data discovery though metadata sub-module, as well. Due to the heterogeneity of the coastal zone, and the need to efficiently incorporate the environmental structure and function, a regional sub-national scale between 1:50,000 and 1:250,000 was selected as appropriate for the Catalan coast (UNEP, 1995). Information system conceptual relationship to specific management objectives and management tools, as well as other external databases are shown on Fig. 34.1.

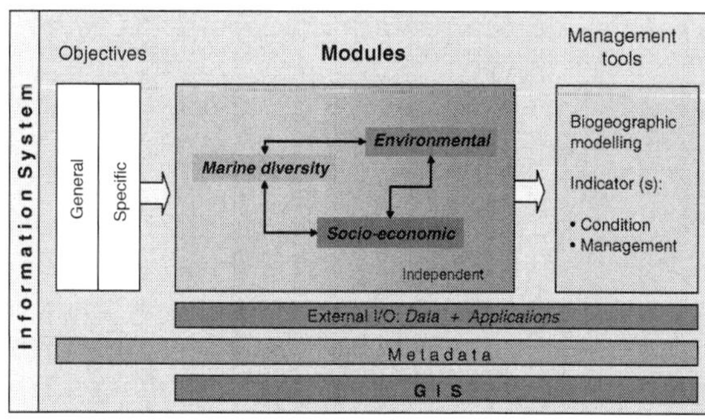

Fig. 34.1 Conceptual model of the Catalan coastal zone GIS (See also Plate 63 on Page 430 in Color Plate Section)

Catalonia has been subject of a limited thematic mapping effort, and thus module variables were selected from the available GIS layers published mainly by the local administration departments (see Tables 34.1 and 34.2 for an example of included layers). Layers have been frequently incorporated into the Catalan Coastal GIS since 2003, as they have been developed by either the Coastal Management Group of the Universitat Politècnica de Catalunya or local management agencies. Selection has been based on the thematic and independent representativeness of the spatial and temporal semi-constant characteristic of the coastal zone. Therefore, themes have been determined according to its conceptual environmental contribution as valuable phenomena of the coastal dynamic sub-system and quality of the available data. Previous to their incorporation, they followed common spatial standardization and data quality assurance process, based on international technical and metadata standards (a data update criteria is also applied).

The sustainable use of biodiversity requires a more holistic, multispecies and ecosystem-based approach, and such level of understanding of the distribution patterns and relations among species and the ecosystem needs to be based on a robust

Table 34.1 Themes by dimension used for the Catalan coastal zone HEMU definition

Dimension	Theme	Scale	Year	Descriptor (s)
Socioeconomic	Population size	Comarca	2004	Inhabitants count[1]
	Population growth	Comarca	2001	Mean anual rate[1]
	Gross National Product	Comarca	1996	Euros at market price[1]
	Accommodation coefficient	Comarca	2002	Hotel beds by population[1]
	Impervious surface	Comarca	2003	Urban area and infrastructure[1,2]
Natural	Natural protected area	25,000	2004	Protected areas and wetlands surface[3]
	Geomorphologic relevance	50,000	2002	Areas surface[3]
	Vegetation condition	25,000	2004	Naturalness, diversity and rarity[4]
	Landscape transformation	Municipality	2004	Environmental degradation[3]
	Running water condition	Municipality	2003	River flow and quality[3,5]

Data source: [1]Institut d'Estadística de Catalunya – Generalitat de Catalunya (GenCat);
[2]Centre d'Estudis Avançats de Blanes – Consejo Superior de Investigación Científica;
[3]Departament de Medi Ambient i Habitatge – GenCat;
[4]Departament de Biologia Vegetal – Universitat de Barcelona;
[5]Agència Catalana de l'Aigua – GenCat.

Table 34.2 Selected land-to-sea pressure indicators of the Catalan coastal zone

Indicator	Pressure descriptor(s)	Impact factor
Industry	Nuclear Plant/other[1]	1–1000 m
Aquaculture	Surface/type/organism/intensity[1]	1000 m
Coastal tourism	Beach length >= 100 m/high use/urban[2]	Beach length
Submarine waste outfalls	Diameter/long/category/status[1]	Outfall length
Ports	Type/surface class[1]	2000 m
Coastal urban pressure (CUP)	Municipal urban surface/municipality coastal length[1]	Coastal length

Data source: [1] *Generalitat de Catalunya;*
[2] *Ministerio de Medio Ambiente.*

spatial biogeographical data model (Tsontos and Kiefer, 2003). We have developed a specific module for Biogeographical Information (BI) to better understand the role of coastal-marine biodiversity in support of an ecosystem management approach of the Catalan coast. The BI module focuses on the integration, analysis visualization and discovery of species and the environmental variables that control their success. Although the module is operational at the moment, only a few capabilities have been implemented. The BI module will be completing in the near future and moving towards the understanding of: (1) the biogeographic complexity; (2) the role of biodiversity function; (3) the ecosystem's ecological condition; and (4) the identification of conservation priorities and gaps.

Spatial features will include behaviours attribution, to model natural features as they are supposed to be found on nature and/or anthropogenic environments. Also, it will be structured by several geo-ecological domains as habitats, sites (conservation) and other tiling forms as distribution (species range), ecological communities and geopolitical boundaries. See module conceptual structure in Fig. 34.2. Basic data model features were selected from other existing and successful implemented ESRI GIS data models as the Marine Data Model (Oregon State, Duke University and ESRI; http://dusk.geo.orst.edu/djl/arcgis/index.html) and the Biodiversity Data Model (NatureServe and ESRI) (see Breman et al., 2002 for a detailed marine GIS feature list).

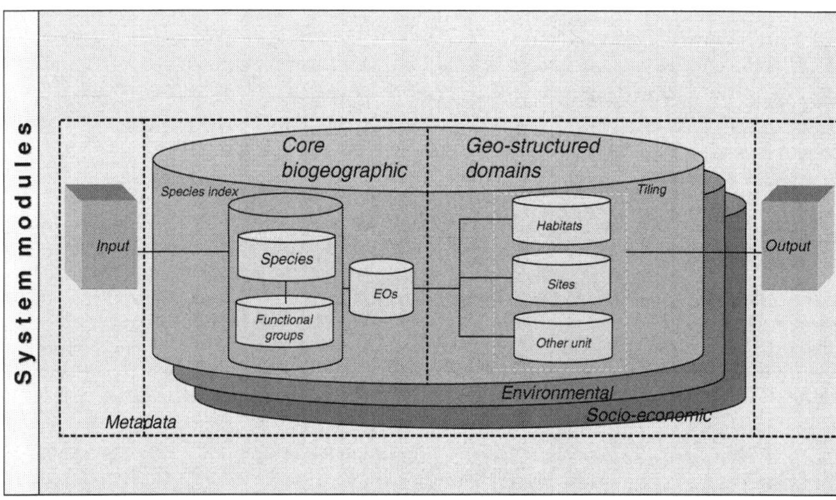

Fig. 34.2 Conceptual entities and relations of the biogeographic module (See also Plate 64 on Page 430 in Color Plate Section)

34.3 Implementation Examples

Two cases of the Catalan coast GIS implementation are presented here. Due to heterogeneity of the coastal zone; the first one consists of an Homogeneous Environmental Management Unit (HEMU) regionalisation of the terrestrial subsystem. In the second case the spatial relationship between human land-based activities and the ichthyofauna is analyzed in order to produce a series of indicators of human pressure and impact along the Catalan coastal zone.

34.3.1 Coastal Zone Environmental Regionalisation

The fundamental purpose of all ICZM initiatives is to maintain, restore or improve specified qualities of coastal zone systems and their associated human societies. A relevant feature of ICZM is that it addresses needs for both socioeconomic development and natural conservation in geographically specific planes, be the local, national or regional supra-national level. Geographic areas constitute the basic implementation locus of ICZM strategies and actions. This exercise offers a general and simple method to regionalize the Catalan coastal zone into HEMUs (Brenner et al., 2006). Discrete spatial units were developed from although Catalan coast specific characteristics (see Table 34.1), but general that methodology could be applied to other coastal cases. The territorial administrative *comarca* unit was chosen as coastal regionalisation spatial planning unit. Regionalisation analysis was developed in a GIS environment. Socioeconomic and natural (biophysical) subsystems were selected as the appropriate dimensions of the regionalisation process.

A final map of four classes of HEMU is provided in Fig. 34.3. Classes range from highly urban and developed to highly natural areas and are intended to constitute a neutral territorial management schema to the coastal environment in Catalonia using Jenks optimization method (Jenks, 1967). The proposed method has been considered a straightforward spatial analytical tool for coastal homogeneous unit's definition in Catalonia. Numbers attached to *comarcas* represent the total richness or direct adding of both dimensions values, which in this case represent the "fuzzy transitional areas" that could pass from natural to developed areas as the ones with 1.5 and 2.0 values. Although resulting HEMU schema depends on specific characteristics, the ones proposed here can be substituted from other site-specific relevant. The proposed HEMU map is expected to constitute an important tool in the implementation of the recent ICZM Plan of Catalonia.

34.3.2 Evaluation of Human Pressure on Ichthyofauna

Along its 699 km the Catalan Coast accounts for 490 out of the 655 fish species of the Mediterranean (Allué et al., 2000). In Catalonia, the ichthyofauna is the best know group among marine biodiversity. Moreover, a potential ecological functional group as environmental bioindicator of the condition of the coastal zone, due to its contribution to spatial resilience. Several of those species are considered rare and deserve special conservation attention. The majority of those species have been extensible studied by taxonomists, ecologists and fisheries biologist in separate efforts. Moreover Catalan coast also accounts the two most important pressures on the Spanish littoral system; the tourist industry, the urbanization, and

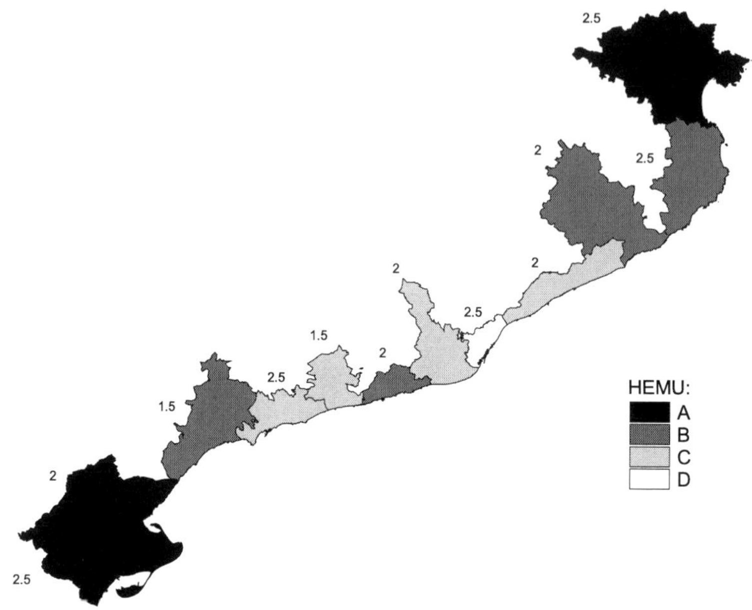

Fig. 34.3 Homogeneous environmental management units of the Catalan coastal zone. Numbers indicate total socioeconomic and natural richness by unit

their associated impacts (Sardá, 2001). In order to alleviate the pressure on coastal environments and its biodiversity the European Community has recently recommended a new strategic and harmonized framework for regional and local implementation (COM/00/547). This new instrument recommends studying in integral manner all the driving forces that generates pressure on the coastal system. In this study we propose a methodology to identify the relationship between the forces of pressure that drive the species diversity and distribution. Some of the impacts are evident and have been studied in great detail as fisheries and some kinds of pollution, but in this case we analyze the effect of inland-coastline generated pressures on this group.

The approach involved the definition of threat categories based on diversity and distribution criteria, and the identification of component sources of pressure that can be mapped. These stressors include simple population and infrastructure features such as cities, ports, and discharge pipelines as well as more complex modelled layers of river inputs. Coast stressors have been previously identified in the Catalan Port Plan (GenCat, 1983), and six were selected for this study (Table 34.2). Once these components have been selected, model rules were developed for translating them into spatial measures of threat. The indicators were designed to highlight areas where, either high diversity or high pressure occurs and provide a regionally consistent indication of human impact on this biodiversity group, which could serve as a proxy guide to fish conditions across the Catalan coast.

Two types of analysis were conducted in order to investigate whether or not there was a relationship between the anthropogenic pressures and coastal fish diversity. The first one consisted on the visual and spatial analysis in the GIS environment. A total of 265 species with 2598 element occurrences (EO; point data) from Fishbase Database were available to conduct the analysis (Froese and Pauly, 2004). As a result, a possible impact area of 479,997 ha was identified in the coastal-marine area. The impact area reached as far as 19 km from the coast and no shoreline was impact-free, but the semi-isolated northern natural protected area Creus Cape. Matching the impact region were 66.7% (177) of the species, where 54.5% (6) had some degree of concern following the IUCN (2004) criteria (species numbers in parenthesis). The 32.8% (854) of the total EOs matched the 10.7% (331) of the hexagon model that was implemented. Impact indicators area and species by hexagon unit can be seen in Fig. 34.4. Mean resilience to fisheries pressure of all impacted species was 2.3 (based on a 1 to 4 scale, being 4 the highest resilience to population perturbation). Among those, the mean of special concern species was 2.18, with a maximum of 3. The most vulnerable species were defined to be those with costal distributions in depths <= 50 m. 66% of the species (150 species in 66 families) were selected as vulnerable to the analyzed pressure and impact indicators, and 83% were of special concern (5). Second types of analysis were the canonical redundancy analysis (RDA) and Mantel's simple correlation between species EOs and independent variables (Legendre and Legendre 1998; Casgrain and Legendre 2001). RDA is the canonical form of the Principal Component Analysis, and the technique selects the linear combination of environmental variables that gives the smallest total residual sum of squares (Jongman et al., 1995).

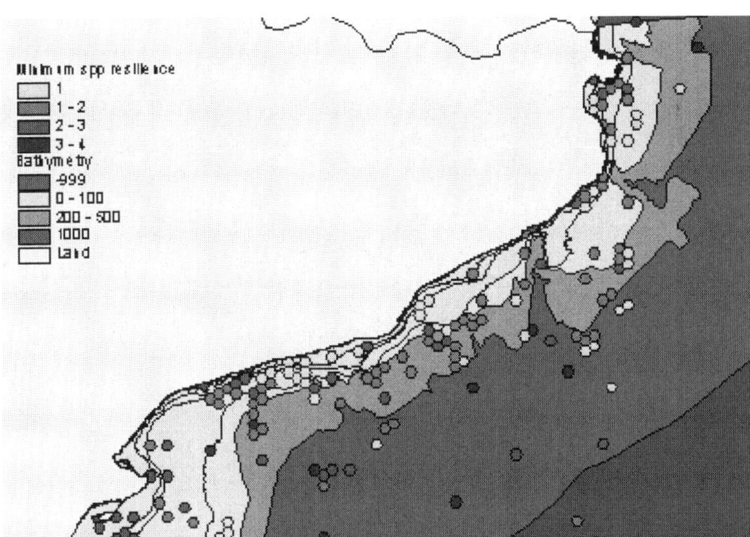

Fig. 34.4 Marine hexagon model of anthropogenic impact indicators and fish species resilience of the Catalan coast (See also Plate 65 on Page 430 in Color Plate Section)

Conclusions based on the analysis performed suggest that species occurrences were not significantly correlated with the impact areas. This was maybe due to an inexistent relation or because present data model needs to be improved. The correlation through RDA permits to identify species groups that could be used as indicators of environmental quality (e.g. Gobiidae and Triglidae families) and finally, using the resilience of the species, the analysis has permitted to identify pressure/impacts able to be "resisted" by fish, on the assumption that those species occur on the coast at present time (impacts). We suggest that future work should move towards the improvement of the marine biophysical model, through adding spatial hydrodynamics at some level, in order to develop a better predictor variable set. Another action should be, to develop a similar approach using presence/absence data of special concern and documented "indicator" species. We recommend that, this type of analysis should be done systematically by coastal managers in order to identify conservation priority elements and sites, since more than 50% of fish families, species occur on littoral waters below 50 m depth. We hope this will constitute a new way of looking at and understanding the effect of human activities on the coastal zone.

Aknowledgements This work was done in the framework of the Mevaplaya Project (REN2003-09029-C03-01/MAR) funded by the Spanish Ministry of Education and Science. The first author thanks the support of the Escola Tècnica Superior d'Enginyers de Camins, Canals i Ports de Barcelona-UPC for its financial support; and through a doctoral studies grant of the National Science and Technology Council of México (CONACyT). Authors thank the organizations and institutions supplying data used in this work, speciallt from Fishbase. Special thanks to Albert Ferré (UB), Ramón Jordana (DARP), Marta Manzanera (ACA), and Xavier Martí (DMAiH). The help of Modest Fluvià (UdG) on the definition of the economic dimension is also acknowledged.

References

Ader, R.R. (1982). "A geographic information system for addressing issues in the coastal zone", *Computers, Environment and Urban systems*, vol. 7, pp. 233–243.

Allué, C., Lloris, D., and Meseguer, S. (2000). *Colecciones biológicas de referencia (1982–1999) del Instituto de Ciencias del Mar (CSIC): Catálogo de Peces*, Instituto de Ciencias del Mar, Barcelona, pp. 200.

Bartlett, D.J. (1990). "Spatial data structure and coastal information systems", *Proceedings of EGIS'90, The First European Conference on GIS*, pp. 30–39.

Bartlett, D.J. (1994). *GIS and the coastal zone: past, present and future*, AGI Publication, London.

Bartlett, D.J. (2000). "Working on the frontiers of science: applying GIS to the coastal zone", in: *Marine and Coastal Geographical Information Systems*, Wright, D.J., and D.J. Bartlett, (eds.), Taylor & Francis, London, pp. 11–24.

Belfiore, S. (2003). "The growth of integrated coastal management and the role of indicators in integrated coastal management: introduction to the special issue", *Ocean & Coastal Management*, vol. 46, pp. 225–234.

Bowen, R.E., and. Riley, C. (2003). "Socio-economic indicators and integrated coastal management", *Ocean & Coastal Management*, vol. 46, pp. 299–312.

Breman, J., Wright, D., and Halpin, P.N. (2002). "The inception of the ArcGIS Marine Data Model", in: *Marine geography: GIS for the oceans and seas*, J. Breman, (ed.), ESRI Press, Redlands, pp. 3–9.

Brenner, J., Jiménez, J.A., and Sardá, R. (2006). "Definition of homogeneous environmental management units for the Catalan coast", *Environmental Management*, vol. 38, pp. 993–1005.

Burbridge, P.R. (1997). "A generic framework for measuring success in integrated coastal management", *Ocean & Coastal Management*, vol. 37, pp. 175–189.

Casgrain, P., and Legendre, P. (2001). *The R Package for multivariate and spatial analysis v4, User's Manual*, Département de Sciencies Biologiques, University of Montréal, Montréal, pp. 136.

Cendrero, A., Frances, E., del Corral, E., Ferman, D., Fisher, J.L., del Rio, D., Camino, L., and Lopez, A. (2003). "Indicators and indices of environmental quality for sustainable development in coastal areas; application to case studies in Europe and the Americas", *Journal of Coastal Research*, vol. 19, pp. 919–933.

Crober, A.M. (1999). *The ecosystem approach to ecosystem management*, Bachelors Thesis, University of Waterloo, pp. 34.

Davis, B.E., and Davis, P.E. (1988). "Marine GIS concepts and considerations", *Proceedings of the GIS/LIS'88 Conference*, AGI Publication, San Antonio, TX, ACSM/ASPRS, pp. 159–168.

Deakin, R., and Miment, R.P. (1994). "Strategic-planning and decision support. The use of GIS in the coastal environment", *Proceedings of the AGI'94 Conference*, 13.3.1–13.3.5.

Ellis, R.H. (1972). "Coastal zone management system: a combination of tools", in: *Tools for coastal management*, Marine Technology Society (eds.), Marine Technology Society, Washington, DC

Fairfield, J. (1987). "Marine planning with Geographic Information Systems", *Proceedings of the Coastal Zone'87 Conference*, pp. 1023–1030.

Fornés, A., Basterretxea, G., Orfila, A., Jordi, T., Moral, F., Alvarez, A., Vizoso, G., Casas, B., Duarte, C., and Tintoré, J. (2003). "The IMEDEA M&CGIS: a GIS based interdisciplinary tool for scientific management of the coastal zone", *Proceedings of the CoastGIS'03 Conference*, Genoa, pp. 7.

GenCat (1983). *Estudi del Pla de Ports Esportius*, Direcció General de Ports i Transports, Departament de Política Territorial i Obres Publiques, Generalitat de Catalunya, 3 volumes, Barcelona, Spain.

GenCat (2004). *Plan estratégico para la gestión integrada de las zonas costeras de Cataluña, Memoria ambiental*, Departamento de Medio Ambiente y Vivienda, Generalitat de Catalunya, Barcelona, Spain, pp. 66.

Froese, R., and Pauly, D. (2004). (eds.) FishBase (online: http://www.fishbase.org), Fishbase Project, Revised in May 2004.

Henocque, Y. (2003) "Development of progress indicators for coastal zone management in France", *Ocean & Coastal Management*, vol. 46, pp. 363–379.

IUCN (2004). *Red list of threatened species* (online: http://www.redlist.org), IUCN Species Survival Commission, Revised in May 2004.

Jenks, G.F. (1967). "The data model concept in statistical mapping", *International Yearbook of Cartography*, vol. 7, pp. 186–190.

Jongman, R.H.G., ter Braak, C.J.F., and van Tongeren, O.F.R. (1995). *Data analysis in community and landscape ecology*, Cambridge University Press, Cambridge, pp. 299.

Jones, A.R. (1995). "GIS in coastal management: a progress review", *Proceedings of the CoastGIS'95 Conference*, University College, Cork, Ireland, pp. 163–178.

Kay, J.J. (1994). *Some notes on: the ecosystem approach, ecosystem as complex systems and state of the environment reporting* (online: http://www.fes.uwaterloo.ca/u/jjkay/pubs/nac/), Revised on March 8, 2005.

King, O.H. (1995). "Estimating the value of marine resources: a marine recreation case", *Ocean & Coastal Management*, vol. 27, pp. 129–141.

Legendre, P., and Legendre, L. (1998). *Numerical ecology, developments in environmental modelling 20*, Elsevier, Amsterdam, pp. 853.

Li, R. (2000). "Data models for marine and coastal Geographic Information Systems", in: *Marine and coastal geographical information systems*, Wright, D.J., and D.J. Bartlett (eds.), Taylor & Francis, London, pp. 25–36.

Longhorn, R.A. (2003). "Coastal/marine GI/GIS – a pan European perspective", in: *Coastal and marine geo-information systems*, D.R. Green, and S.D. King (eds.), Kluwer Academic Publisher, Dordrecht, The Netherlands, pp. 35–60.

OECD (1993). *OECD core set of indicators for environmental performance reviews*, Organization for Economic Co-operation and Development, Paris.

Olsen, S.B. (2003). "Frameworks and indicators for assessing process in integrated coastal management initiatives", *Ocean & Coastal Management*, vol. 46, pp. 347–361.

Pickaver, A.H., Gilbert, C., and Breton, F. (2004). "An indicator set to measure the progress in the implementation of integrated coastal zone management in Europe", *Ocean & Coastal Management*, vol. 47, pp. 449–462.

Ricketts, P.J. (1992). "Current approaches in Geographic Information Systems for coastal management", *Marine Pollution Bulletin*, vol. 25, pp. 82–87.

Riddel, K. (1992). "Geographic Information and GIS: keys to coastal zone management", *GIS Europe*, vol. 1, pp. 22–25.

Sardá, R. (2001). "Shoreline development on the Spanish Coast", in: *Science and integrated coastal management*, Von Bodungen, B., and R.K. Turner (eds.), Dahlem University Press, Berlin, pp. 149–163.

Sardá, R. (2003). *Caracterizació del litoral Català i l'impacte de las activitats per a una gestió integral, Sìntesi general i recomanacions d'actuació*, Technical report, Departament de Medi Ambient i Habitatge, Generalitat de Catalunya, Barcelona, Spain.

Stanbury, K.B., and Starr, R.M. (1999). "Applications of Geographic Information Systems (GIS) to habitat assessment and marine resource management", *Oceanologica Acta*, vol. 22, pp. 699–703.

Tsontos, V.M., and Kiefer, D.A. (2003). "The Gulf of Maine biogeographical information system project: developing of a spatial data management framework in support of OBIS", *Oceanologica Acta,* vol. 25, pp. 199–206.

Turner, R.K., Lorenzoni, I., Beaumont, N., Bateman, I.J., Langford, I.H., and McDonald, A.L. (1998). "Coastal management for sustainable development: analysing environmental and socio-economic changes on the UK coast", *The Geographical Journal*, vol. 3, pp. 164.

Townend, I.H. (1990). "The application of GIS to coastal zone management", *Proceedings of EGIS'90, First European Conference on Geographic Information Systems*, pp. 1096–1107.

UNEP, 1995, "The development and implementation of ICAM", in: *Guidelines for Integrated Management of Coastal and Marine Areas, With Special Reference to the Mediterranean Basin*, UNEP Regional Seas Reports and Studies, PAP/RAC (MAP-UNEP), No. 161, Split, Croatia, pp. 19–33.

Vallega, A. (2003). "From Rio to Johannesburg: the role of coastal GIS", Conference Preliminary Text, Integrating Information in Coastal Zone Management, *Proceedings of the CoastGIS'03 Conference*, Genoa, pp. 28.

Watts, P.A.G. (2003). "Plans for the coastal zone", in: *Coastal and marine geo-information systems*, D.R. Green, and S.D. King (eds.), Kluwer Academic Publisher, Dordrecht, The Netherlands, pp. 61–74.

Welch, R., Remillard, M., and Alberts, J. (1992). "Integration of GPS, remote sensing, and GIS techniques for coastal resource management", *Photogrammetric Engineering and Remote Sensing*, vol. 58, pp. 1571–1578.

Zeng, T.Q., Zhou, Q., Cowell, P., and Huang, H. (2001). "Coastal GIS: functionality versus applications", *Journal of Geospatial Engineering*, vol. 3, pp. 109–126.

Chapter 35
Geohazards and Geographic Information in the Coast of Tarragona (Spain)

Isabel Montoya, Inmaculada Rodríguez, María José Sánchez, and Francisco Carreño

Abstract The increasing amount of people that every year visits the coastal areas makes necessary to carry out a detailed analysis of all the processes that occur in these areas, in order to avoid the damages that the natural processes can cause.
To this respect, the creation of a database, containing geographical, numerical, and alphanumerical information in various digital formats, including vector and raster maps, terrestrial, aerial and satellite imagery, is essential. This information set constitutes the basis to develop a risk assessment working method in a pilot area, and apply it later on along the Mediterranean Coast, considering it as the starting point to get a comprehensive risk mapping in the Mediterranean littoral.

Keywords GIS · Hazard · Littoral

35.1 Introduction

The increase of the concentration of population in littoral areas interferes and somehow modifies the normal development and evolution of some littoral ecosystems. This is due, on the one hand, to the presence of the littoral users, which will take part in the alteration, degradation and contamination of natural areas such as dune systems, lagoons or wooded areas, and even could affect the water net; on the other hand, these modifications will rise as a response to the diverse interventions performed on these spaces, such as beach nourishment or construction of jetties, harbours or promenades.

The project being developed at the Biology and Geology Department of the Rey Juan Carlos University in Madrid aims to carry out a complete study on vulnerability and hazard in a determined area of the Catalonian littoral.

A risk assessment is an evaluation of the potential adverse impact of a given activity, or a lack of it, upon the well being of an individual, a population, a community,

I. Montoya (✉)
Universidad Rey Juan Carlos, ESCET, C/Tulipán s/n, 28933, Móstoles, Madrid, Spain
e-mail: isabel.montoya@urjc.es

D.R. Green (ed.), *Coastal and Marine Geospatial Technologies*, Coastal Systems and Continental Margins 13, DOI 10.1007/978-1-4020-9720-1_35,
© Springer Science+Business Media B.V. 2010

or an organization. It is a process by which information or experience concerning the cause and effect under a set of circumstances (exposure) is integrated with the extent of exposure in order to assess risk (Department of the Army: US Army Corps of Engineers 1996).

35.1.1 Geographical and Social-Environmental Setting

The chosen study area sites in the Baix Camp zone, located in the southern part of the province of Tarragona (Spain) (Fig. 35.1). It has 22 km of coast line, half of which show medium or high occupation rate (Table 35.1), with breadths between 15 and 80 m.

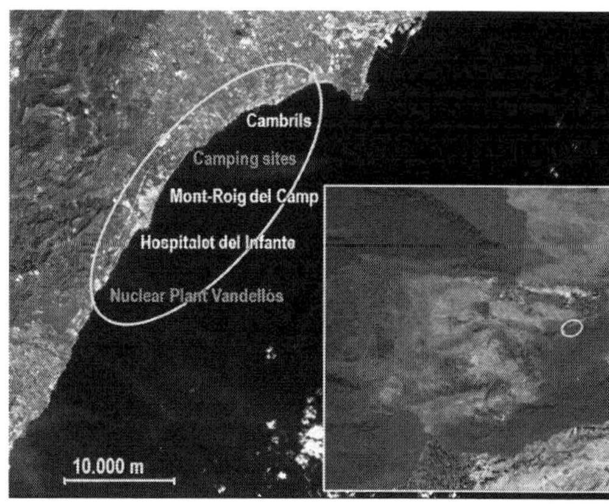

Fig. 35.1 Study area map location, including the municipalities and infrastructures that are susceptible to be affected by hazards

When talking about human settlements continuity in the littoral, we can make a difference between two different concentration stretches. The first one, where the urban area is comprised by houses, apartments, camp sites or leisure or holiday centres extends from Cambrils to Hospitalet del Infante. Due to the high population density, this coastal stretch constitutes a very fragile environmental strip. A second littoral stretch begins at Hospitalet and spreads southwards, where urban settlements become quite scarce. Besides tourism, the economic activity in the area is quite industrial because of the proximity of the nuclear power station of Vandellós and a thermal power station.

35.1.2 Geological and Geomorphological Setting

The Baix Camp region is part of the Reus Depressed Area, a Tertiary basin filled up with recent, fine-grained sedimentary detrital materials, with irregularly distributed coarse pebbles, which have an alluvial origin, sometimes displaying torrential characteristics as well.

Table 35.1 Beaches in the study area

Beach (municipalitie)	Length (m)	Width (m)	Occupation	Housing development
La Almadraba (a)	1200	35	**Medium**	**Semi-urban**
Cala Justell (a)	170	40	Low	Isolated
El Torn (a)	1400	70	Low	Isolated
El Arenal (a)	2200	80	Low	**Semi-urban**
Punta del Riu (a) (b)	320	67	Low	**Semi-urban**
Playa de Cristal (b)	1150	60	Low	**Semi-urban**
Cala de los Ángeles (b)	130	30	Low	**Urban**
Cala Bot (b)	160	30	**High**	**Urban**
Cala Misteri (b)	210	35	Low	**Urban**
Cala del Solitari (b)	120	25	Low	**Urban**
Cala de las Sirenas (b)	190	45	Low	**Urban**
Cala Vieneses (b)	75	35	Low	**Urban**
L'Estany Gelat (b)	2450	60	Low	Isolated
La Porquerola (b)	2600	30	Low	Isolated
Rifà (b)	600	15	Low	Isolated
La Pixerota (b)	1900	20	**Medium**	**Semi-urban**
L'Ardiaca (c)	1470	20	**High**	**Urban**
La Llosa (c)	1100	20	**High**	**Urban**
L'Horta de Santa María (c)	360	15	**Medium**	**Urban**
Prat d'En Fores i Regueral (c)	830	30	**High**	**Urban**
El Cavet (c)	590	20	**Medium**	**Semi-urban**
L'Esquirol (c)	750	20	**High**	**Semi-urban**
Vilafortuny (c)	1760	40	**High**	**Urban**
Cap de Sant Pere (c)	590	20	**High**	**Urban**

Data from Dirección General de Puertos y Costas, Ministerio de Medio Ambiente.
(a) Vandellós and Hospitalet del Infante; (b) Mont-Roig del Camp; (c) Cambrils.

The coastal waterfront is made up by sandy beaches with smooth slopes displaying river pebbles in some areas. Moreover, it is necessary to highlight the presence of cliffs surrounding coves, composed of pebbles included in sand-clay matrix, covered by a calcareous crust. These materials are easily weathered under the action of swell storms, and thus potentially unstable, being prone to landslide and falling over the beach.

In the southern part of the coast, before reaching the Vandellós nuclear power station, there is a rocky coast stretch made up of Jurassic dolomites showing signs of karstification (IGME 1980).

Along the shoreline, and transversally to it, a whole set of streams and torrents contribute recent detrital materials to the beaches. These deposits are related to the autumn torrential rains, these streams being active only in that season and occasionally in spring.

35.1.3 Littoral Dynamics

The oncoming swell is characterized by very short periods, the main incidence component of which has a northwest direction (Jiménez et al. 2000). Observing the wave height and wave period roses it is possible to determine that the waves which are liable to reach the coast have E to S components.
In the Mediterranean Sea, general littoral drift follows the N-S direction. At a local scale and near the coast, due to the coastline morphology, littoral drift could change its direction, but in this case that does not occur and it keeps the normal way.

35.2 Potential Geohazards

Once the factors that describe the physical medium have been analyzed, it is possible to identify the potential hazards to which the working area will be exposed. Thus, processes such as coastal erosion, fall of blocks, landslides and floods caused by the usual heavy torrential rains in the area will have to be thoroughly studied.

35.2.1 Coastal Erosion

One of its characteristics is that it is a high-energy coast, where the littoral dynamics plays an important role on the shaping of the morphologies located there. This coast has an erosive character. The erosion-accumulation values for this area from 1947 to 1994 were calculated by Jiménez et al. (2000) who obtained the variation of sediments volume from the shoreline evolution rates. These authors in their study established erosion rates in –0.6 m/year^{-1} for 1947/1977 period, and –0.4 m/year^{-1} for 1977/1994.

35.2.2 Rock Falls

The cliffs that nowadays surround the coves in Mont-Roig del Camp are fractured, and their current state is quite poor, constituting a potential danger of rocky blocks falls upon the beaches and therefore upon the population that use to frequent them, which supposes quite a special concentration during the summer. A thorough and accurate study requires knowledge of the following data: lithology, slope, vegetation, structural conditions, morphological indicators, drainage, climate, fluvial and coastal erosion, seismicity and human action. Having collected all these details, it will be possible to analyze them and map the potentially unstable areas.

35.2.3 Floods

Streams, torrents and gullies in the area constitute a potential flood hazard during torrential rains. Characterizing of this type of processes starts establishing the bound of areas susceptible to flood in a 1:5000 scale (Fig. 35.2). It is necessary to perform a later classification of hazardous areas, as well as a vulnerability assessment taking into account the damages suffered by people, roads, railway and infrastructures.

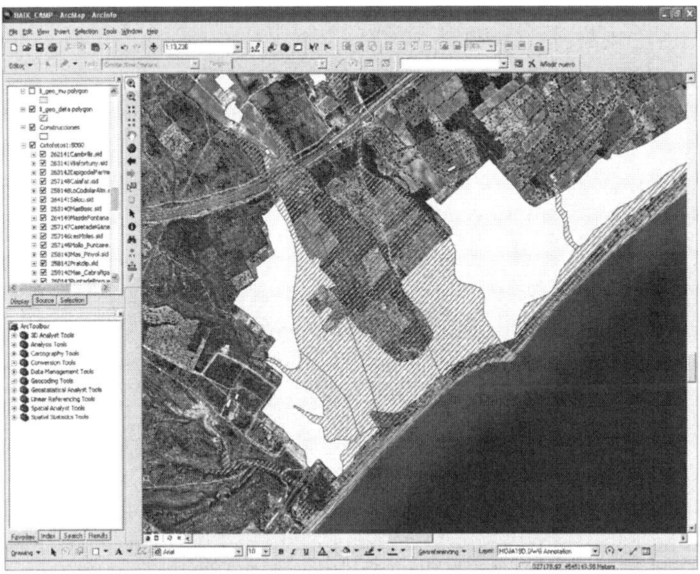

Fig. 35.2 Example of GI management showing human settlements (*white area*) probably affected by flood hazard during torrential rains (*striped area*)

35.3 Geographic Information Management

Mapping and Geographic Information Systems (GIS) are techniques clearly directed towards the improvement of territorial management. When working on risk assessment, these tools will be especially necessary in: (1) natural disasters prevention, (2) improvement on environmental planning and (3) sustainable use of natural resources. GIS are very useful tools to: incorporate geographic data in order to analyse very different variables, obtain detailed hazard maps, create prediction models and simulate future settings.

To process and analyze such amount of information in a correct way, one of the current commercial GIS, the ESRI software, ArcGIS®, will be used. The main target consists in collecting and reviewing all the information useful to characterize the study area, and then homogenize, refine and integrate it in different layers to analyze it later, obtaining a database which will include the following information: (a) Raster format data: aerial photographs, orthophotomaps, digital elevation models, satellite imagery; (b) Vector format data: topographic information, bathymetry, geology, vegetation, land use, hydrology and infrastructures data; (c) Database table format: meteorological, hydrodynamics and socio-economical data. In order to describe the processes acting in the area, some previous studies, such as: geomorphologic and slope maps, hydrological description, shoreline evolution analysis and swell behaviour simulation, are needed. (Fig. 35.3). All of them are obtained from this previous integrated information (raster, vector, and database).

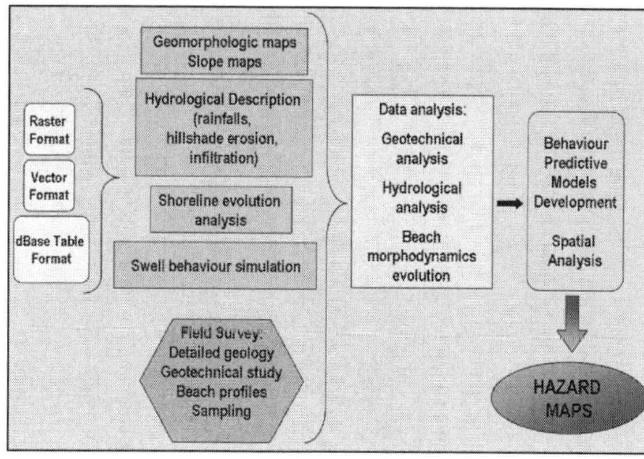

Fig. 35.3 Flow chart describing the methodology developed to obtain hazards maps through the Geographic Information Management

Data regarding to geomechanical properties of the materials and beach topography will be obtained through out an exhaustive field survey (Montoya, 2008).

In a fourth step, after information integration, previous studies and field survey, the sampled sediments, the geotechnical and hydrological properties must be analyzed. Furthermore, the beach morphodynamic evolution will be determined.

Thus, by using spatial analysis tools of GIS, behaviour predictive models to obtain finally hazard maps for each process will be developed.

GIS help people to analyze changes occurred in an area, as a whole, as well as in specific stretches, with accuracy. They are also useful because they allow to process in an instantaneous way any new configuration in order to analyze such changes (Ministerio de Medio Ambiente 2001). The coupling between database and georreferenced graphic information is suitable to carry out logical and arithmetic calculations with the different layers obtained with the GIS. (Ministerio de Medio Ambiente 2002). These calculations lead to the creation of a coastal hazard map.

35.4 Conclusions

The knowledge of the sensitive areas and the risk map elaboration, help to manage preventive securities, oriented to territorial management.

The obtained database and the creation of this Risk Assessment GIS, will allow the knowledge of all the parameters that have influence and modify the littoral. Moreover, it will be the starting point to apply this methodology to other littoral places with potential risk occurrence.

It will allow the checking of current land uses and determining the areas where the government legislation will need to be more demanding.

References

Department of the Army: U.S. Army Corps of Engineers (1996) Environmental Quality Risk Assessment Handbook. Volume II: Environmental Evaluation. Washington, DC 20314-1000. 409 pp.

Instituto Geológico y Minero de España (IGME) (1980) Mapa Geológico de España: Hoja 472, 33–18: Reus. Escala 1:50000.

Jiménez JA, Valdemoro HI, Gracia V, Sánchez-Arcilla A (2000) Estudio sobre la situación del tramo costero Cambrils-Vandellós (T.M. Mont-Roig del Camp). Evolución costera y dinámica litoral. Laboratori d'Enginyeria marítima. E.T.S.E.C.C.P.B. Universitat Politécnica de Catalunya. Barcelona.

Ministerio de Medio Ambiente. Dirección General de Puertos y Costas. Guía de playas españolas. http://www.mma.es/costas/guia_playas/index.htm

Ministerio de Medio Ambiente (2001) Proyecto de actuación medioambiental en el entorno de la Península del Fangar, playa de La Marquesa y playa de Pal en el Delta del Ebro. Estudio de Impacto Ambiental. CETYSIG SL.

Ministerio de Medio Ambiente (2002) Proyecto de paseo peatonal, accesos y adecuación medioambiental de las calas situadas en Mont-Roig del Camp, T.M. Mont-Roig del Camp (Tarragona). CETYSIG SL.

Montoya, I. (2008) Análisis de susceptibilidad a los movimientos de ladera en los acantilados de la franja costera de Mont-Roig del Camp (Tarragona). PhD Thesis. Universidad Rey Juan Carlos, Madrid. 194 pp.

Chapter 36
Assessing the Vulnerability of Asian Megadeltas to Climate Change Using GIS

Colin D. Woodroffe

Abstract Susceptibility of Asian megadeltas to climate change, including sea-level rise, is investigated using GIS. The Indus, Ganges-Brahmaputra-Meghna, Irrawaddy, Chao Phraya, Mekong, Red, Pearl, Changjiang, and Huanghe deltas began to form around 6000 years ago and have prograded since. The surface topography of active and abandoned delta plains is examined using digital terrain models derived from Shuttle Radar Topography Mission data and channel morphology is investigated using radar imagery. After delta plains are abandoned they become increasingly dominated by tidal processes. Population density is estimated using gridded world population data but highly variable local microtopography and uncertainty regarding future climate changes preclude detailed vulnerability analysis.

Keywords: GIS · Deltas · Topography · Population · Climate change

36.1 Introduction

Coasts appear to be particularly susceptible to impacts as a result of human-induced climate change. Amongst the most vulnerable coastlines are the low-lying plains associated with deltas and estuaries (McLean and Tsyban, 2001; Nicholls et al., 2007). In order to better assess the relative vulnerability of different coastlines and provide quantitative information on the impacts, there have recently been a series of global vulnerability assessments (e.g. Hoozemans et al., 1993). For example, a global database of world coastlines has been developed as a part of the European Union DINAS-COAST project, integrating information on physical, ecological and socio-economic characteristics (Vafeidis et al., 2005). Although the project included a flexible, interactive assessment tool, termed DIVA, it is based on decomposition

C.D. Woodroffe (✉)
School of Earth and Environmental Sciences, University of Wollongong, Wollongong, NSW 2522, Australia
e-mail: colin@uow.edu.au

D.R. Green (ed.), *Coastal and Marine Geospatial Technologies*, Coastal Systems and Continental Margins 13, DOI 10.1007/978-1-4020-9720-1_36,
© Springer Science+Business Media B.V. 2010

of the world's shoreline into a series of 1-dimensional coastal segments and does not therefore capture the multidimensional complexity of extensive low-lying areas such as deltas (David et al., 2008). This chapter describes a standardised approach to describe the surface topography and population distribution for the 9 major Asian megadeltas using GIS.

36.2 Physical Characteristics

Deltas occur at the mouths of large rivers where substantial sediment loads are brought to the coast. The Asian megadeltas are a series of extensive, productive and heavily populated delta plains at the mouths of the Indus, Ganges-Brahamputra-Meghna, Irrawaddy, Chao Phraya, Mekong, Red (Song Hong), Pearl (Zhujiang), Chiangjiang (Yangtze) and Huanghe (Yellow) Rivers (Fig. 36.1). These rivers are fed by runoff, snowmelt and sediments from the uplifting Himalayan system and are influenced seasonally by the monsoon. In addition to intensive agriculture, aquaculture and silviculture, they are becoming rapidly urbanised and contain some of the world's largest, and most rapidly growing cities, many of them already megacities (cities of more than 8 million people).

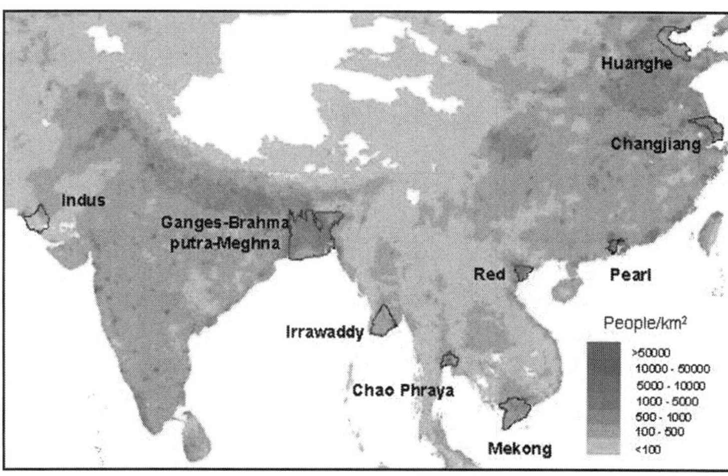

Fig. 36.1 Gridded population density of Asia (based on CIESIN, GPW3 2000), showing concentration in river valleys and on megadeltas (See also Plate 66 on Page 431 in Color Plate Section)

36.2.1 Holocene Delta Evolution

The Holocene evolution of the extensive deltaic plains associated with Asian megadeltas provides the context within which to understand modern sedimentation patterns and deltaic processes and is fundamental to planning and sustainable development in these heavily populated areas. At the millennial scale, there has been a broadly similar history of development for each delta characterised by transgression during the final stages of postglacial sea-level rise up to around 6000 years ago, with deposition of intertidal or shallow marine sediments unconformably over an eroded Pleistocene alluvial surface (Tanabe et al., 2003a, b).

Each delta has followed an increasingly distinctive and individual evolutionary pathway during the past 5–6 millennia as a regressive sequence of sediments has prograded seaward (Woodroffe, 2000). Different deltas can be placed within a framework in relation to the physical processes that exert most influence on their development, river, wave and tide processes. River-domination can be inferred where there are prominent levées and evidence of channel meandering or avulsion, such as paleochannels. Wave-domination is characterised by shore-parallel sand ridges, often coalescing into beach-ridge plains. Tide-domination is indicated by exponentially tapering channels and complexly meandering, mangrove-lined tidal creeks (Woodroffe et al., 2006).

The millennial-scale evolution of these deltas has been subject to recent investigation and short-term sedimentary processes are understood at the microscale, but the behaviour of individual deltas at decade-to-century scale remains poorly understood by comparison. The large rivers are dynamic and adjust their course; the surface of the Holocene delta plains is often shaped by a cycle of distributary activity and there are numerous examples of river capture or avulsion. The most remarkable of these has been the Huanghe, which has switched its course completely in historical time. The Ganges-Brahmaputra-Meghna has undergone changes; there has been a major recent shift in the course of the Brahmaputra, resulting in the sequestration of sediment in the Sylhet basin and there is evidence of the progressive migration of the course of the Ganges eastward, with the abandonment of successive distributaries.

36.2.2 Dominant Processes

Fluvial activity is focused on the active delta plain that is shaped by the river channel and comprises the floodplains that are subject to inundation when the river floods. Former distributaries in the abandoned delta plain carry much diminished flows and infill with sediment. Whereas subsidence is offset by accretion of new alluvial sediment in the active delta plain, the surface of the abandoned delta plain appears to become progressively lowered with supply of sediment unable to keep pace with regional subsidence. In contrast to the prominent delta lobes of the Mississippi, which become reworked by low-energy wave activity in the Gulf of Mexico, Asian megadeltas are subject to much greater wave or tidal energy. The Indus experiences considerably higher-energy waves, and the abandoned delta front of many of the deltas becomes increasingly dominated by tidal processes. For example, the Mekong delta underwent a transition from embayed estuary to lobate delta and the rate of

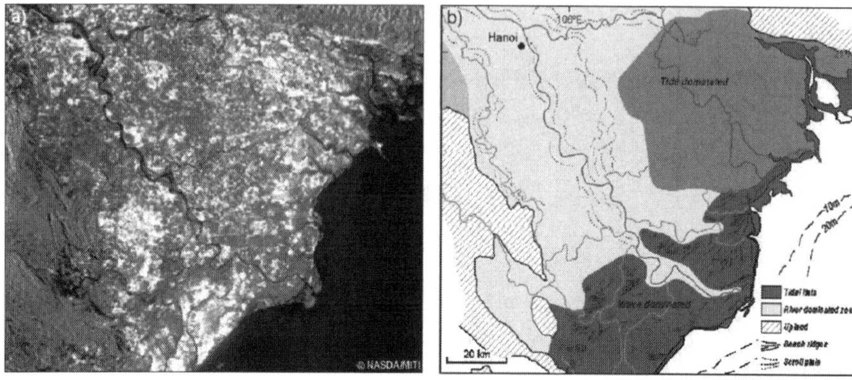

Fig. 36.2 The Red River Delta; (**a**) JERS radar imagery of the delta showing the complexity of land-use, and (**b**) an interpretation of subaerial delta geomorphology that shows distinct sectors of the delta dominated by different processes (based on Mathers and Zalasiewicz, 1999) (See also Plate 67 on Page 431 in Color Plate Section)

seaward progradation slowed after 4000 years ago from 30 to 35 m/year to an average of around 11 m/year (Ta et al., 2002). The large tidal range in the Bay of Bengal means that the abandoned delta plain of the Ganges-Brahamputra-Meghna is tide-dominated, and behaves in a different way to the river-dominated active delta distributaries that carry large and highly variable discharges and substantial sediment loads.

The separation of processes within a delta is most clearly seen on the plains of the Red River delta (Mathers and Zalasiewicz, 1999). The active floodplain flanking the main channel is river-dominated, a series of meander scroll plains flank the distributaries, and the location of formerly-active channels can be seen from the higher ground that has been formed by the deposition of levées. The southwest sector is wave-dominated, and contains a series of shore-parallel beach ridges. The eastern sector is tide-dominated; there are extensive tidal flats and tapering and meandering creek systems (Fig. 36.2).

36.2.3 Delta Plain Morphology

The surface geomorphology of the 9 Asian megadeltas has been investigated using Shuttle Radar Topography Mission (SRTM) elevation data. This comprises single pass synthetic aperture radar (SAR) interferometry (C and X band) at 3 arc second resolution, corresponding to a cell size of around 90 m on the ground. The SRTM elevations are reported to the nearest metre, with the sea surface set to zero, and are available from the National Aeronautics and Space Administration (NASA – http://edcsns17.cr.usgs.gov/srtm/index.html).

The largest delta in the region is the combined Ganges-Brahmaputra-Meghna (GBM) delta, covering an area of more than 100,000 km^2 (Fig. 36.3). The Meghna,

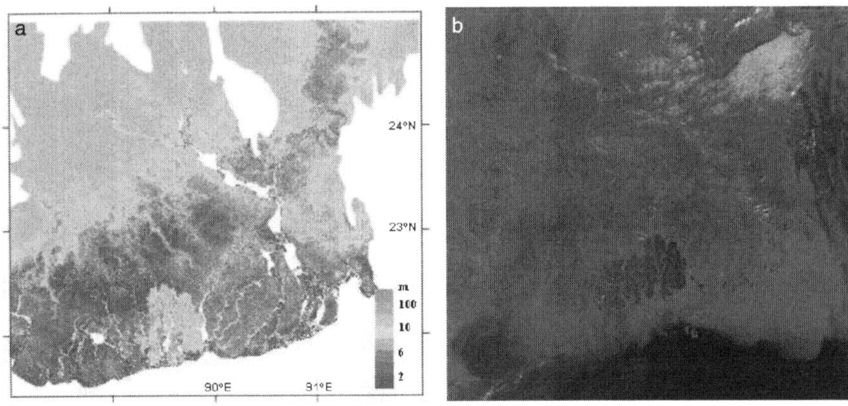

Fig. 36.3 The delta of the Ganges-Brahmaputra-Meghna delta; (**a**) DTM determined from SRTM, and (**b**) MODIS image of delta (See also Plate 68 on Page 432 in Color Plate Section)

to the east of this delta, is strongly river-dominated as the combined flow of several rivers (primarily the Ganges and Brahmaputra) reaches the Bay of Bengal. However, the western section of coastline is tide-dominated. Termed the Gangetic Tidal plain, and covered by the extensive mangrove forests of the Sundarbans, it contains a network of tapering tidal channels, and channel bank erosion and retreat of the coast are typical. It appears that, whereas the contributing rivers bring a large volume of sediment to the coast, it is the strong tidal currents associated with the large tidal range that re-suspend sediment and ensure that waters remain turbid.

Figure 36.3 shows the topography of the GBM with low-lying areas flanked by natural levées marking existing and former distributaries of the Ganges. These are important in channelling flow, are significant for human activities, and influence the extent and duration of flooding (Umitsu, 1985). These levées coalesce over much of the northern part of the delta resulting in considerably higher elevation. By contrast, the particularly low-lying, actively-subsiding Syhlet basin, which in places is only 2 m above sea level, is clear in the DTM extending northeastwards inland from the coast; this can be flooded by water that is up to 6 m deep in the wet season. Less than 4% of the flow of the Ganges feeds into the Sunderbans. The DTM indicates the former distributaries of the Ganges, such as the Hooghly and the Gorai. The location of delta deposition has been influenced by the progressive migration of the course of the Ganges eastward, and with the abandonment of successive distributaries. Loss of land in the order of 100 km^2 over the past 30 years has occurred along this coast, and islands such as Sagar Island have experienced retreat on the seaward side (Allison et al., 2003). This example from the Ganges-Brahmaputra-Meghna system demonstrates that deltas are dynamic, changes in their morphology are driven by sediment supply and accommodation space, which are in turn a function of the morphodynamics of the delta (Brammer, 1993).

Figure 36.4 shows the topography of the Irrawaddy delta determined from SRTM and the morphology of the delta as seen by JERS radar imagery. Whereas the

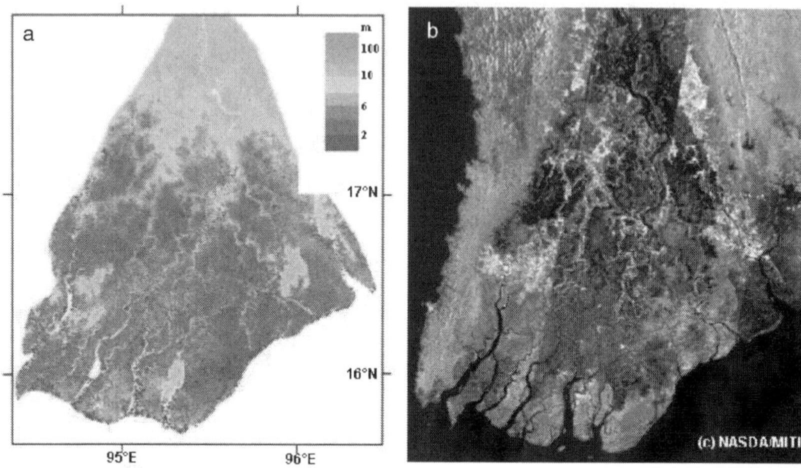

Fig. 36.4 The delta of the Irrawaddy River, Myanmar; (**a**) Digital terrain model (DTM) derived from Shuttle Radar Topography Mission (SRTM) data; (**b**) JERS radar imagery, showing the tide-dominated "Mouths of the Irrawaddy" (See also Plate 69 on Page 432 in Color Plate Section)

river has been structurally constrained between the Araken mountains and the Shan plateau, the delta has switched between distributaries. River-dominated channels are characteristically braided, or marked by meander scroll plains. The western portion of the coastline comprises a series of tide-dominated channels, the "mouths of the Irrawaddy". These show negative exponential changes in width along their course, and mangrove-lined tidal creeks. Whereas deposition may occur at the active locus of river supply of sediment, erosion is likely to be the main process in the tide-dominated parts of the abandoned delta plain.

Figure 36.5 presents the SRTM-derived DTM for the other Asian megadeltas. The Indus is characterised by an active delta plain that is river-dominated, and abandoned delta plains that are tide-dominated and covered by extensive mangrove forests. The delta shows a progressive increase in elevation inland, composed of several abandoned delta lobes to the east of the modern river course. The Chao Phraya has formed the extensive low-lying central plain of Thailand. SRTM data indicate that the Mekong is especially low-lying, with more than 70% of the delta plain less than 4 m above sea level. More than 50% of the plains of the Pearl, Changjiang and Huanghe deltas are also below 4 m above sea level. The abandoned former Huanghe mouth in Jiangsu is below 4 m above sea level over more than 80% of its area and is subject to rapid coastal erosion (Li et al., 2004).

Although the megadeltas are low-lying, projected rates of sea-level rise are only gradual and inundation of most of the area of the delta plains does not appear imminent. However, vulnerability of these deltas to flooding is determined not just by elevation, but by wet season river flood levels over the active floodplain, and the upper levels reached by the tide in the tide-dominated abandoned delta plain. Sea-level rise and its effect on the Vietnamese Mekong Delta has been modelled in terms

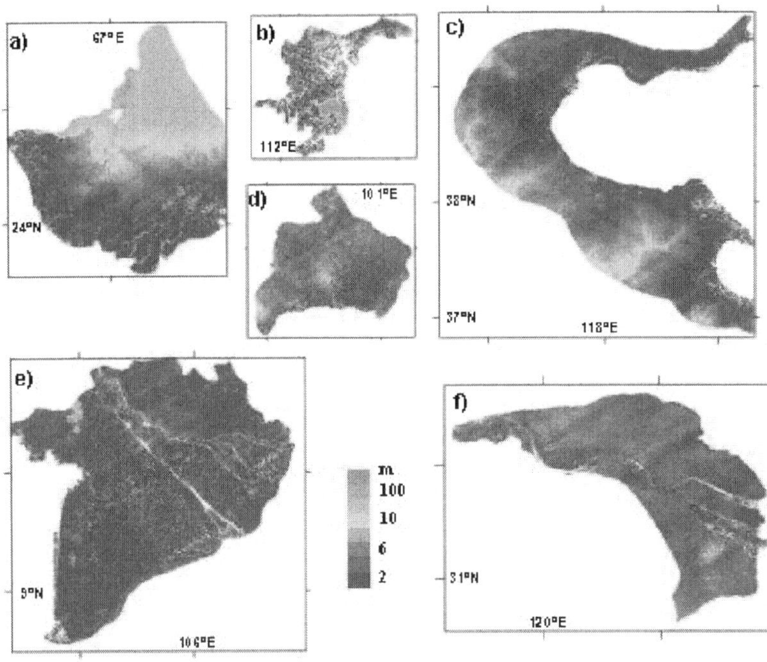

Fig. 36.5 DTM of (**a**) Indus, (**b**) Pearl, (**c**) Huanghe, (**d**) Chao Phraya, (**e**) Mekong, and (**f**) Changjiang deltas based on Shuttle Radar Topography Mission (SRTM) data (after Woodroffe et al., 2006) (See also Plate 70 on Page 433 in Color Plate Section)

of water elevation in the flood season, and the implications for rice production have been examined (Wassmann et al., 2004).

36.3 Variability of Socio-Economic Factors Across Deltas

The natural dynamic behaviour of Asian megadeltas is subject to alteration by human influence including indirect effects and direct impacts. Land-use change in the catchment has been one of the most significant indirect influences. In the case of the Huanghe, land-use change in the loess plateau has meant substantially increased sediment loads in the river (Saito et al., 2001). By contrast the construction of dams on rivers has resulted in the reduction of sediment loads through retention of sediment in dams (Ericson et al., 2006). For example, the Indus is now extensively dammed and sediment supply to the coast has been reduced to low levels (Syvitski et al., 2005).

36.3.1 Delta Populations

Figure 36.1 shows gridded population densities for Asia based on the updated and expanded GPW-3 (beta version), produced by the Center for International Earth

Table 36.1 Principal characteristics and population estimates of the Asian megadeltas

Megadelta	Catchment area $km^2 \times 10^3$	Mean annual discharge $m^3\ s^{-1}$	Population 2000 (GPW3) millions	Population 2015 (GPW3) millions	Pre-human sediment discharge $kg\ s^{-1}$	Post-human sediment discharge $kg\ s^{-1}$
Indus	1082	6564	3.1	4.4	9593	3686
G-B-M	1667	22102	129.9	166.2	40534	46287
Irrawaddy	414	11953	10.6	12.1	16331	8239
Chao Phraya	179	961	11.5	16.4	452	256
Mekong	806	15900	15.7	19.0	2551	2531
Red (Hong)	171	3900	13.3	16.1	1039	4119
Pearl	409	10700	9.8	27.1	1547	1427
Changjiang	1722	29460	25.9	33.1	109444	49504
Huanghe	945	1990	14.0	16.6	3237	931

Science Information Network (CIESIN – http://sedac.ciesin.columbia.edu/gpw), adjusted to UN estimates of country total populations. It is clear that people are disproportionately concentrated within the river valleys and on the Asian megadeltas (Small and Nicholls, 2003). The populations of the megadeltas have been calculated using ArcGIS raster calculator with the delta margins defined by the maximum Holocene shoreline (Table 36.1). The gridded population datasets are based on census returns distributed within administrative units, using a mass-conserving algorithm for 1990, 1995 and 2000, and an estimated population for 2015 using country-specific growth rates. Data are gridded using 2.5 arc minute cells (4.6 km at equator). Spatial uncertainties in population distributions are generally larger than boundary uncertainties.

The population of the GBM delta is the largest; almost the entire nation of Bangladesh occurs on the delta, although those higher areas such as the Barind and Madhupur tracts have been excluded, but West Bengal (India) has been included. The population of the Changjiang delta consists of nearly 26 million. The population on the Holocene plains deposited at the present mouth of the Huanghe consists of around 14 million, but the coast of Jiangsu on the abandoned Huanghe, adjacent to the Changjiang, has nearly 20 million. The Indus has the smallest population with around 3 million, and the other deltas each have 10–15 million.

The gridded population data are only an approximation of the spatial distribution of people within the deltas; the spatial resolution of census data poses a fundamental limitation on calculating population in the near-coastal zone. Extrapolations of populations based on GPW2015 are also available. Figure 36.6 shows the gridded estimates for the Red River delta, determined assuming continuation of recent geometric demographic patterns of growth between 1990 and 2000 censuses, adjusted at national level to UN 2015 population projections. Urbanisation is the most significant contributor to growth; rural populations are generally distributed in highly nucleated settlements, and the spatial resolution of the gridded cells does not capture the concentration of settlements along higher ground, such as levées and beach ridges that can be seen in the SRTM-derived DTM of the Red River delta.

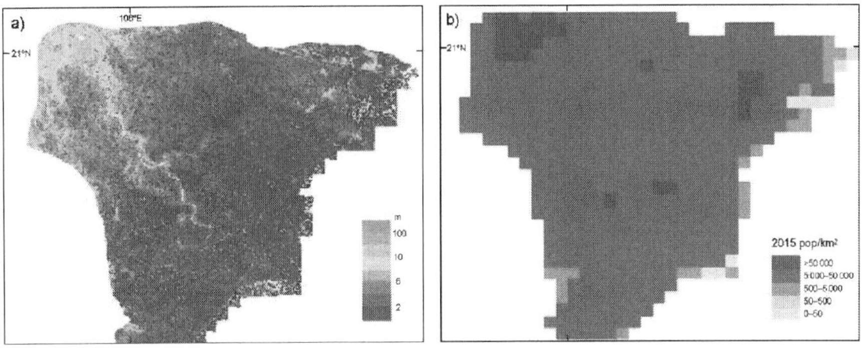

Fig. 36.6 The Red River Delta; (**a**) DTM derived from SRTM; (**b**) gridded population density for 2015 using estimates of population increase (based on GPW3, CIESIN) (See also Plate 71 on Page 433 in Color Plate Section)

The population data are not at sufficient resolution for detailed hazard analysis, or local vulnerability assessment. Such vulnerability analyses should be focused on detailed local topography and integration with other variables such as flood levels, land use, and other relevant factors. The population data does not adequately represent urbanisation and growing rural-urban imbalance with growth of enormous cities in the megadeltas.

36.3.2 Flooding and Flood Management

Natural levées are frequently built up as a component of flood control, decreasing the frequency with which the plains are overtopped but increasing the magnitude of flood when overtopping does occur. Such activities have ramifications for the supply of sediment and nutrients to the delta plain surface. Drainage and irrigation works impact the movement of water, but have more subtle effects on sediment and sediment geochemistry. A widespread problem is the acidification of potential acid sulphate soils with oxidation of pyrite.

There has been a long history of interference with water in the deltas, from flood mitigation and groundwater extraction to irrigation and channelisation. The natural systems which underlie the Asian megadeltas have been increasingly influenced by human alterations, and the incorporation of human factors into a systems approach remains challenging (Milliman et al., 1989; Syvitski and Saito, 2007). The spatial variability of elevation and hydrodynamics within a delta implies that management of deltaic environments needs to be based on an understanding of the geomorphological dynamics of each system. Widespread land-use change to agriculture, silviculture and aquaculture in the rural areas, and more recent urbanisation can lead to irreversible effects, such as compaction, subsidence and acid sulphate soil development. There is a need to understand these processes more completely and to reinforce the resilience of the natural systems.

GIS provides a geoscientific framework within which to assess the balance of natural processes, including subsidence, compaction, sediment supply, sea-level change and changes in tidal amplitude. An integrated assessment framework needs to be based on an appreciation of the role of sediment delivery and sequestration within suites of landforms and the socio-economic factors such as population and land use, emphasising spatial variability and the interconnectedness of the components of a delta.

Erosion is a part of the natural dynamics of some sectors of the delta, becoming progressively accentuated as the delta has built seaward during the late Holocene. It is an essential phase of a delta cycle, when subsidence or compaction exceeds the rate of supply of new sediment, or when it is accentuated by human dewatering through groundwater extraction. Abandonment of a delta lobe is followed by self-cannibalisation. The contrasting dominance of different physical processes along the front of the delta results in differing trends of coastal erosion or deposition.

Inundation of low-lying areas is already a problem around the world's shorelines, with on average around 10 million people experiencing flooding per year. Megadeltas are subject to a series of hazards, such as cyclones and storm surges, and extreme events of this kind have resulted in massive death tolls, particularly along the shores of the Bay of Bengal (Ali, 1996). Although the number of people at risk might be expected to increase as a result of future sea-level rise, the principal factor influencing future flood vulnerability appears to be the socio-economic pathway that is followed, including population growth rates and the economic constraints on the extent of flood protection (Nicholls, 2004).

36.4 Discussion

Despite the fact that deltas represent some of the largest sedimentary deposits in the world, the shorelines of many of the world's deltas, and those of Asian megadeltas in particular, are highly variable and in many places are already undergoing erosion. Geomorphological variability across delta plains means that different parts of a delta are likely to respond differently to global climate change, with erosion accentuated in those parts of a delta already subsiding and no longer receiving fresh sediment inputs.

This natural geomorphological variability is further complicated by variability in human settlement and land use. Water extraction and water/flood management has been spatially variable across deltas, ranging from levée enhancement to installation and maintenance of dikes and canals. A corollary of water control has been a rerouting of sediment pathways, and human modification of natural patterns of sedimentation and erosion (Thanh et al., 2004). Further use of groundwater, and the subtle adjustments of ground level as a consequence, appear to have exacerbated problems such as compaction associated with the natural dynamics of deltas. Subsidence as a result of groundwater extraction is a serious problem, especially for Bangkok (Phienwej and Nutalaya, 2005). Wise sustainable management of these systems requires more attention to sediment pathways and consideration of the consequences of elevation change.

Long-term sustainability of these populated deltas is frequently affected by large-scale engineering projects with impacts that often outweigh those likely to be associated with global warming and sea-level rise (Syvitski and Saito, 2007). The rate of relative eustatic sea-level rise is often exceeded by the rate of isostatic-controlled subsidence and of similar magnitude to natural sediment compaction. Accelerated compaction may result from petroleum and groundwater extraction, exacerbated by reduction in sediment delivery due to trapping behind dams and the construction and enhancement of levées (Syvitski and Saito, 2007).

36.5 Conclusion

GIS provides a valuable tool for assessment of the impacts of climate change. Whereas the topographic variability of delta plains is partially captured at the scale of the SRTM radar altimetry, for many of the deltas the socio-economic data is not available at a suitable scale. Still more problematic is the nature of probable impacts as a result of climate change. Although sea-level rise has received much attention, rates of sea-level rise have been revised downward in successive IPCC assessments, and it is becoming clear that other aspects of climate change, such as rainfall and runoff change may play an equally important role. In the case of these deltas the extent of human influence has been overwhelming. Anthropogenic impacts on river flow and sediment delivery have had, and seem likely to continue to have, just as far-reaching impacts as those related to climate change.

References

Ali, A. (1996). "Vulnerability of Bangladesh to climate change and sea level rise through tropical cyclones and storm surges", *Water, Air and Soil Pollution,* 92, 171–179.

Allison, M.A., Khan, S.R., Goodbred, S.L., and Kuehl, S.A. (2003). "Stratigraphic evolution of the late Holocene Ganges-Brahmaputra lower delta plain", *Sedimentary Geology,* 155, 317–342.

Brammer, H. (1993). Geographical complexities of detailed impact assessment for the Ganges-Brahmaputra-Meghna delta of Bangladesh. in: Warrick, R.A. Barrow, E.M., and Wigley, T.M.L. (eds) *Climate and Sea Level Change: observations, projections and implications.* Cambridge University Press, Cambridge, pp. 246–262.

David, L.T., Maneja, R., Goh, B., Lansigan, F., Sereywath, P., Radjawane, I.M., Matsumoto, B.M.M., Tantichodok, P., Snidvong, A., Tri, N.H., Nguyen, K.A.T., Saito, Y., and Hinkel, J. (2008). "Sea level rise vulnerability of southeast Asian coasts". *Land-Ocean Interactions in the Coastal Zone, Inprint,* 2008, 3, 3–6. (See http://www.loicz.org/products/publication/newsletter/index.html.en)

Ericson, J.P., Vorosmarty, C.J., Dingman, S.L., Ward, L.G., and Meybeck, M. (2006). "Effective sea-level rise and deltas: causes of change and human dimension implications". *Global and Planetary Change,* 50, 63–82.

Hoozemans, F.M.J., Marchand, M., and Pennekamp, H.A. (1993). "*A global vulnerability analysis: vulnerability assessment for population, coastal wetlands and rice production on a global scale*", 2nd edition, Delft Hydraulics, The Netherlands.

Li, C., Fan, D., Deng, B., and Korotaev, V. (2004). "The coasts of China and issues of sea level rise", *Journal of Coastal Research, Special Issue,* 43, 36–49.

Mathers, S., and Zalasiewicz, J. (1999). "Holocene sedimentary architecture of the Red River Delta, Vietnam", *Journal of Coastal Research,* 15, 314–325.

McLean, R.F., and Tsyban, A. (2001). "Coastal zones and marine ecosystems", in: McCarthy, J.J. Canziani, O.F., Leary, N.A., Dokken, D.J., and White, K.S. (eds), *Climate change 2001: impacts, adaptation, and vulnerability*, Cambridge University Press, Cambridge, 343–379.

Milliman, J.D., Broadus, J.M., and Gable, F. (1989). "Environmental and economic implications of rising sea level and subsiding deltas: the Nile and Bengal examples", *Ambio,* 18, 340–345.

Nicholls, R.J. (2004). "Coastal flooding and wetland loss in the 21st century: changes under the SRES climate and socio-economic scenarios", *Global Environmental Change,* 14, 69–86.

Nicholls, R.J., Wong P.P., Burkett V.R., Codignotto J.O., Hay J.E., McLean R.F., Ragoonaden, S., and Woodroffe, C.D. (2007). Coastal systems and low-lying areas. In: Parry, M.L. Canziani, O.F., Palutikof, J.P., van der Linden, P.J., and Hanson, C.E., (eds), *Climate Change 2007: impacts, adaptation and vulnerability.* Contribution of Working Group II to the Fourth Assessment Report of the Intergovernmental Panel on Climate Change, Cambridge University Press, Cambridge, UK, 315-356.

Phienwej, N., and Nutalaya, P. (2005). "Subsidence and flooding in Bangkok", in: Gupta, A (ed) *The Physical Geography of Southeast Asia,* Oxford University Press, Oxford, pp. 358–378.

Saito, Y., Yang, Z.S., and Hori, K. (2001). "The Huanghe (Yellow River) and Changjiang (Yangtze River) deltas: a review on their characteristics, evolution and sediment discharge during the Holocene", *Geomorphology,* 41, 219–231.

Small, C., and Nicholls, R.J. (2003). "A global analysis of human settlement in coastal zones", *Journal of Coastal Research,* 19, 584–599.

Syvitski, J.P.M., Vörösmarty, C.J., Kettner, A.J., and Green, P. (2005). "Impact of humans on the flux of terrestrial sediment to the global coastal ocean", *Science,* 308, 376–380.

Syvitski, J.P.M., and Saito, Y. (2007). "Morphodynamics of Deltas under the influence of humans". *Global and Planetary Change*, 57, 261–282

Ta, T.K.O., Nguyen, V.L., Tateishi, M., Kobayashi, I., Tanabe, S., and Saito, Y. (2002). "Holocene delta evolution and sediment discharge of the Mekong River, southern Vietnam", *Quaternary Science Reviews,* 21, 1807–1819.

Tanabe, S., Ta, T.K.O., Nguyen, V.L., Tateishi, M., Kobayashi, I., and Saito, Y. (2003a). Delta evolution model inferred from the Holocene Mekong Delta, southern Vietnam, in: Sidi, F.H., Nummedal, D., Imbert, P., Darman, H., and Posamentier, H.W. (eds), *Tropical deltas of southeast Asia: sedimentology, stratigraphy, and petroleum geology*, SEPM Special Publication, Tulsa, OK, 76, 175–188.

Tanabe, S., Saito, Y., Sato, Y., Suzuki, Y., Sinsakul, S., Tiyapairach, S., and Chaimanee, N. (2003b). "Stratigraphy and Holocene evolution of the mud-dominated Chao Phraya delta, Thailand", *Quaternary Science Reviews,* 22, 789–807.

Thanh, T.D., Saito, Y., Huy, D.V., Nguyen, V.L., Ta, T.K., and Tateishi, M. (2004). "Regimes of human and climate impacts on coastal changes in Vietnam", *Regional Environmental Change,* 4, 49–62.

Umitsu, M. (1985). "Natural levees and landform evolutions in the Bengal lowlands", *Geographical Review of Japan,* 58 (Ser. B), 149–164.

Vafeidis, A.T., Nicholls, R.J., McFadden, L., Hinkel, J., and Grashoff, P.S. (2005). "Developing a global database for coastal vulnerability analysis: design issues and challenges", *The International Archives of the Photogrammetry, Remote Sensing and Spatial Information Sciences*, 34, 5 pp.

Wassmann, R., Hien, N.X., Hoanh, C.T., and Tuong, T.P. (2004). "Sea level rise affecting the Vietnamese Mekong Delta: water elevation in the flood season and implications for rice production", *Climatic Change,* 66, 89–107.

Woodroffe, C. D. (2000). "Deltaic and estuarine environments and their Late Quaternary dynamics on the Sunda and Sahul shelves", *Journal of Asian Earth Sciences,* 18, 393–413.

Woodroffe, C.D., Nicholls, R.J., Saito, Y., Chen, Z., and Goodbred, S.L. (2006). "Landscape variability and the response of Asian megadeltas to environmental change", In Harvey, N. (ed) *Global Change and Integrated Coastal Management: The Asia-Pacific Region*, Springer, Berlin, pp. 277–314.

Color Plates

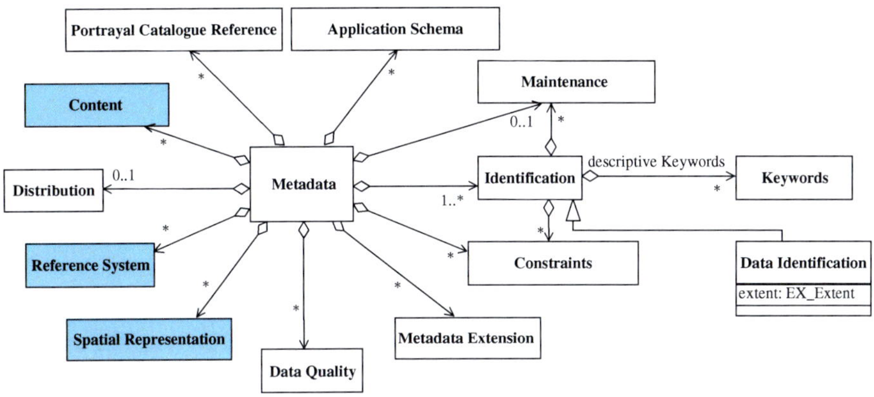

Plate 1 The different heads of ISO 19115 (See also Fig. 1.1 on Page 6)

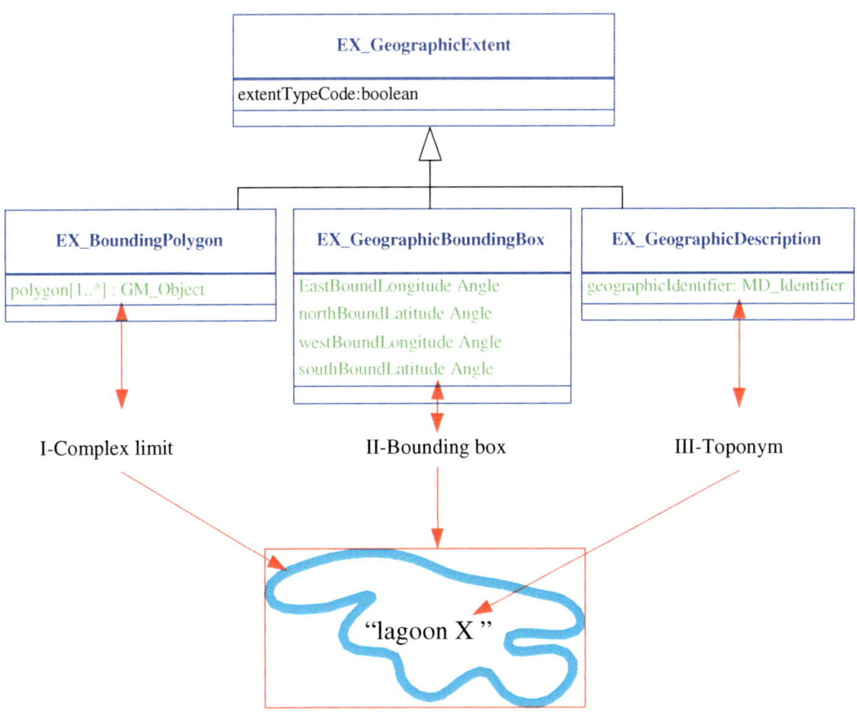

Plate 2 Illustration of the three descriptor types of ISO 19115 (See also Fig. 1.2 on Page 7)

D.R. Green (ed.), *Coastal and Marine Geospatial Technologies*, Coastal Systems and Continental Margins 13, DOI 10.1007/978-1-4020-9720-1,
© Springer Science+Business Media B.V. 2010

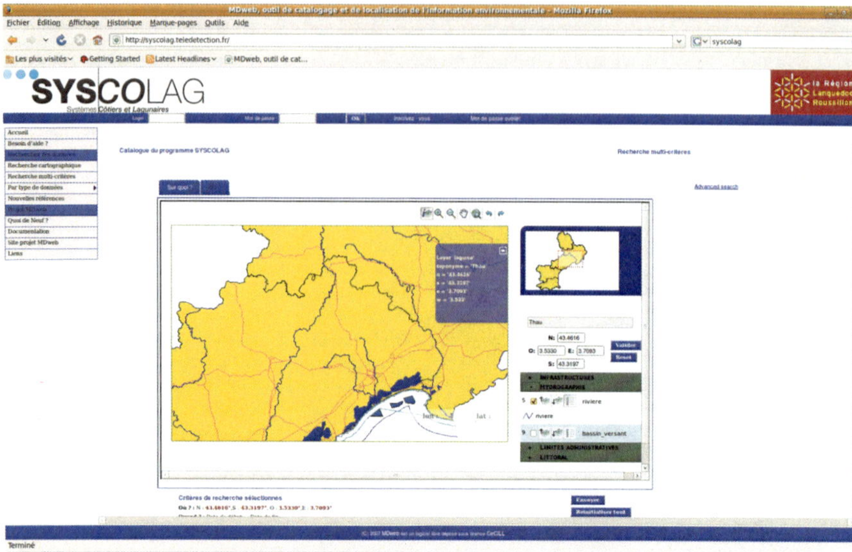

Plate 3 Geographical description of a resource in ISO 19115 (See also Fig. 1.4 on Page 10)

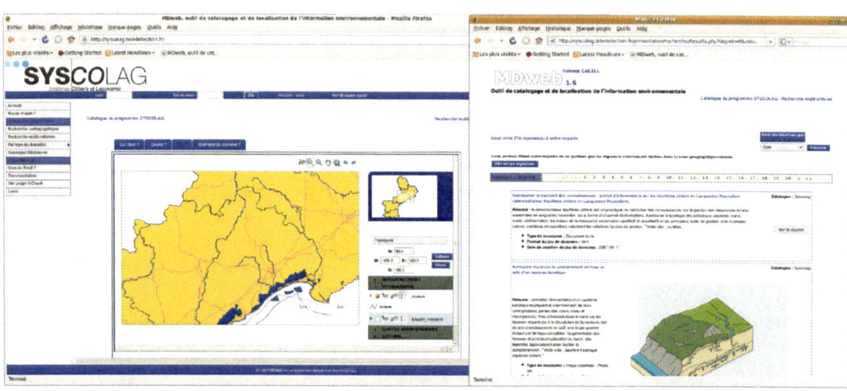

Plate 4 Textual and cartographical responses to a multicriteria request (See also Fig. 1.5 on Page 12)

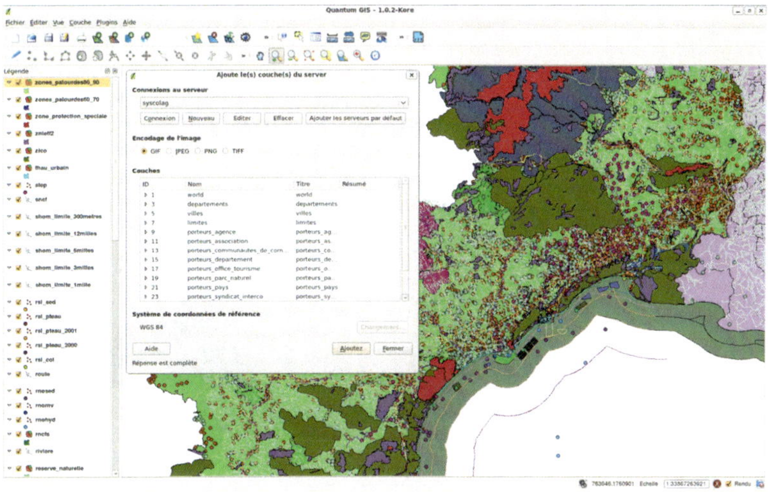

Plate 5 Example of a connection to a Postgis database with the free software application *Quantum GIS* (See also Fig. 1.6 on Page 14)

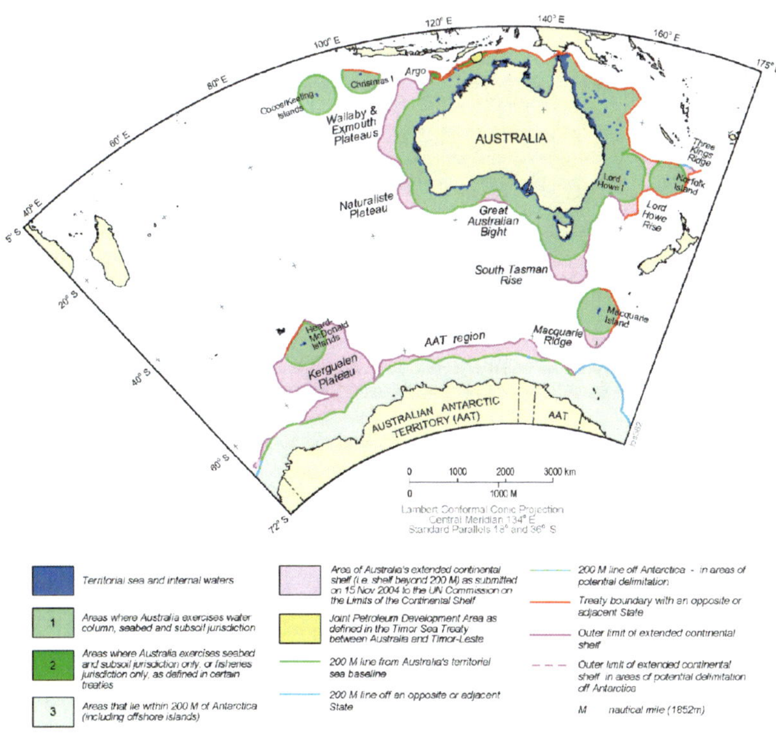

Plate 6 Australia's marine jurisdiction (See also Fig. 2.1 on Page 18)

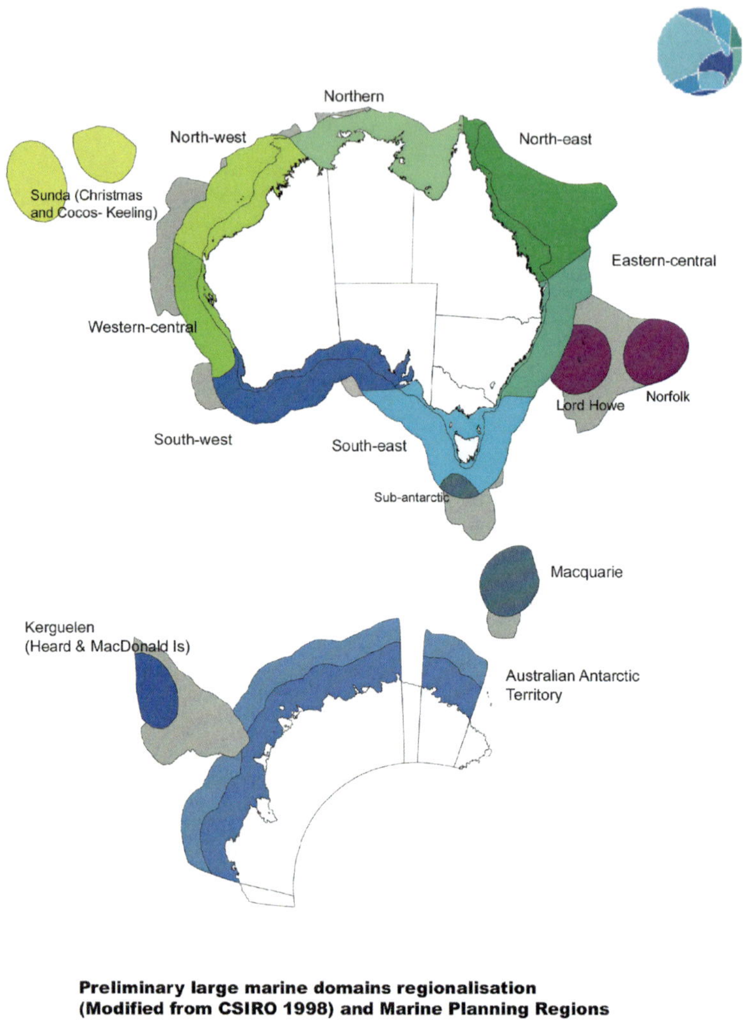

Plate 7 Regional marine planning is being undertaken over a number of regions (See also Fig. 2.2 on Page 21)

Plate 8 Map produced to show port limits under the customs act – Fremantle, Western Australia (See also Fig. 2.4 on Page 24)

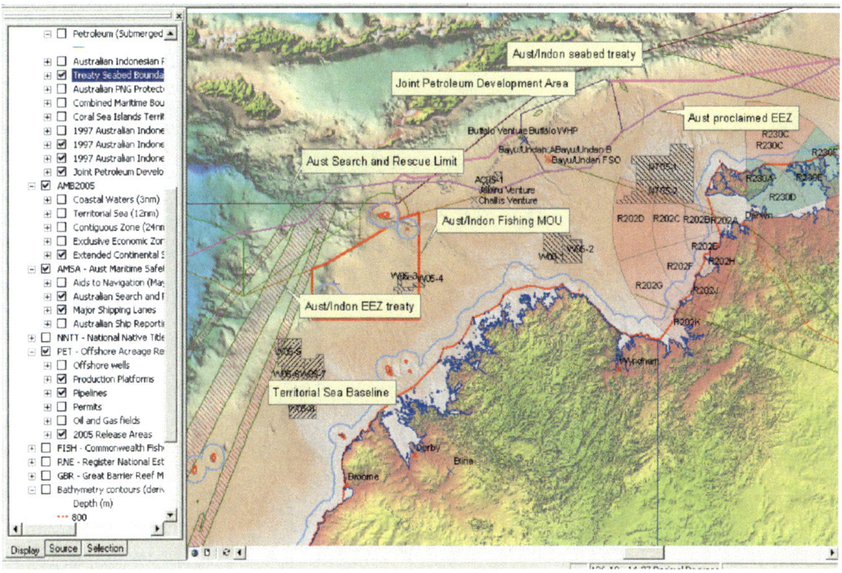

Plate 9 Extent of overlapping interests and treaty boundaries in the Timor Sea (See also Fig. 2.5 on Page 25)

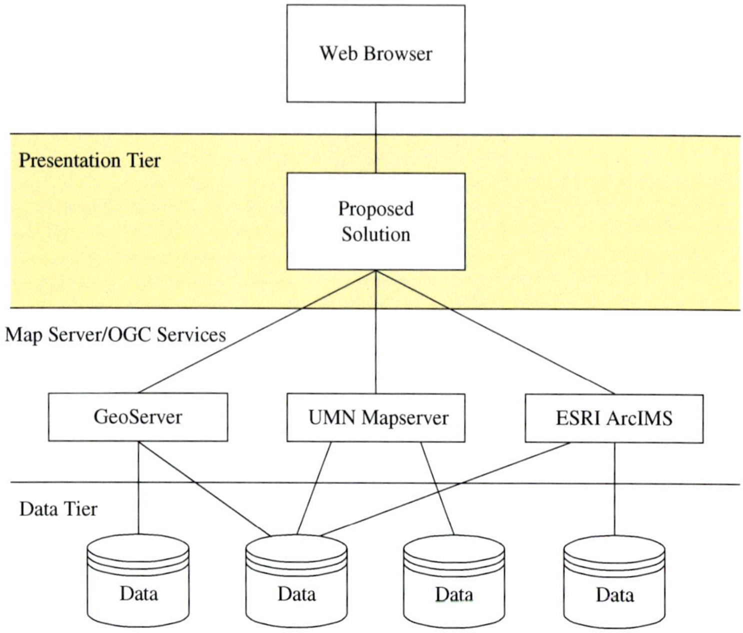

Plate 10 Proposed web mapping architecture for AMSIS (See also Fig. 2.6 on Page 26)

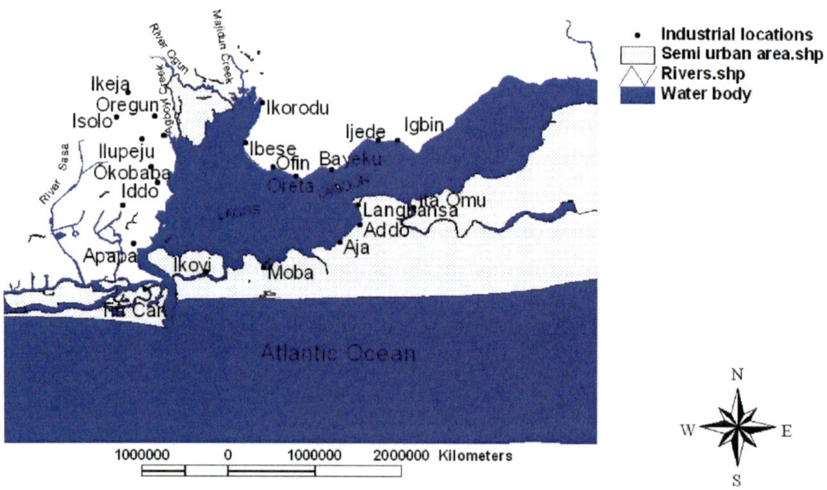

Plate 11 Map of Lagos Lagoon with the rivers and creeks (See also Fig. 3.1 on Page 31)

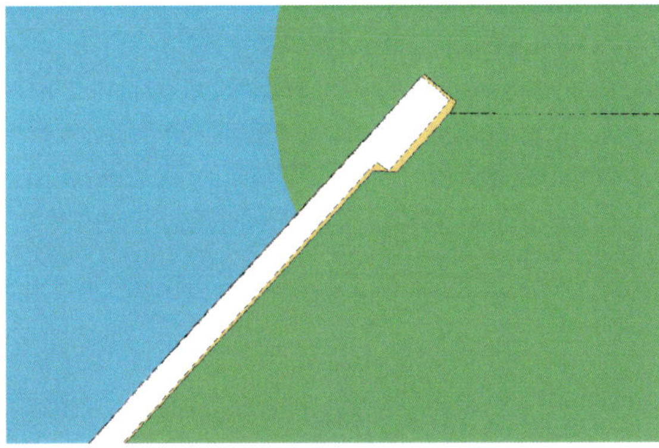

Plate 12 Example of the cartographic differences being identified by SeaZone as follow up to the ICZMap pilot project (See also Fig. 5.1 on Page 53)

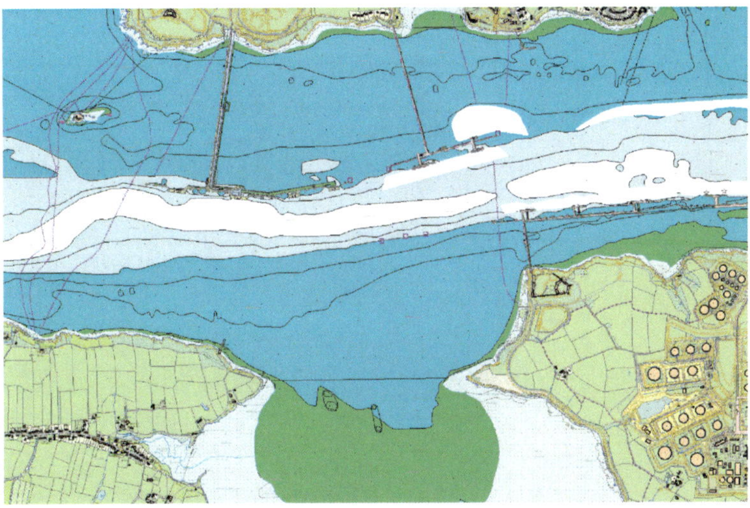

Plate 13 Example of the joined marine geographic information created by SeaZone (land data courtesy of Ordnance Survey) (See also Fig. 5.2 on Page 54)

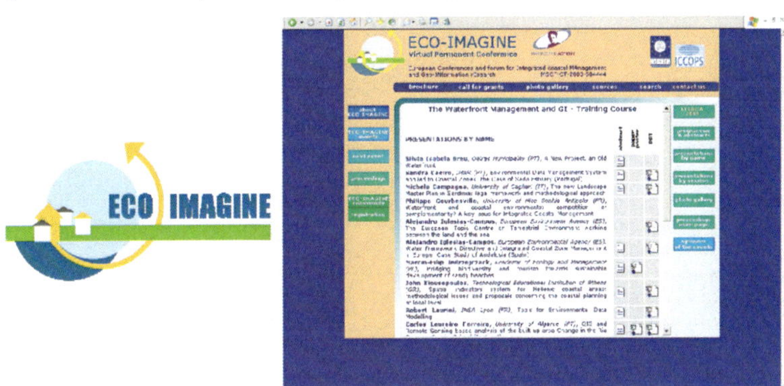

Plate 14 The project logo (**a**) and a page of the virtual permanent conference (**b**) (See also Fig. 7.1 on Page 66)

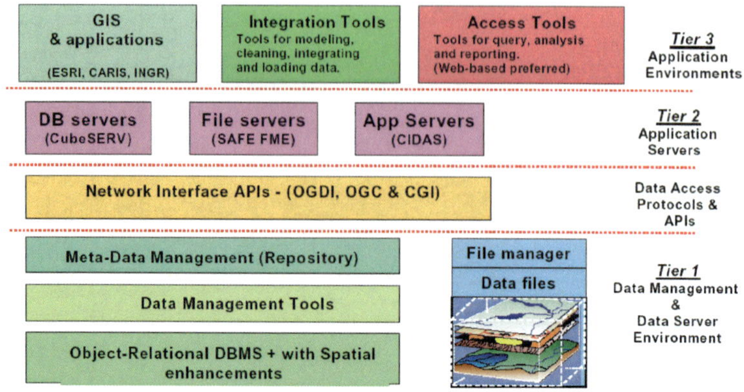

Plate 15 Technical architecture of the marine geospatial data infrastructure (See also Fig. 8.1 on Page 80)

Plate 16 Map generated by the COINAtlantic web mapping application (www.coinatlantic.ca) (See also Fig. 8.2 on Page 81)

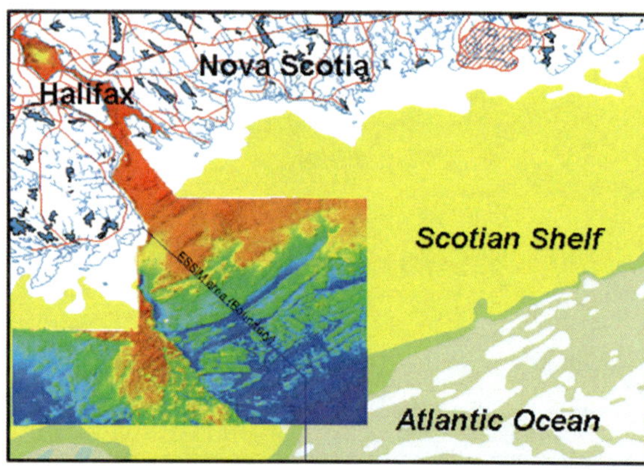

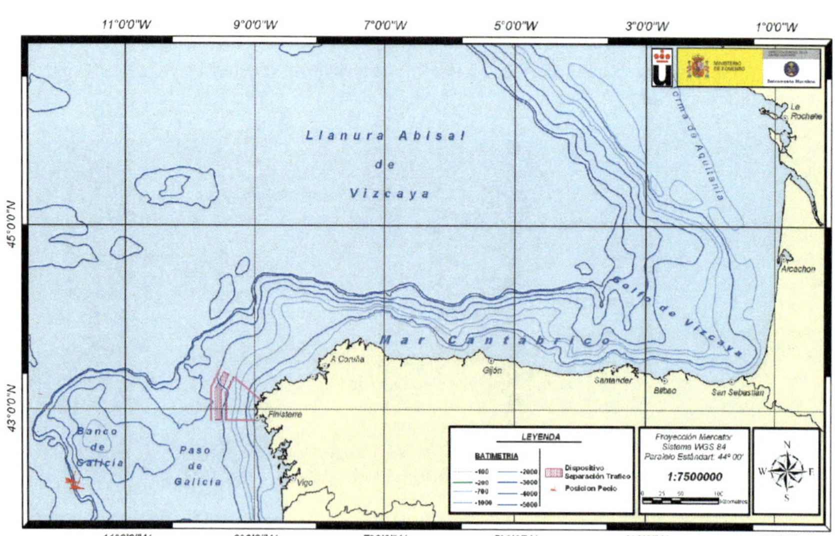

Plate 17 Cartography elements and base map (See also Fig. 11.1 on Page 111)

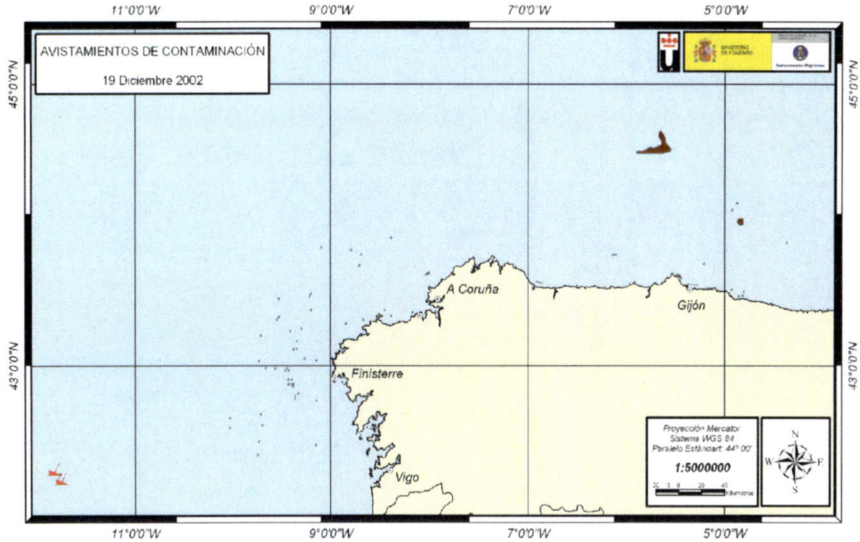

Plate 18 Example of daily map with the oil spill distribution (See also Fig. 11.4 on Page 114)

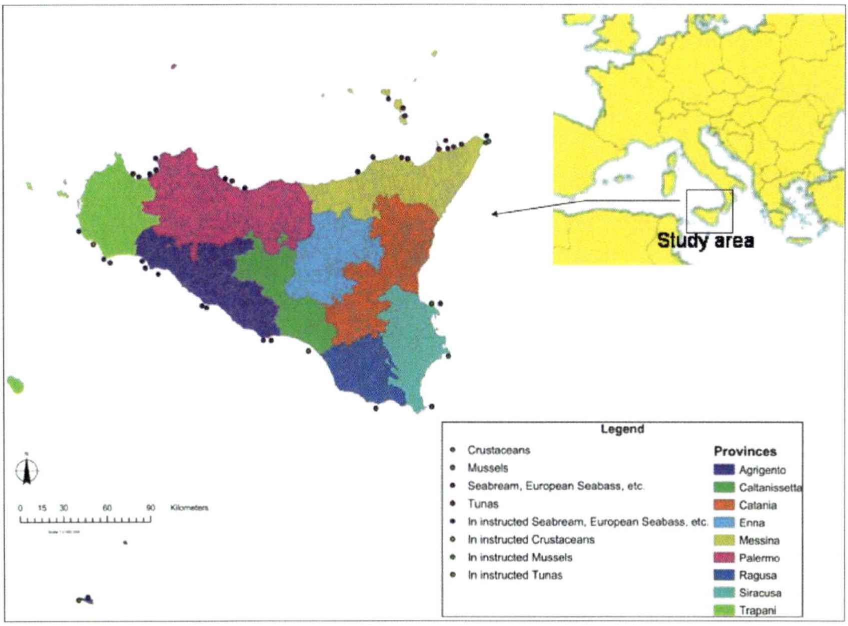

Plate 19 Position of the existing and planned plants along the Sicilian coasts and the distribution of species and provinces (See also Fig. 12.1 on Page 119)

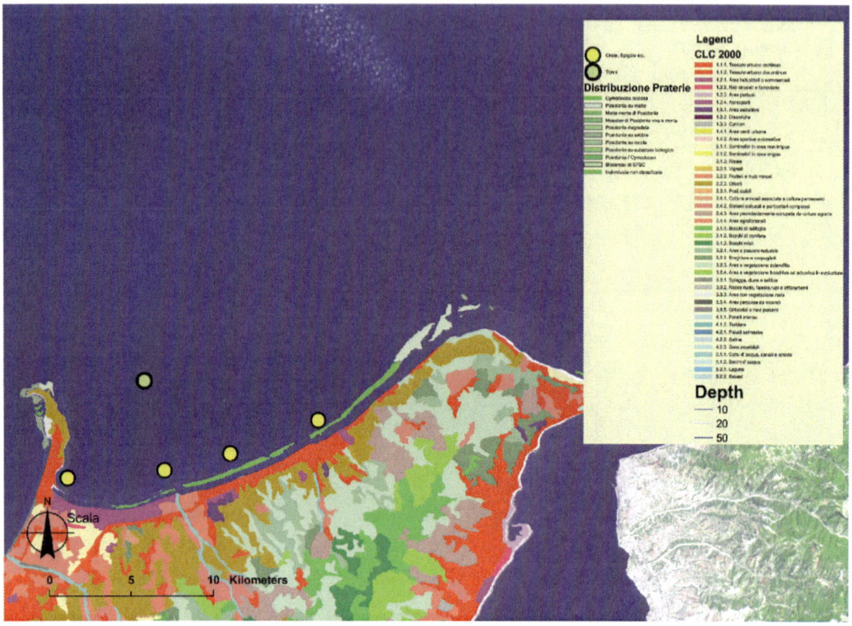

Plate 20 Detail of plants and biotopes (circled in *yellow-green*) distributed along the Tyrrhenian coast in the province of Messina (brown in Fig. 12.1.) (See also Fig. 12.2 on Page 124)

Plate 21 Nine examples of coastal geomorphology in Mexico (*left to right* and *top to bottom*): Cancún, Quintana Roo; Progreso, Yucatán; Huatulco, Oaxaca; Acapulco, Guerrero; Puerto Vallarta, Jalisco; Manzanillo, Colima; San José de los Cabos, Baja California Sur; Mazatlán, Sinaloa; and Hecelchakán, Campeche (See also Fig. 14.1 on Page 140)

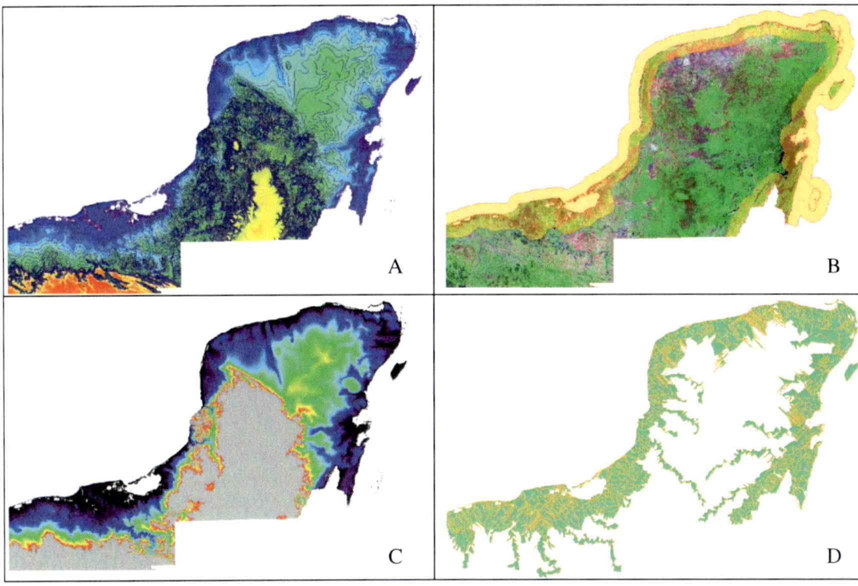

Plate 22 (**A**) Digital elevation map of the Yucatan Peninsula within Mexico; (**B**) Proximity, shoreline buffer 20 km offshore and inland; (**C**) Area below 50 m elevation; (**D**) Hydrological subwatersheds totally or partially within the 20 km buffer (See also Fig. 14.2 on Page 147)

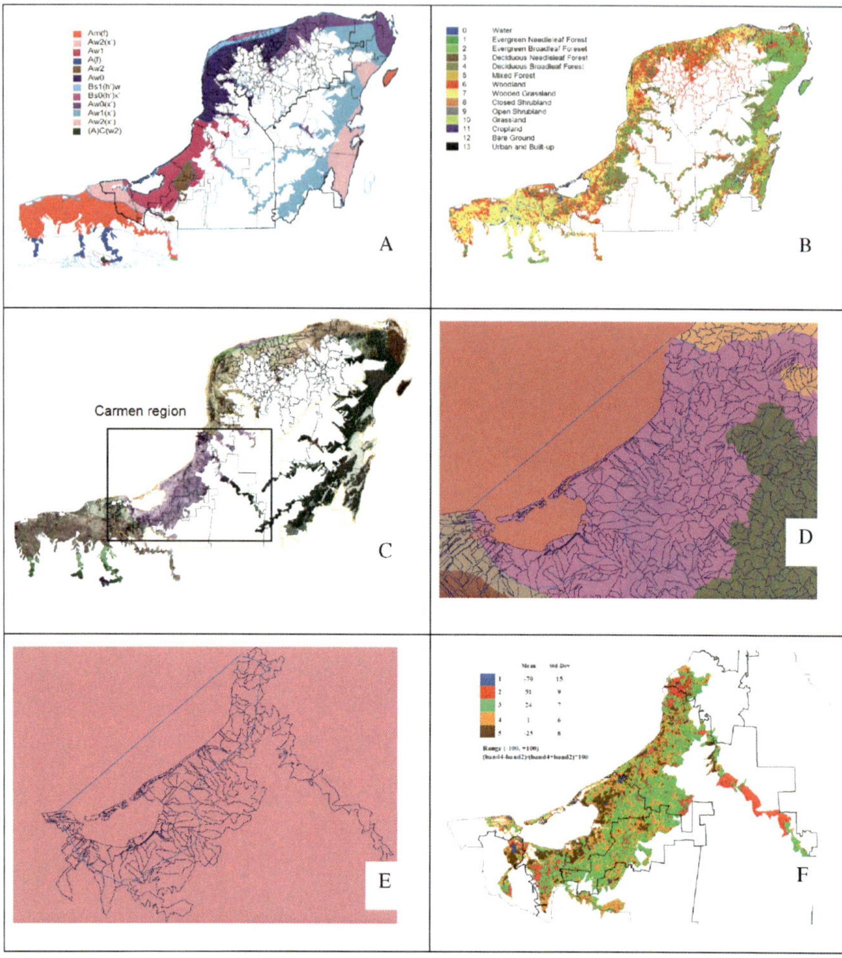

Plate 23 (**A**) Shoreline and government administrative boundaries overlaid with climatic regions contained within watershed boundaries; (**B**) Shoreline and government administrative boundaries overlaid with land uses contained within climatic regions and watershed boundaries; (**C**) Shoreline and government administrative boundaries overlaid with NDVI values contained within climatic regions and watershed boundaries; (**D**) Close-up of Carmen climatic region (Aw1), including sub-watersheds; (**E**) Carmen climatic region showing selected watersheds totally or partially within 20 km SL buffer; (**F**) Carmen climatic region overlaid with five NDVI classes within watershed boundaries, as well as municipal and state boundaries (See also Fig. 14.3 on Page 148)

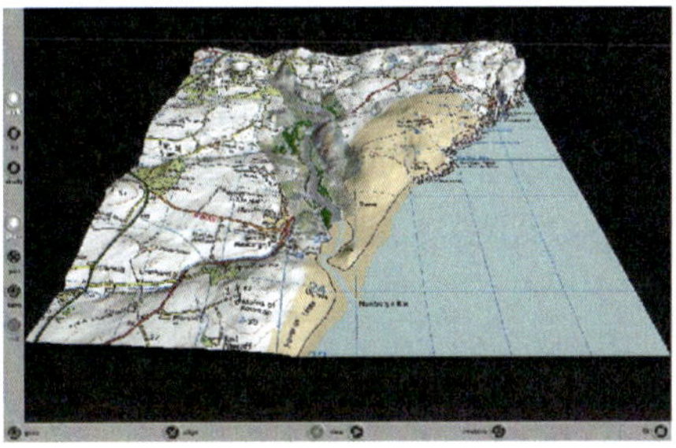

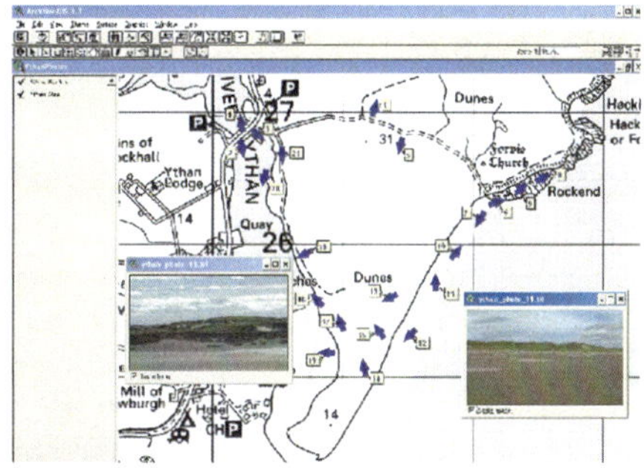

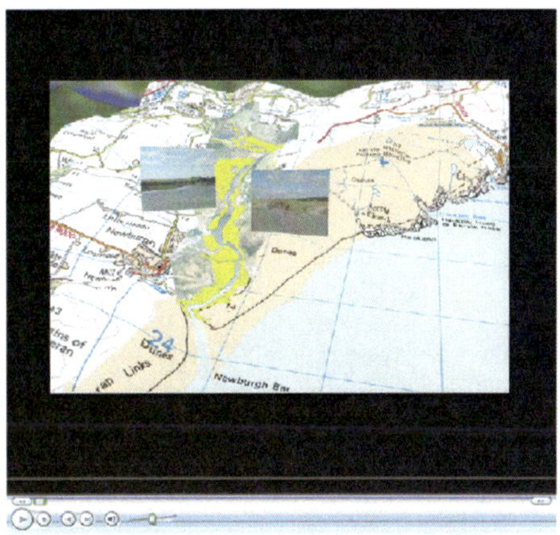

Plate 24 (**a**) An interactive hypermedia animation with several forms of interactivity generated with Erdas Imagine 8.5 using the ParallelGraphics CortonaR VRML Client used to view VRML model. (**b**)An ArcView 3.3. project using raster Ordnance Survey (OS) map data at a scale of 1:50,000 with hotlinks to 21 ground truth colour photographs at 1280 × 960 pixels resolution. (**c**) The third example is a video clip (11.4 MB file size, with a duration of 25 s, comprising 375 frames, at 15 frames per second, using the DivX High Definition Profile, at a resolution of 768 × 512) generated by the Bryce 5 software (See also Fig. 16.2 on Page 181)

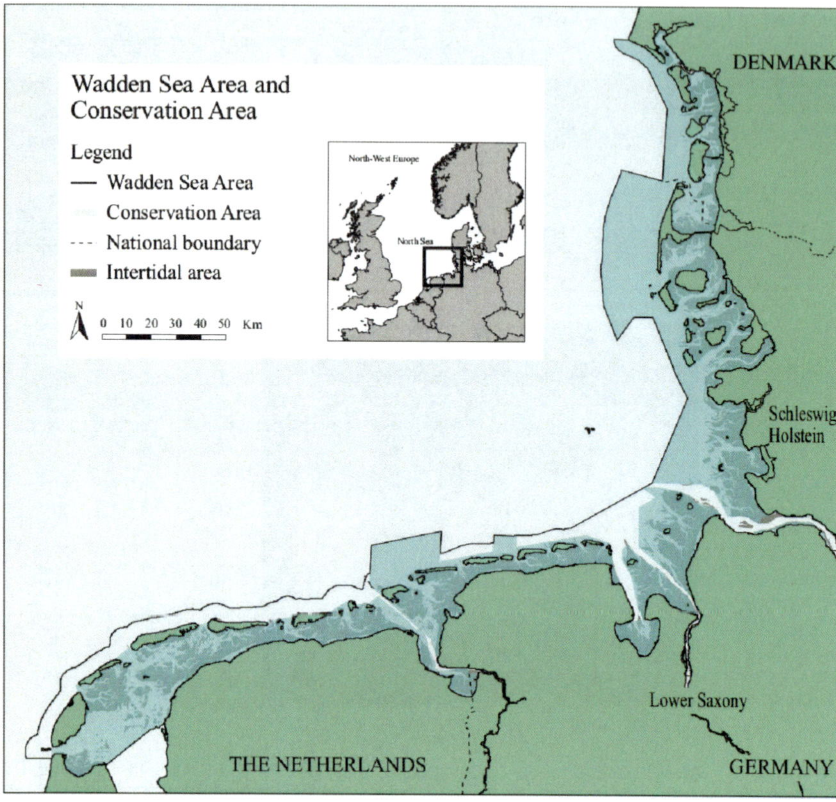

Plate 25 The Wadden Sea area (Trilateral cooperation area) (See also Fig. 18.1 on Page 198)

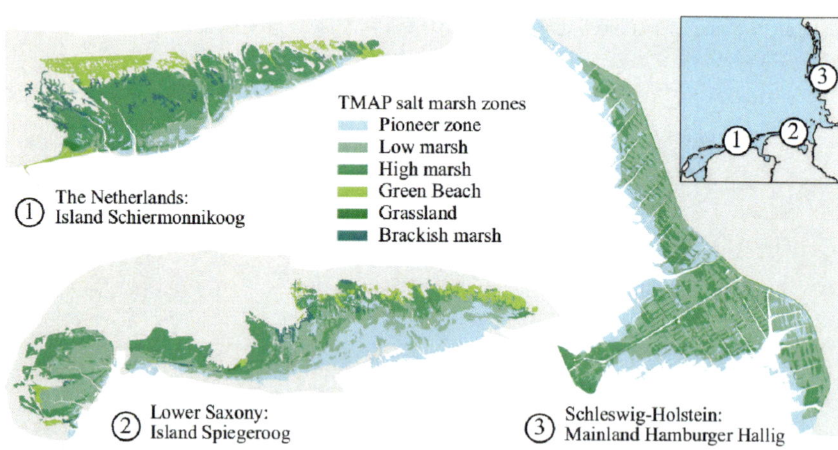

Plate 26 Salt marsh zones in the Netherlands and Germany. The used geographical and attribute data is directly comparable (Essink et al. 2005) (See also Fig. 18.2 on Page 201)

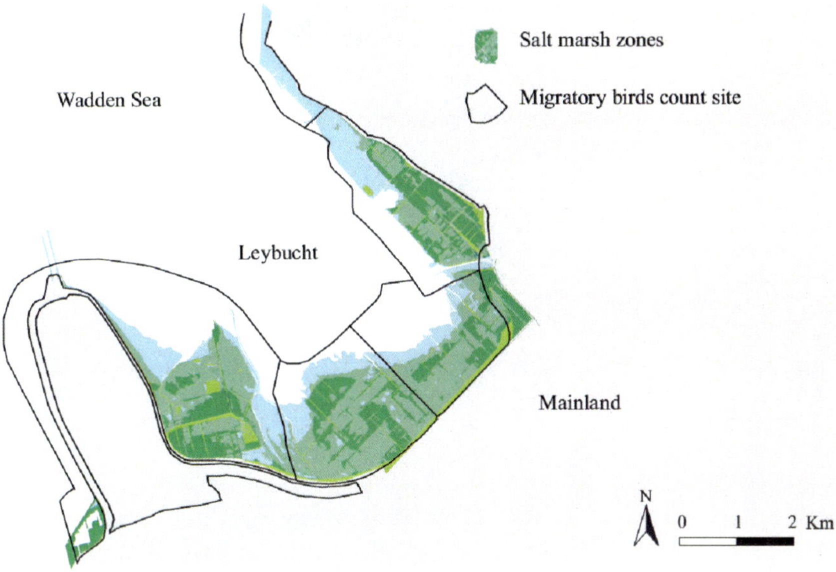

Plate 27 Salt marsh zones (NLPV 1997) combined with migratory bird count sites (NLWKN 2004) in the Leybucht Lower Saxony, Germany (See also Fig. 18.3 on Page 202)

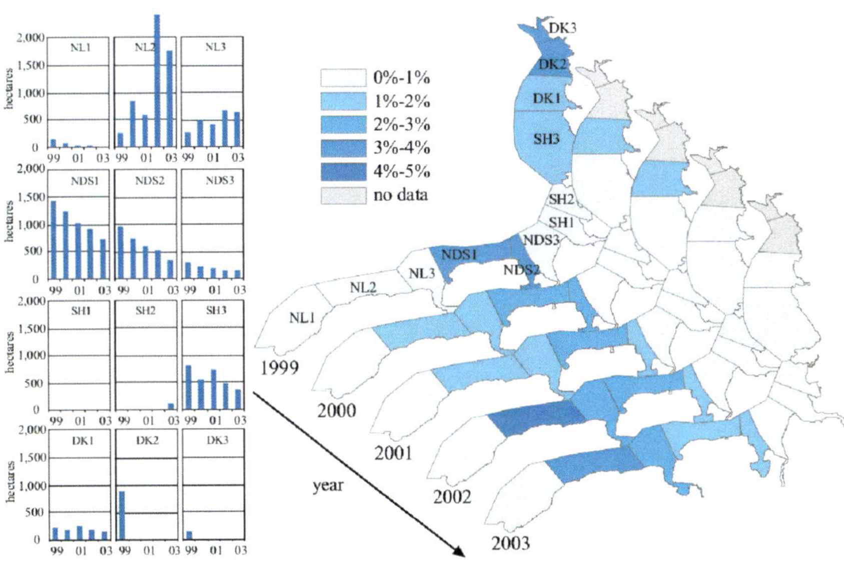

Plate 28 Distribution of intertidal blue mussel beds in the Wadden Sea. Absolute (ha) and relative (%) of intertidal area covered by blue mussel beds (Essink et al. 2005) (See also Fig. 18.4 on Page 203)

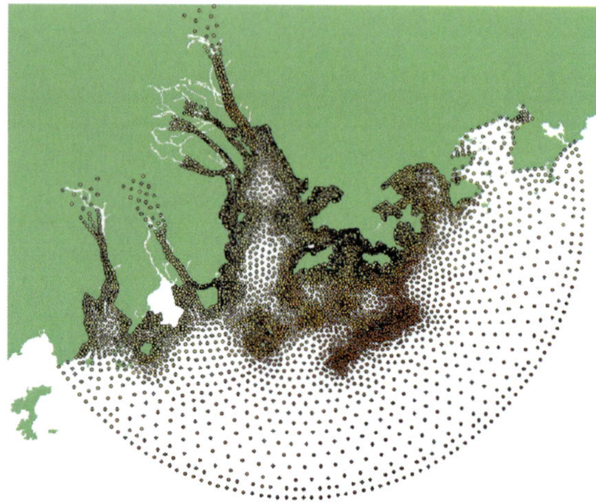

Plate 29 Display of water depth from model input in GIS (After Ng (2004)) (See also Fig. 19.5 on Page 215)

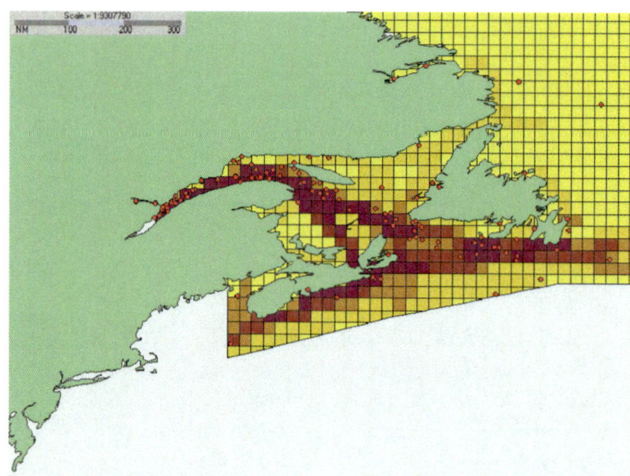

Plate 30 Shipping traffic density distribution through gridding (See also Fig. 21.1 on Page 233)

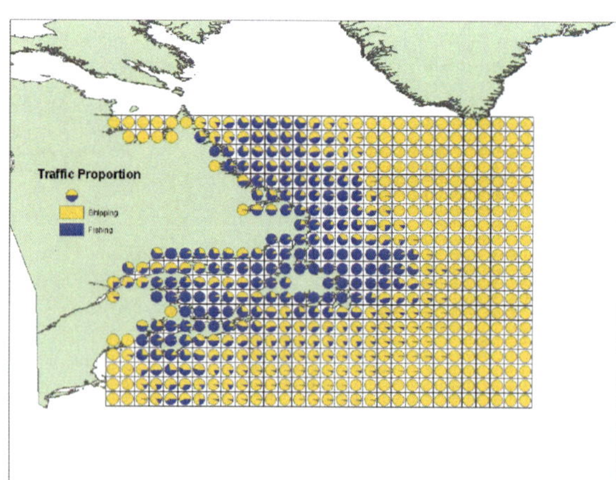

Plate 31 Fishing versus shipping traffic proportion distribution in Atlantic Canada (See also Fig. 21.2 on Page 233)

Plate 32 Single Kernel density analysis of fishing incidents (See also Fig. 21.3 on Page 236)

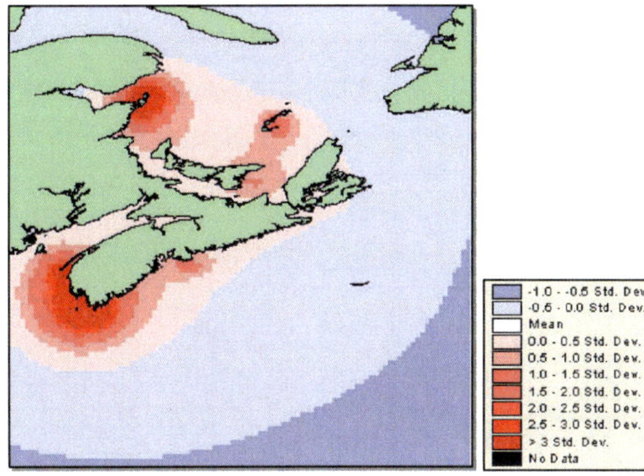

Plate 33 Fishing incident rate distribution using Dual Kernel density interpolation (See also Fig. 21.4 on Page 236)

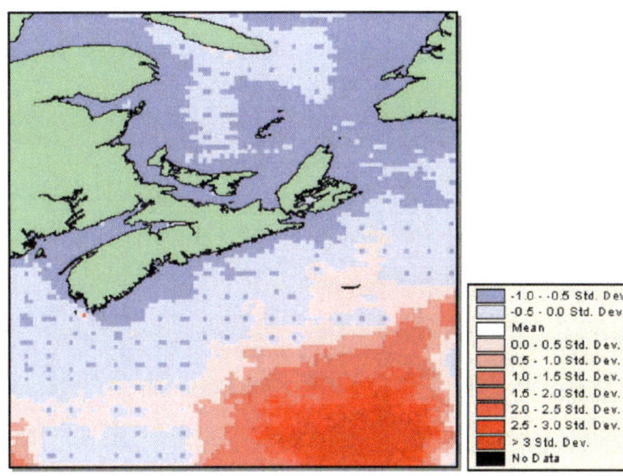

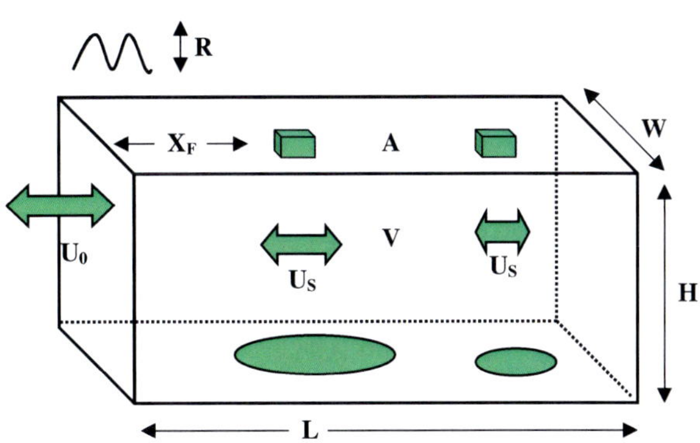

Plate 34 Schematic diagram representing the simplifications used by the carbon deposition model. The sea loch is represented by a rectangular basin with identical physical characteristics to the real system. The tidal current amplitude decreases from U_0 at the mouth to zero at the head, with values at each site, U_S, calculated accordingly. Fish farms are represented by the filled rectangular block, with the areas of impacted seabed denoted by the filled ellipses. All other parameters and variables are described in the text (See also Fig. 22.1 on Page 242)

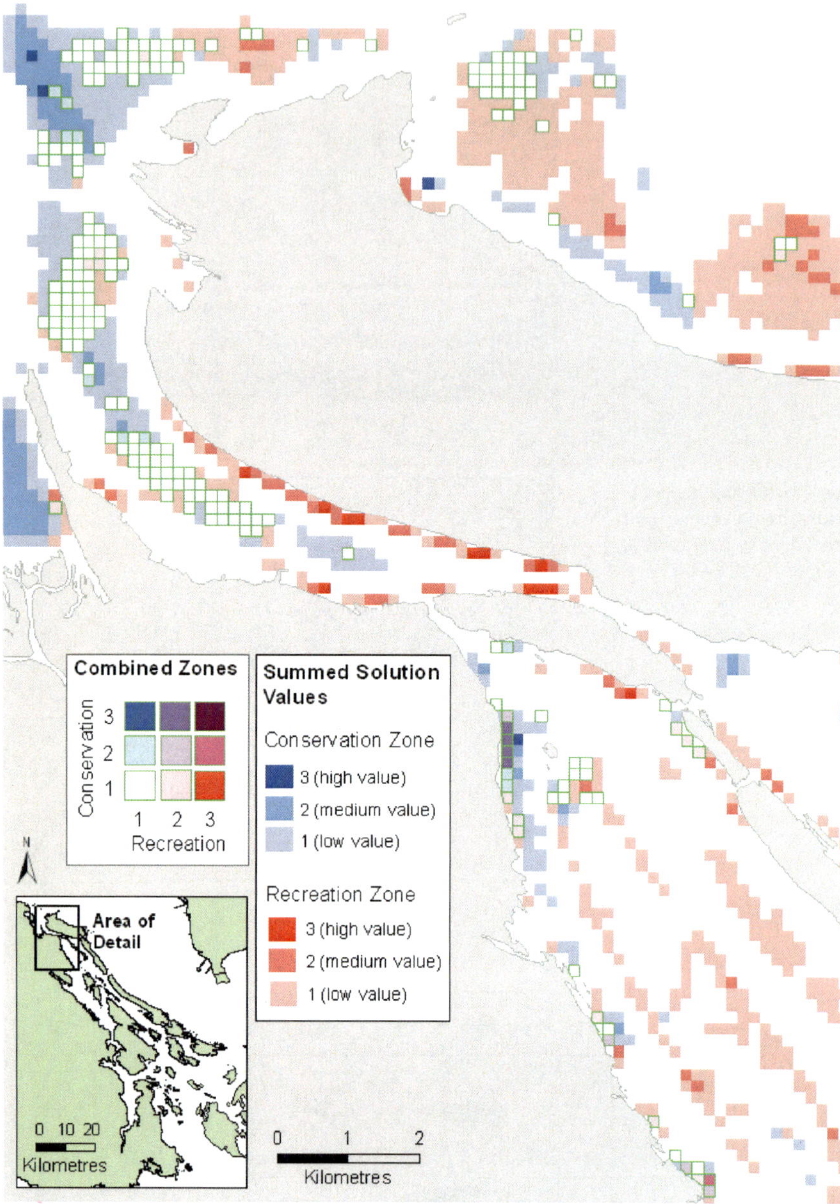

Plate 35 Area of detail from method three combined zones (See also Fig. 23.2 on Page 250)

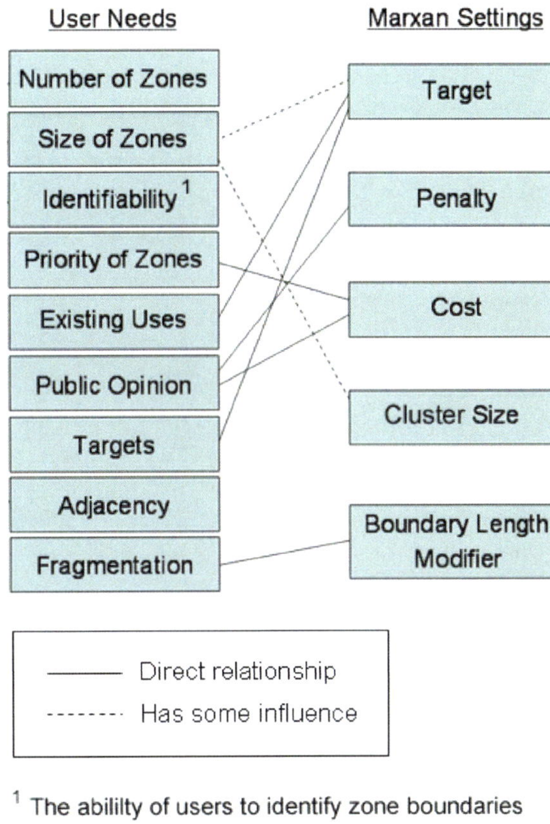

Plate 36 Functionality of Marxan compared with user needs for zoning (See also Fig. 23.3 on Page 252)

Plate 37 Coastal zone in Slovenia (See also Fig. 24.1 on Page 256)

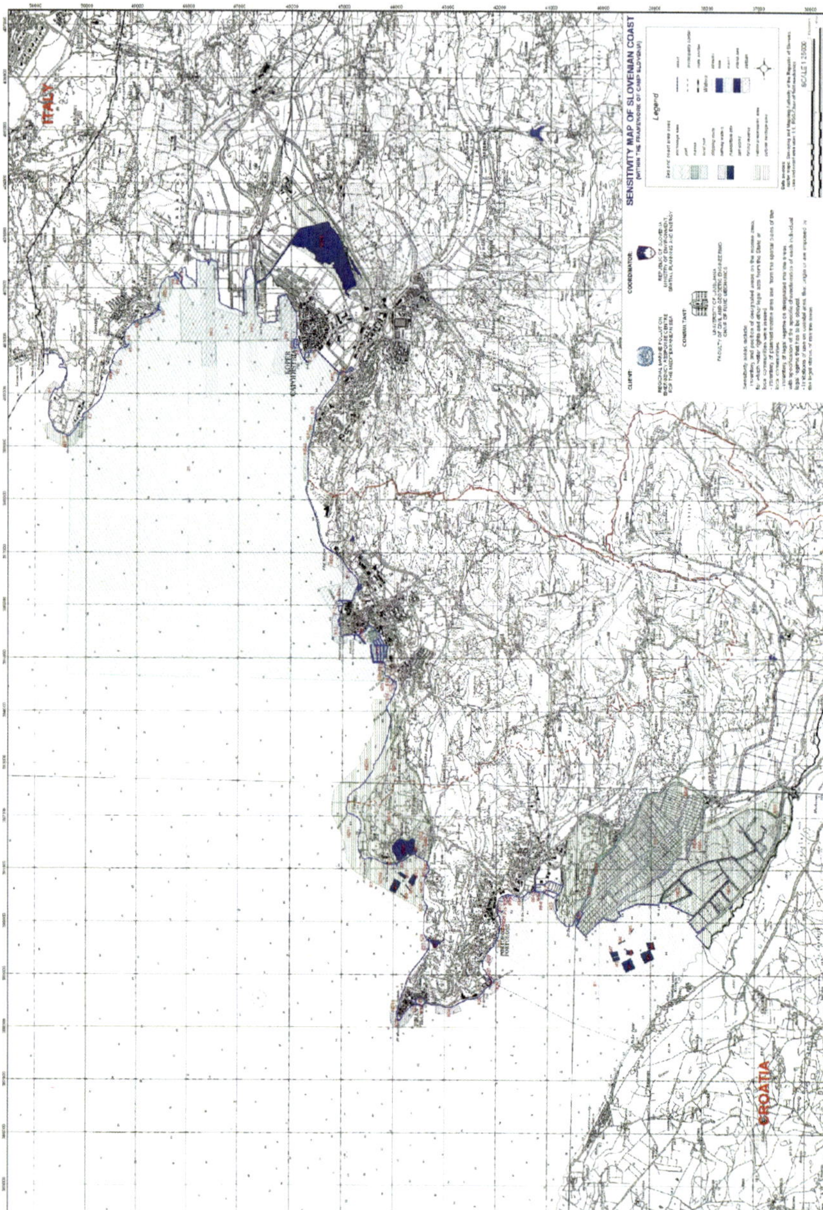

Plate 38 Existing legal regimes as decision support tool for managing sensitivity areas for national contingency plan (See also Fig. 24.2 on Page 260)

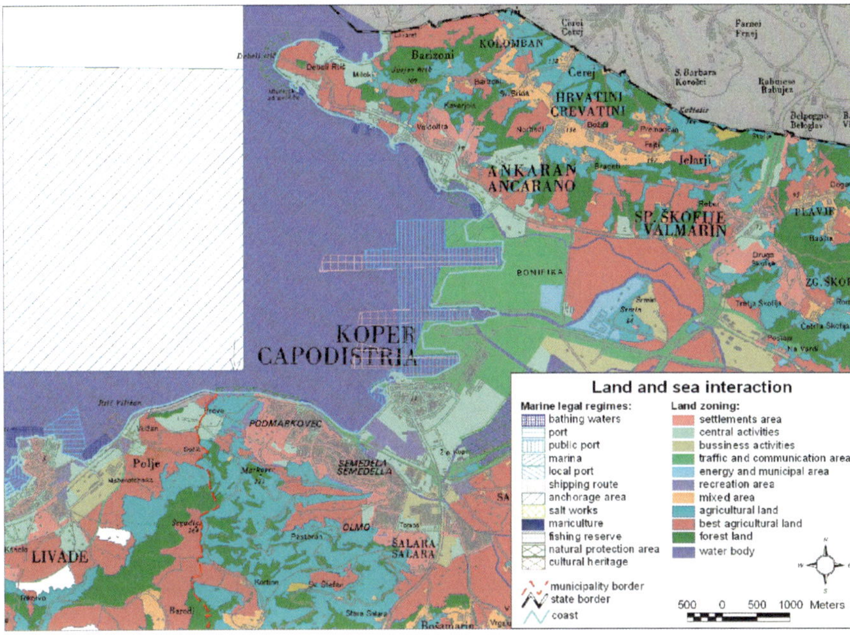

Plate 39 Spatial representation of land and sea interaction as functional connected areas (See also Fig. 24.3 on Page 261)

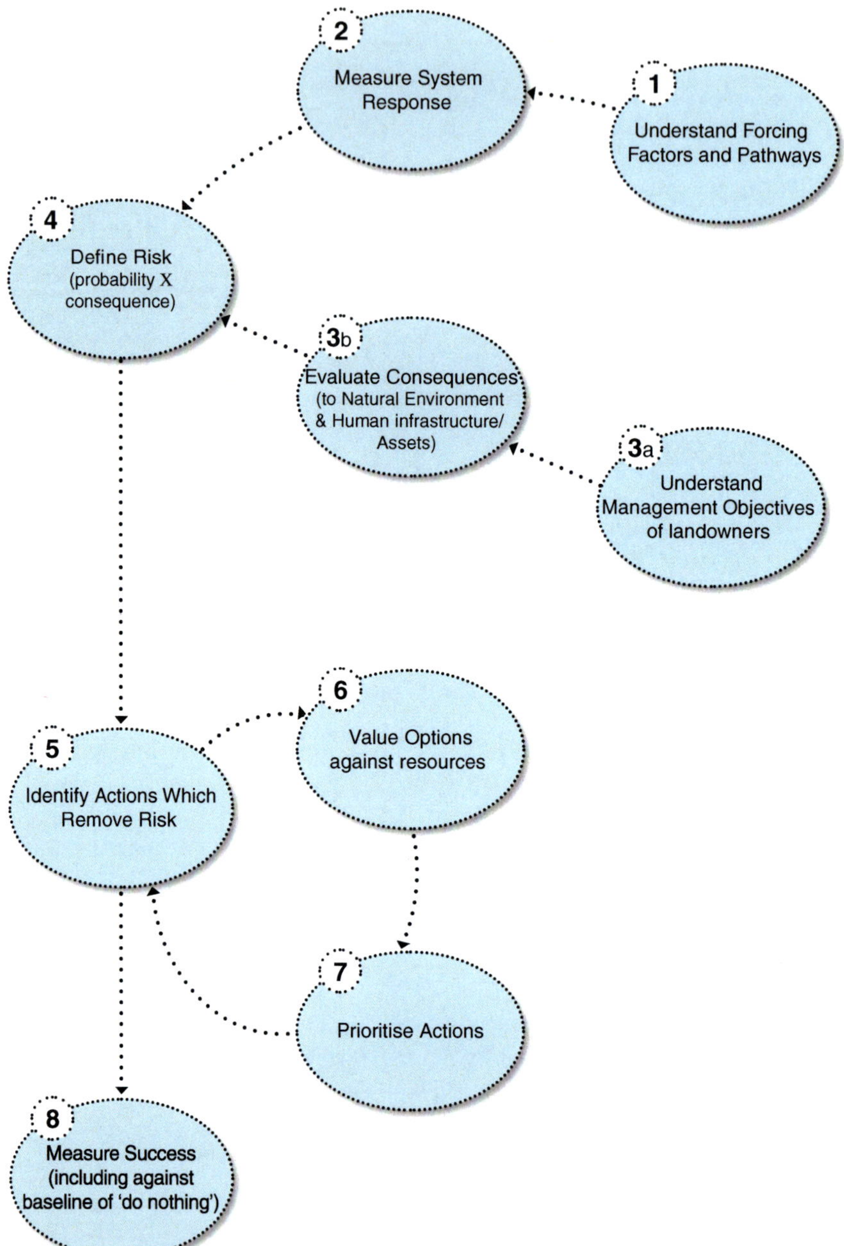

Plate 40 Group model of decision stages in the management of coastal geohazards (See also Fig. 25.1 on Page 268)

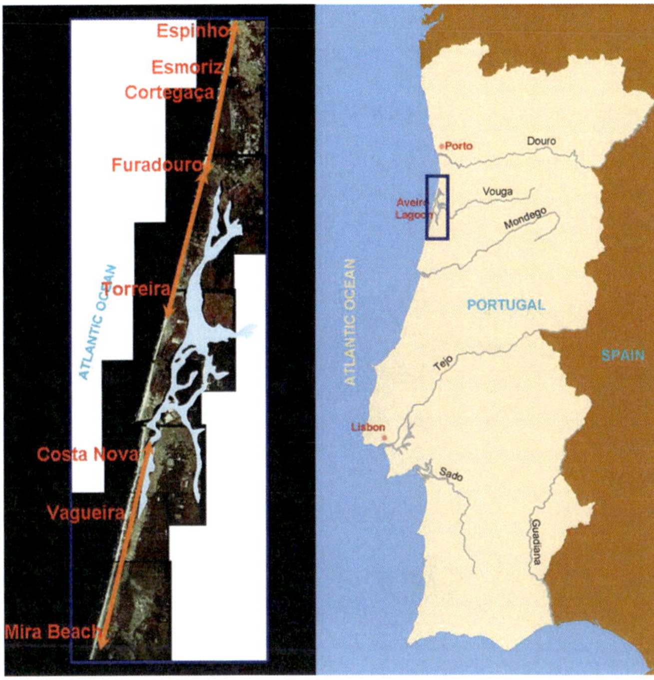

Plate 41 Study area location (See also Fig. 26.1 on Page 276)

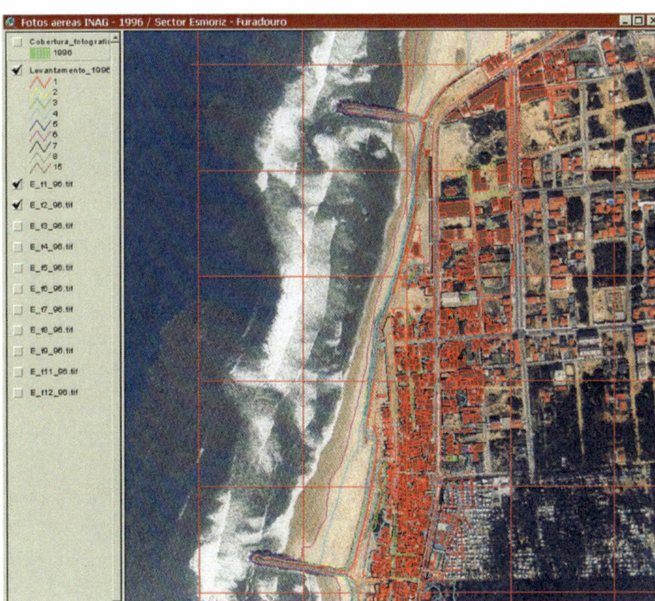

Plate 42 1996 aerial survey after georeferencing (See also Fig. 26.2 on Page 280)

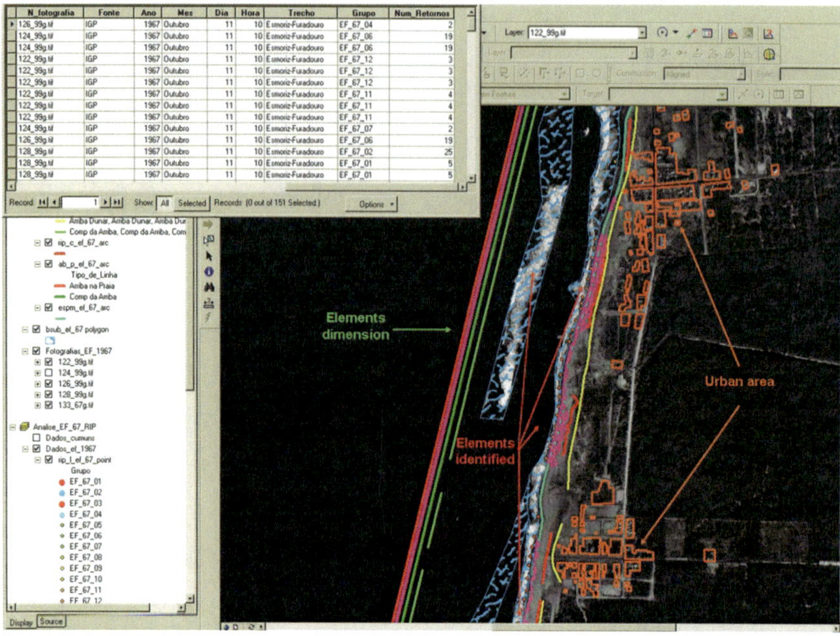

Plate 43 Visual analysis of the 1965 survey (See also Fig. 26.3 on Page 281)

Plate 44 Structure of Matlab analysis tool (See also Fig. 26.4 on Page 282)

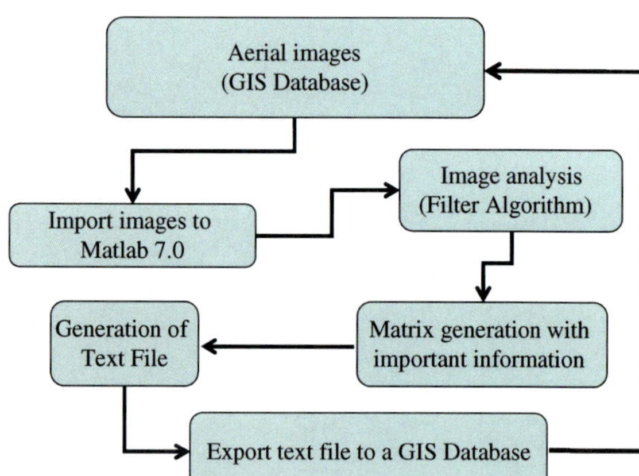

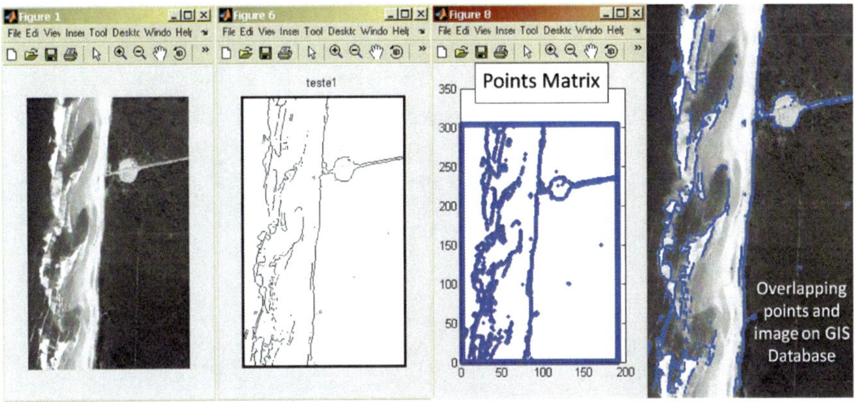

Plate 45 Preliminary results of Matlab analysis and overlapping in GIS database (See also Fig. 26.5 on Page 282)

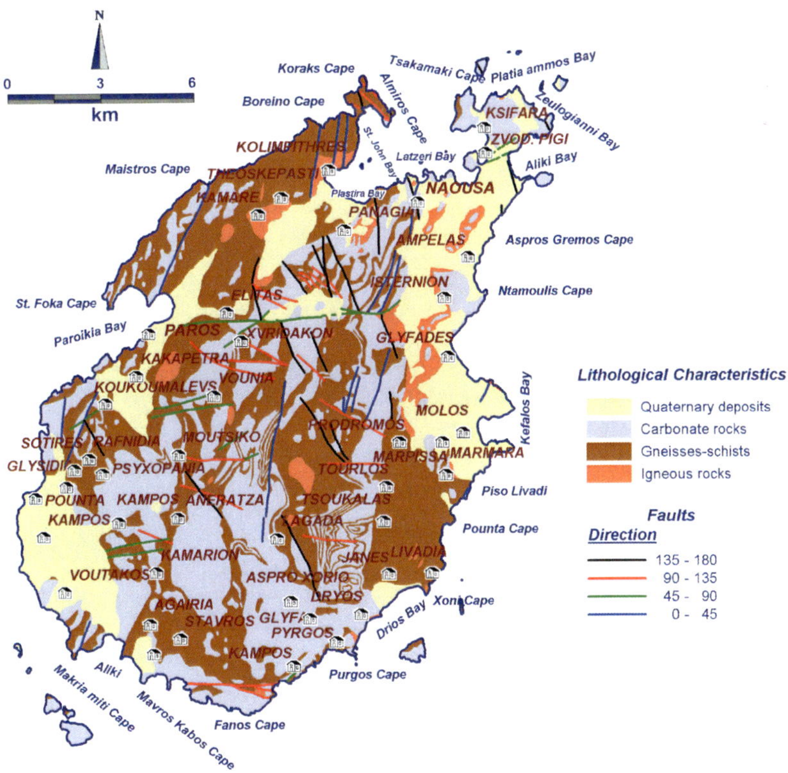

Plate 46 Geological (i.e. lithology, fault lines) characteristics of Paros Island (See also Fig. 27.1 on Page 287)

Plate 47 Geographical and morphological characteristics of Paros Island (See also Fig. 27.2 on Page 288)

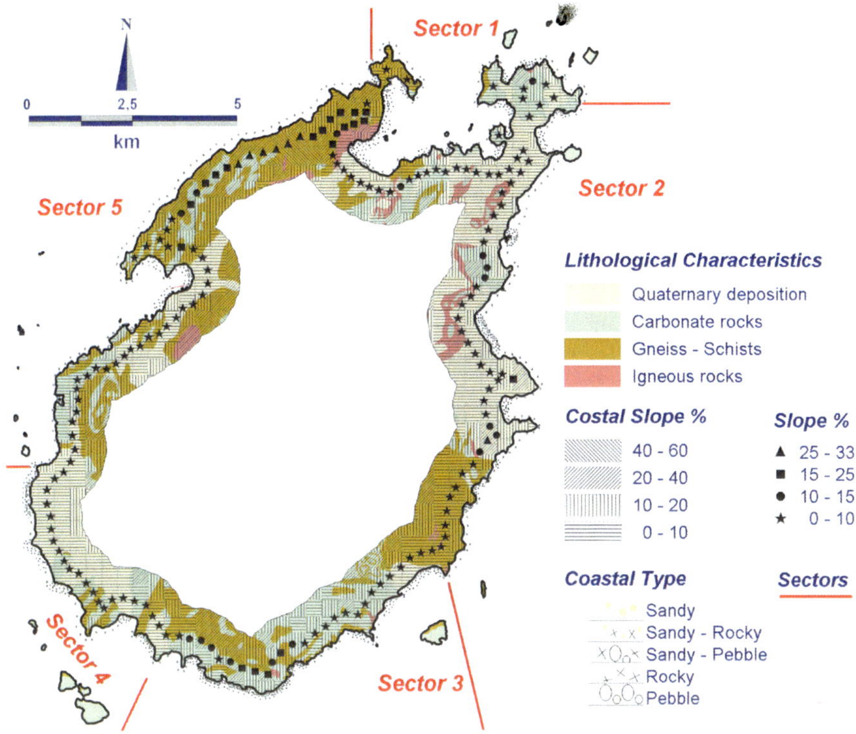

Plate 48 Coastal slopes, coastal types and the sectors defined according to effective fetch of incoming wind-generated waves (See also Fig. 27.3 on Page 292)

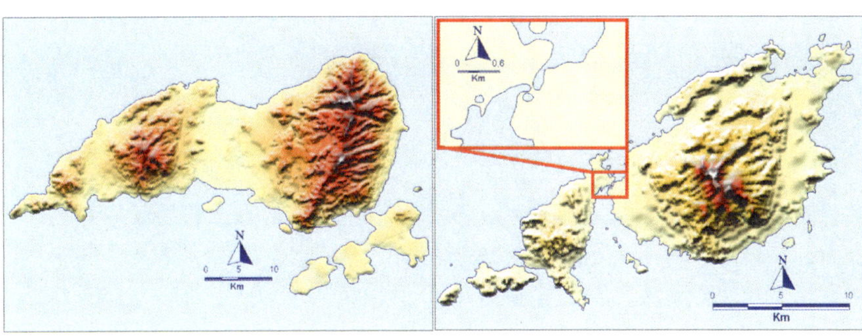

Plate 49 The morphology of Paros Island at (**a**) 10,000 yr BP, when sea level was ca. 54 m lower than its present stage (*left*) and (**b**) 6,000 yr BP, when sea level was ca. 6 m lower than its present stage (*right*) (See also Fig. 27.4 on Page 294)

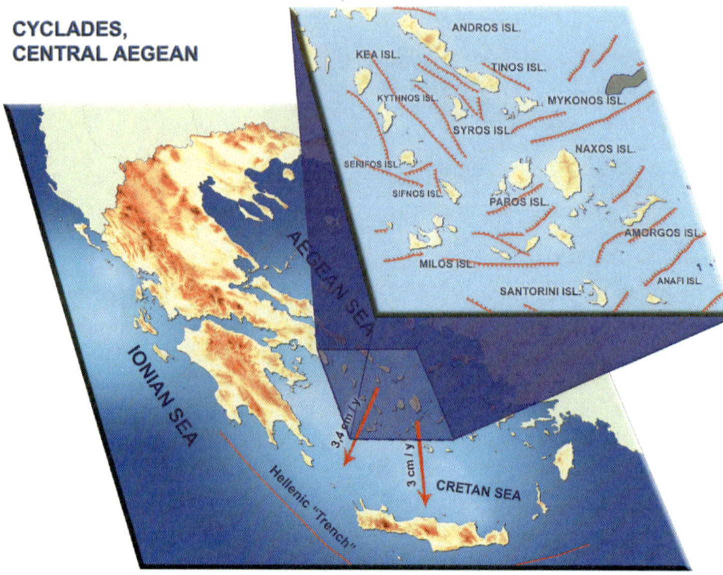

Plate 50 The study area situated in the Central Aegean, showing the rates and direction of motion of Cyclades during the Quaternary and the major submarine fault zones (See also Fig. 28.1 on Page 298)

Plate 51 From left to right: (**a**) Eroding beachrock in Syros. (**b**) Submerged Roman fishtanks in Naxos. (**c**) Eroding Myceanean wall at sea level in the city of Naxos (See also Fig. 28.2 on Page 301)

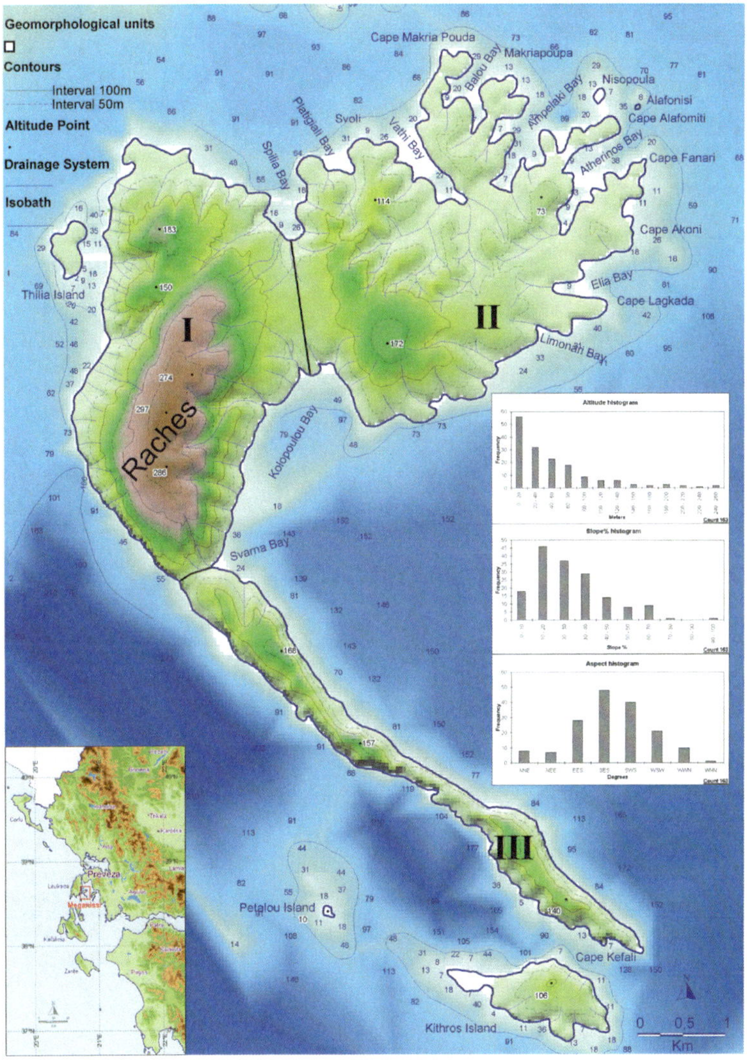

Plate 52 The topography, the drainage system and the geomorphological units of Meganisi island (See also Fig. 29.1 on Page 306)

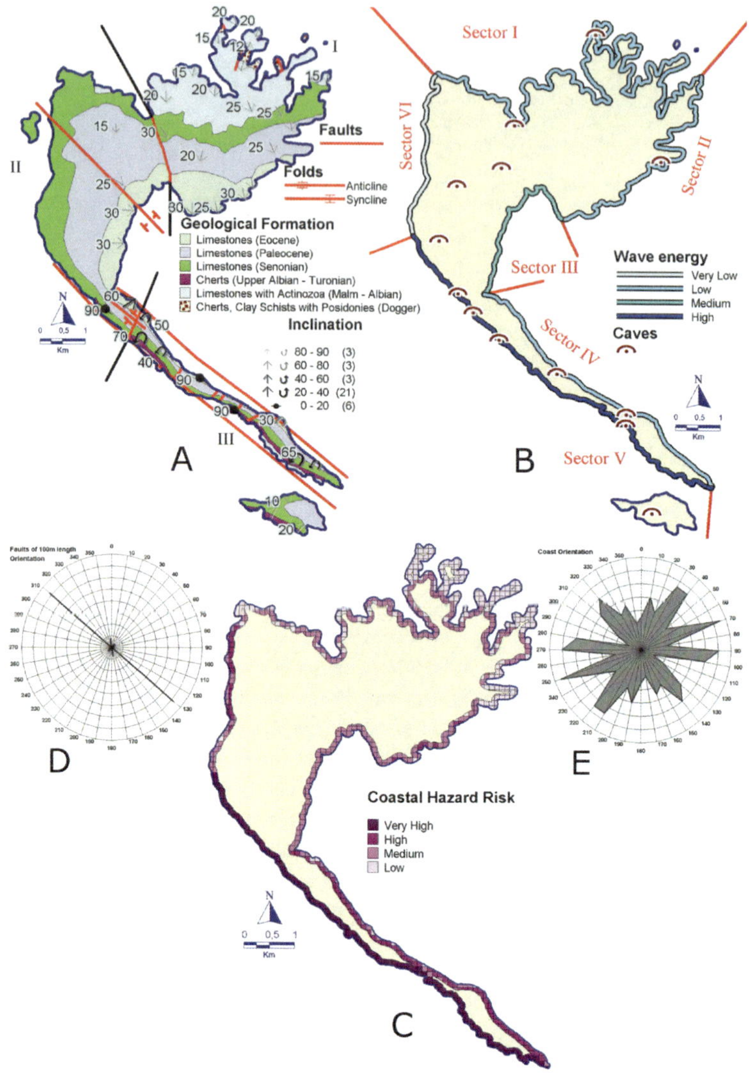

Plate 53 (**A**) The geological map of the Meganisi island (I.G.M.E., 1963, 1994), modified by the authors. (**B**) The wave energy map, (**C**) The final coastal hazard risk map of Meganisi island, (**D**) A rodograme of faults orientation, (**E**) A rodograme of coast orientation (See also Fig. 29.3 on Page 311)

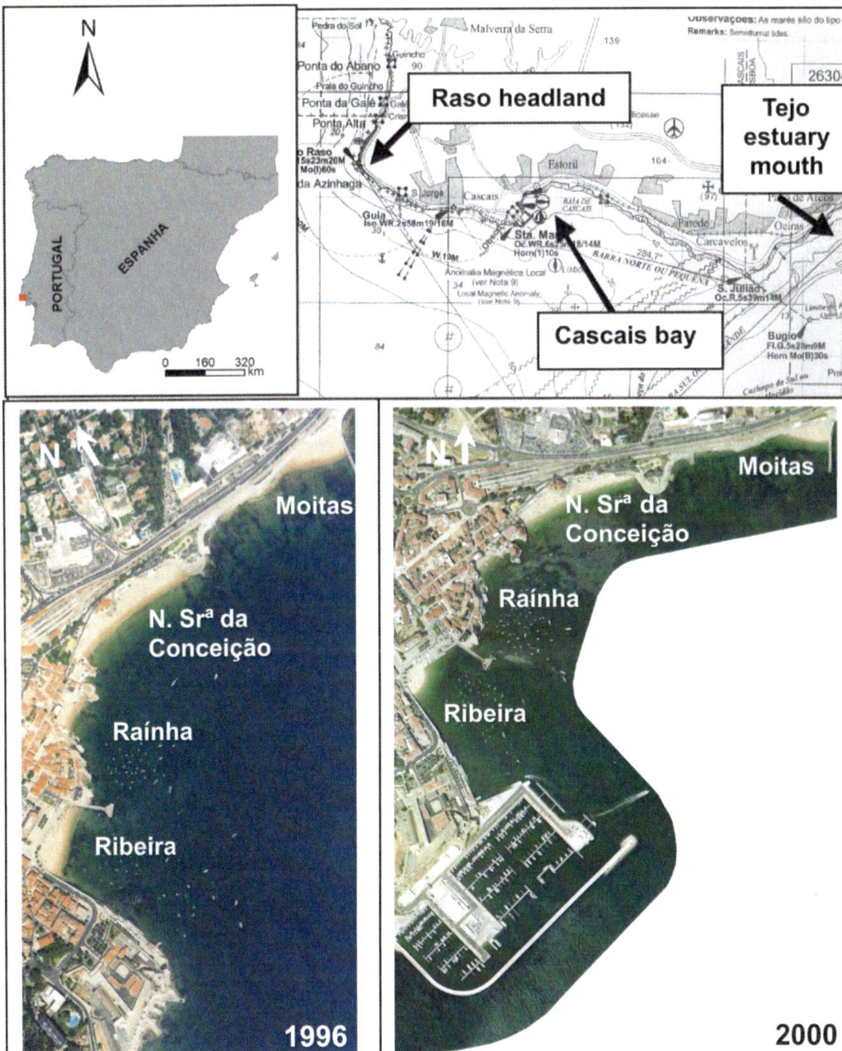

Plate 54 Location of the study area and aerial photographs of Cascais bay beaches before and after the construction of the marina (See also Fig. 30.1 on Page 316)

a)

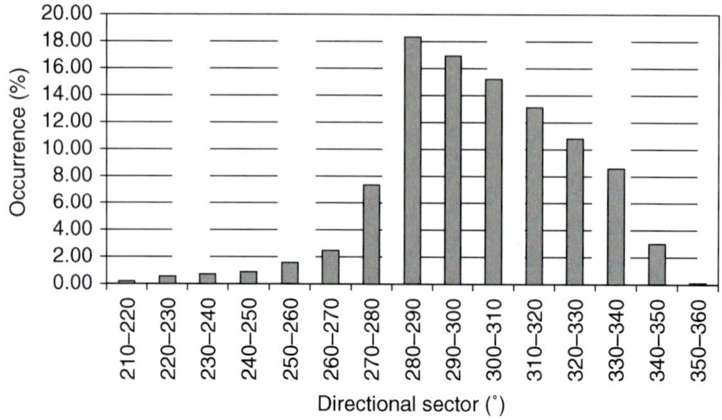

b)

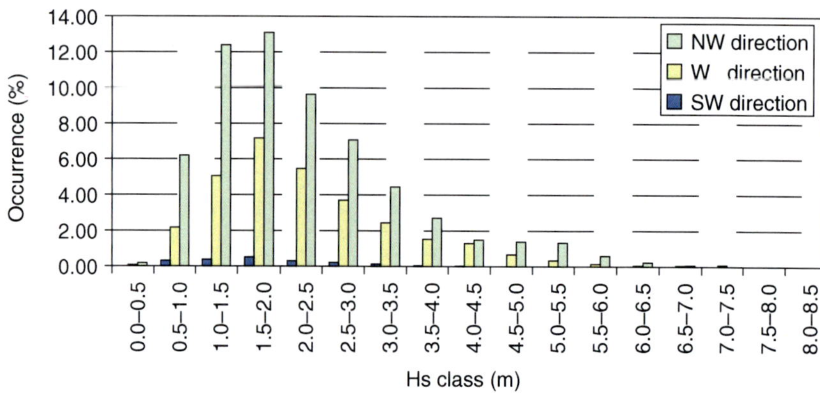

c)

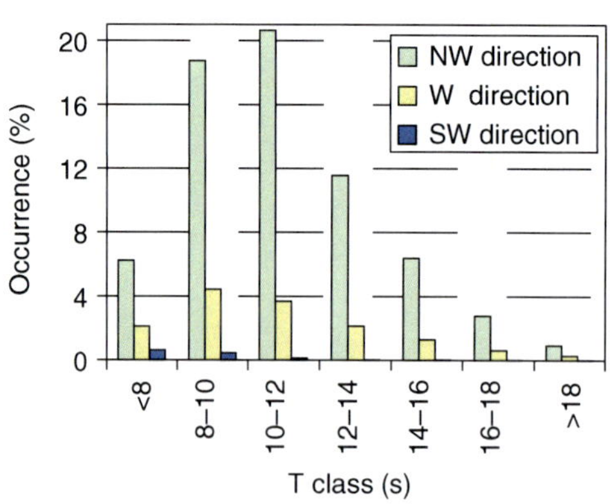

Plate 55 Histograms of occurrence for the offshore wave regime parameters: **a**) direction; **b**) significant height; and **c**) period (See also Fig. 30.2 on Page 318)

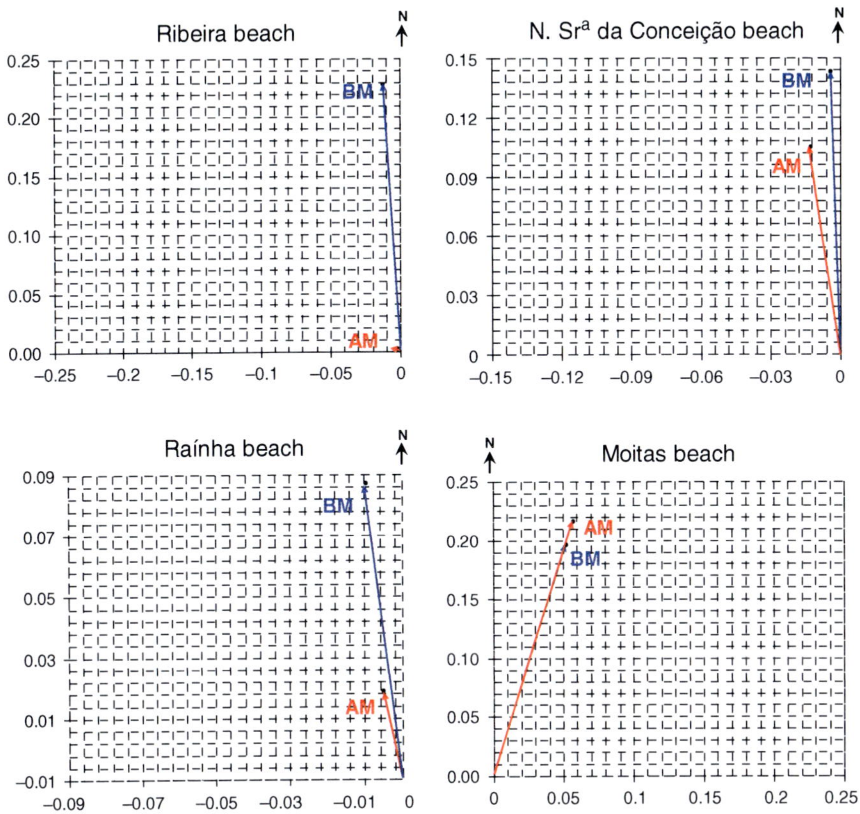

Plate 56 Variable factor of the resultant wave power, for the situations BM and AM, in front of the beaches: Ribeira, Raínha, N. Srª da Conceição and Moitas (See also Fig. 30.4 on Page 323)

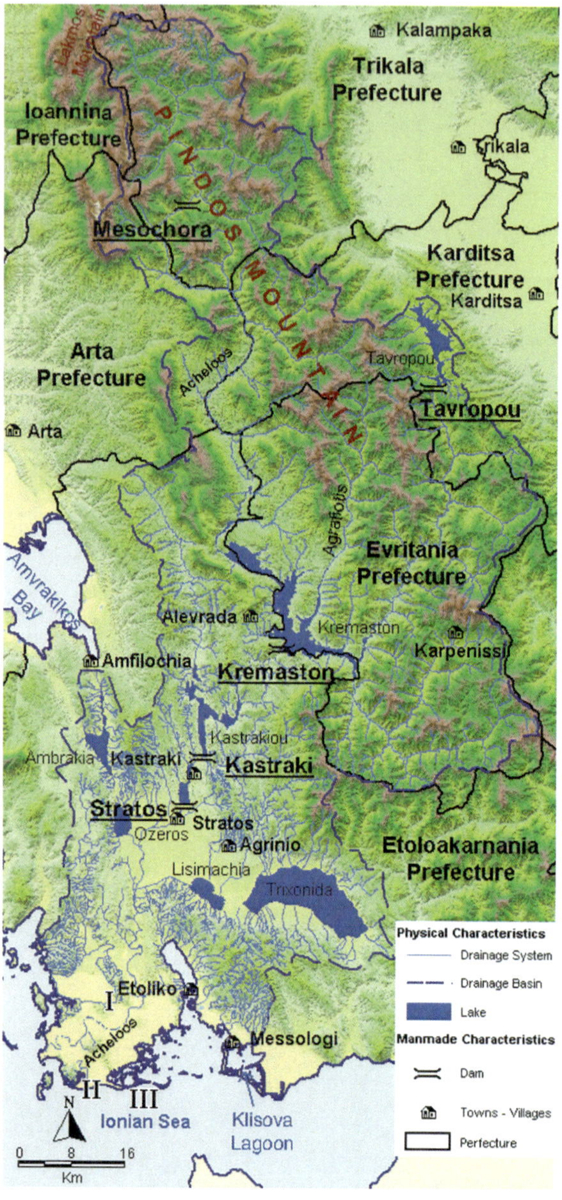

Plate 57 The location of the study area, the development of drainage basin and the Dams of Acheloos basin (See also Fig. 31.1 on Page 326)

Plate 58 The delta plain. Geomorphological and land use changes from 1960 (Fig. 2A) to 2004 (Fig. 2B), based on aerial photos and satellite images interpretation, and field mapping. Fig. 2C: Digital model of the submarine environment (See also Fig. 31.2 on Page 329)

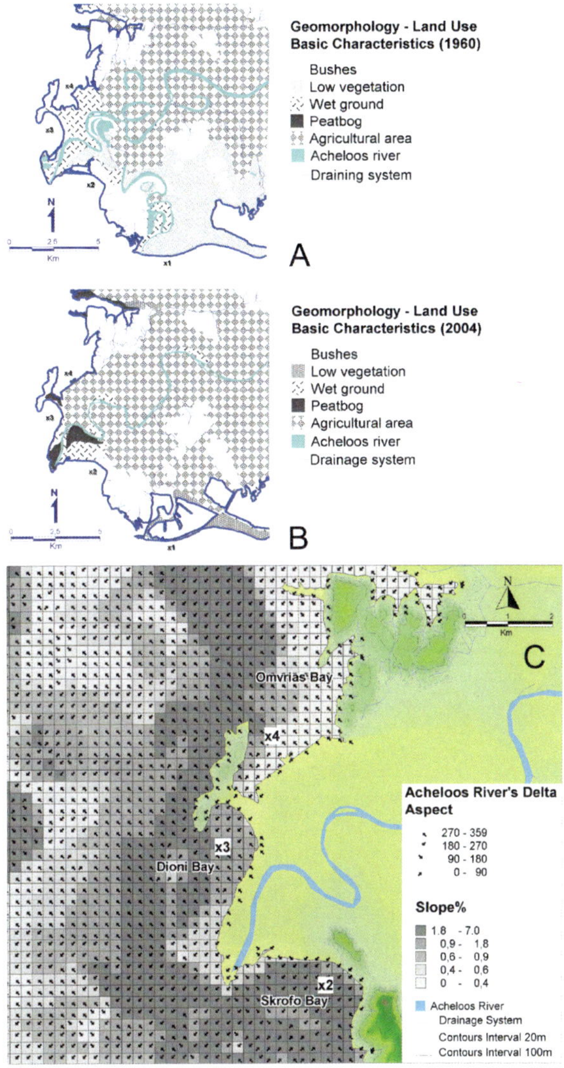

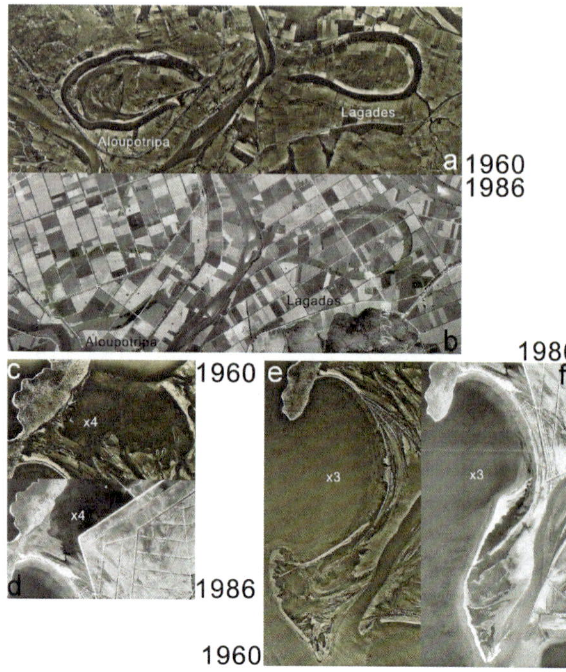

Plate 59 Zooming on geomorphological and land use changes as have been observed by aerial photos and satellite images (1960–2000), in delta plain (See also Fig. 31.3 on Page 330)

Plate 60 Current land use and land cover for the Blakeney-Cley section of the study area (See also Fig. 32.2 on Page 341)

Plate 61 Simulated 2050 land cover/land use for Policy Option 3 and the high sea level rise scenario: same area as Fig. 2 (See also Fig. 32.3 on Page 341)

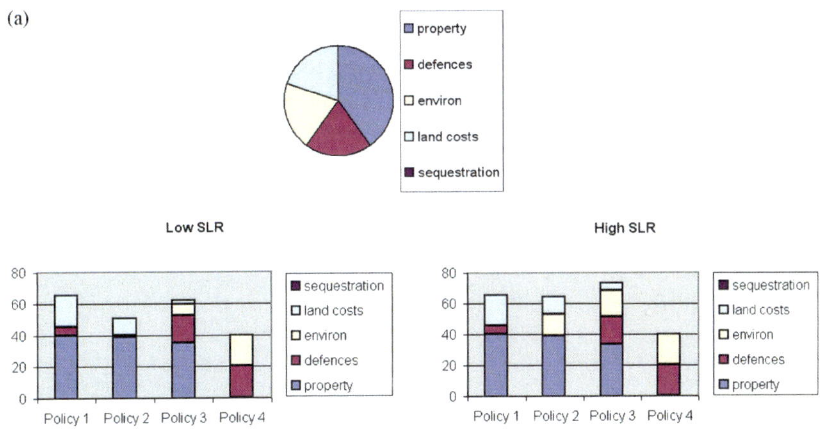

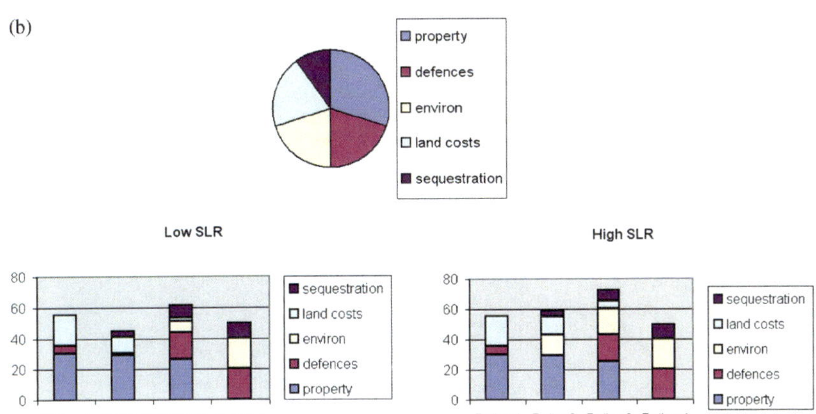

Plate 62 Sample results from the MCA with different criteria weighting systems (**a**) and (**b**). The weighting systems are illustrated in the pie-chart and the resultant scores for each policy shown for two different sea level rise (SLR) scenarios (See also Fig. 32.4 on Page 344)

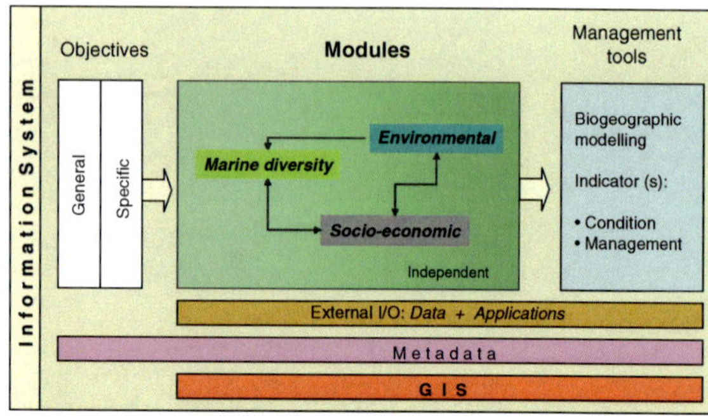

Plate 63 Conceptual model of the Catalan coastal zone GIS (See also Fig. 34.1 on Page 362)

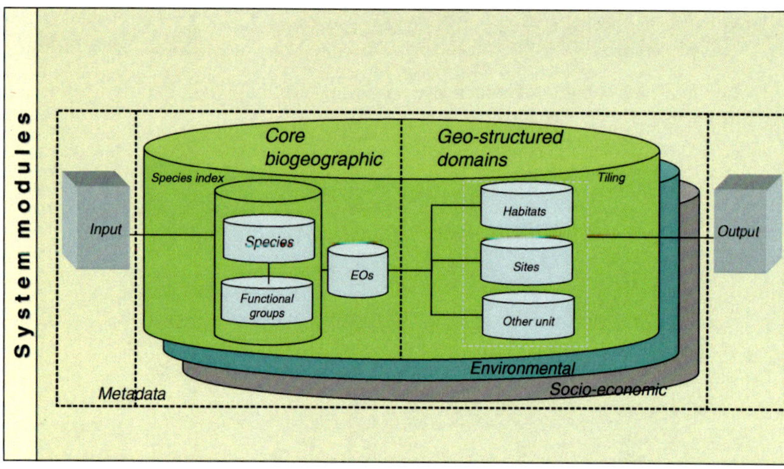

Plate 64 Conceptual entities and relations of the biogeographic module (See also Fig. 34.2 on Page 364)

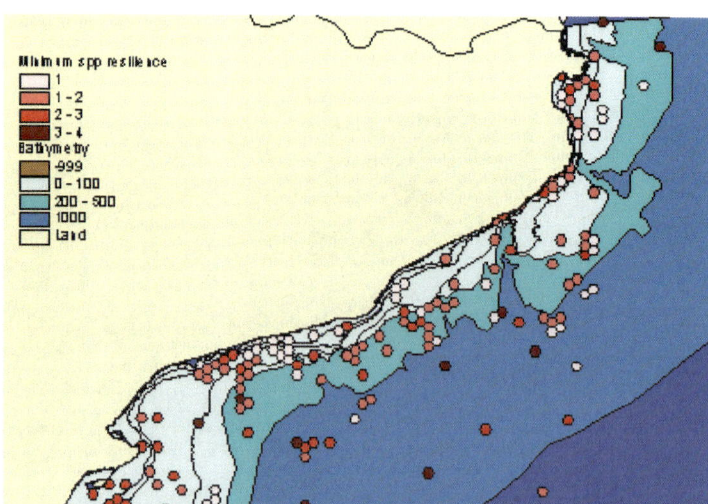

Plate 65 Marine hexagon model of anthropogenic impact indicators and fish species resilience of the Catalan coast (See also Fig. 34.4 on Page 367)

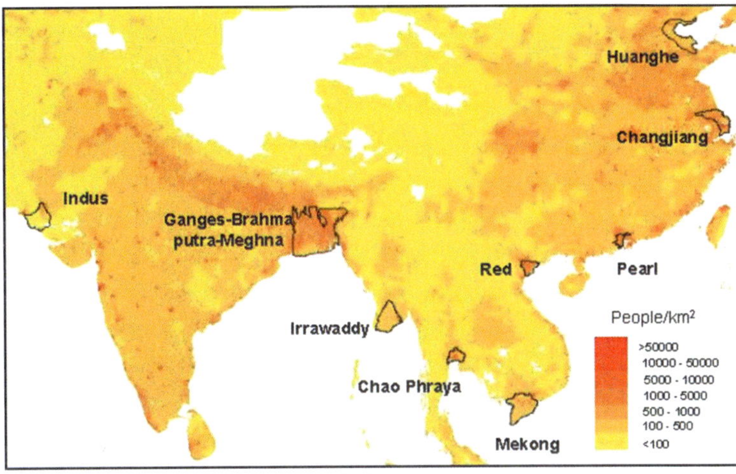

Plate 66 Gridded population density of Asia (based on CIESIN, GPW3 2000), showing concentration in river valleys and on megadeltas (See also Fig. 36.1 on Page 380)

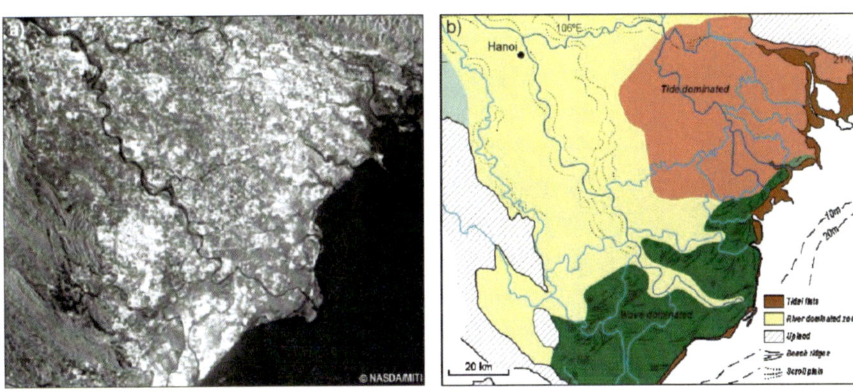

Plate 67 The Red River Delta; (**a**) JERS radar imagery of the delta showing the complexity of land-use, and (**b**) an interpretation of subaerial delta geomorphology that shows distinct sectors of the delta dominated by different processes (based on Mathers and Zalasiewicz, 1999) (See also Fig. 36.2 on Page 382)

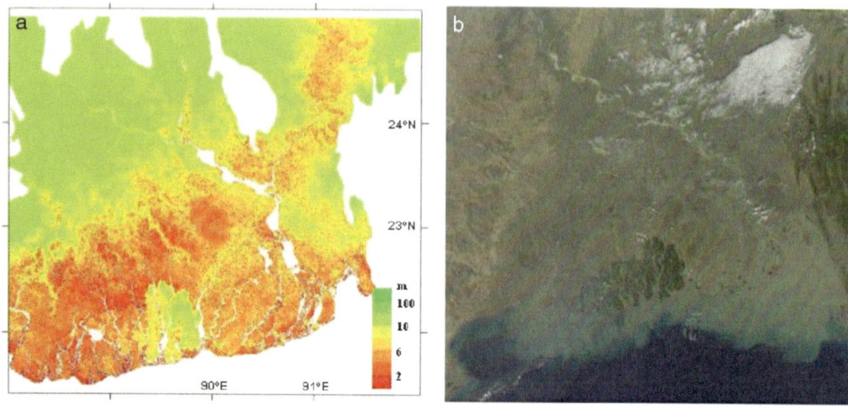

Plate 68 The delta of the Ganges-Brahmaputra-Meghna delta; (**a**) DTM determined from SRTM, and (**b**) MODIS image of delta (See also Fig. 36.3 on Page 383)

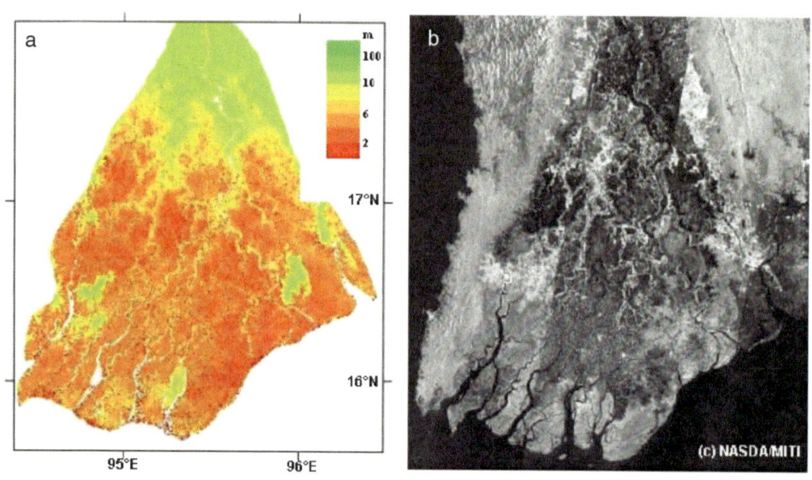

Plate 69 The delta of the Irrawaddy River, Myanmar; (**a**) Digital terrain model (DTM) derived from Shuttle Radar Topography Mission (SRTM) data; (**b**) JERS radar imagery, showing the tide-dominated "Mouths of the Irrawaddy" (See also Fig. 36.4 on Page 384)

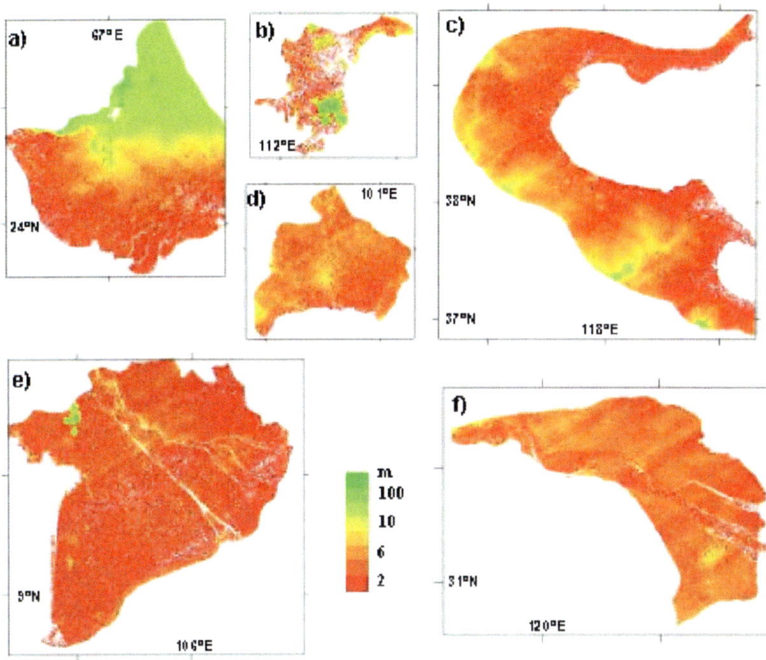

Plate 70 DTM of (**a**) Indus, (**b**) Pearl, (**c**) Huanghe, (**d**) Chao Phraya, (**e**) Mekong, and (**f**) Changjiang deltas based on Shuttle Radar Topography Mission (SRTM) data (after Woodroffe et al., 2006) (See also Fig. 36.5 on Page 385)

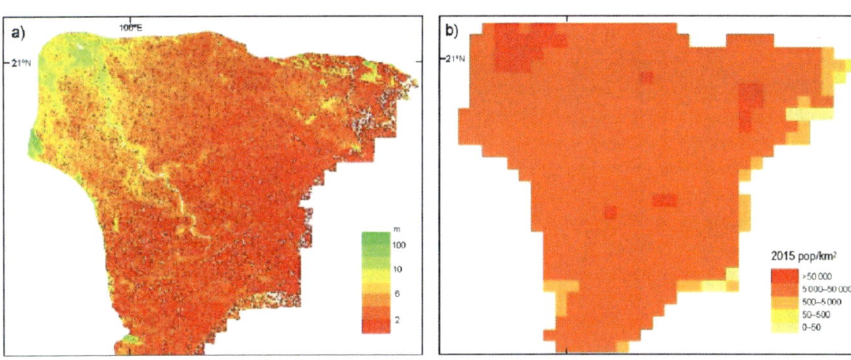

Plate 71 The Red River Delta; (**a**) DTM derived from SRTM; (**b**) gridded population density for 2015 using estimates of population increase (based on GPW3, CIESIN) (See also Fig. 36.6 on Page 387)

Index